Springer Series in Information Sciences 4

Editor: T. S. Huang

Springer Series in Information Sciences

Editors: Thomas S. Huang   Teuvo Kohonen   Manfred R. Schroeder
Managing Editor: H.K.V. Lotsch

1 **Content-Addressable Memories**
   By T. Kohonen   2nd Edition

2 **Fast Fourier Transform and Convolution Algorithms**
   By H. J. Nussbaumer   2nd Edition

3 **Pitch Determination of Speech Signals**
   Algorithms and Devices   By W. Hess

4 **Pattern Analysis and Understanding**
   By H. Niemann   2nd Edition

5 **Image Sequence Analysis**
   Editor: T.S. Huang

6 **Picture Engineering**
   Editors: King-sun Fu and T.L. Kunii

7 **Number Theory in Science and Communication**   With Applications in Cryptography, Physics, Digital Information, Computing, and Self-Similarity   By M.R. Schroeder   2nd Edition

8 **Self-Organization and Associative Memory**   By T. Kohonen   3rd Edition

9 **Digital Picture Processing**
   An Introduction   By L.P. Yaroslavsky

10 **Probability, Statistical Optics and Data Testing**   A Problem Solving Approach   By B.R. Frieden

11 **Physical and Biological Processing of Images**   Editors: O.J. Braddick and A.C. Sleigh

12 **Multiresolution Image Processing and Analysis**   Editor: A. Rosenfeld

13 **VLSI for Pattern Recognition and Image Processing**   Editor: King-sun Fu

14 **Mathematics of Kalman-Bucy Filtering**
   By P.A. Ruymgaart and T.T. Soong   2nd Edition

15 **Fundamentals of Electronic Imaging Systems**   Some Aspects of Image Processing   By W.F. Schreiber

16 **Radon and Projection Transform-Based Computer Vision**   Algorithms, A Pipeline Architecture, and Industrial Applications   By J.L.C. Sanz, E.B. Hinkle, and A.K. Jain

17 **Kalman Filtering** with Real-Time Applications   By C.K. Chui and G. Chen

18 **Linear Systems and Optimal Control**
   By C.K. Chui and G. Chen

19 **Harmony: A Psychoacoustical Approach**   By R. Parncutt

20 **Group Theoretical Methods in Image Understanding**   By Ken-ichi Kanatani

21 **Linear Prediction Theory**
   A Mathematical Basis for Adaptive Systems   By P. Strobach

Heinrich Niemann

Pattern Analysis and Understanding

Second Edition

With 160 Figures

Springer-Verlag Berlin Heidelberg New York
London Paris Tokyo Hong Kong

Professor Dr.-Ing. Heinrich Niemann

Lehrstuhl für Informatik 5 (Mustererkennung),
Friedrich-Alexander-Universität Erlangen-Nürnberg, and

Forschungsgruppe Wissensverarbeitung, Bayerisches Forschungszentrum
für Wissensbasierte Systeme,
D-8520 Erlangen, Fed. Rep. of Germany

Series Editors:

Professor Thomas S. Huang

Department of Electrical Engineering and Coordinated Science Laboratory,
University of Illinois, Urbana IL 61801, USA

Professor Teuvo Kohonen

Laboratory of Computer and Information Sciences, Helsinki University of Technology,
SF-02150 Espoo 15, Finland

Professor Dr. Manfred R. Schroeder

Drittes Physikalisches Institut, Universität Göttingen, Bürgerstraße 42–44,
D-3400 Göttingen, Fed. Rep. of Germany

Managing Editor: Helmut K.V. Lotsch

Springer-Verlag, Tiergartenstrasse 17,
D-6900 Heidelberg, Fed. Rep. of Germany

ISBN 3-540-51378-7 Springer-Verlag Berlin Heidelberg New York
ISBN 0-387-51378-7 Springer-Verlag New York Berlin Heidelberg

ISBN 3-540-10792-4 Springer-Verlag Berlin Heidelberg New York
ISBN 0-387-10792-4 Springer-Verlag New York Berlin Heidelberg

Library of Congress Cataloging-in-Publication Data. Niemann, Heinrich. Pattern analysis and understanding /
Heinrich Niemann. – 2nd ed. p. cm. – (Springer series in information sciences ; 4) First ed. published under title:
Pattern analysis. 1981. Includes bibliographical references. ISBN 0-387-51378-7 (U.S.). 1. Pattern perception.
I. Niemann, Heinrich. Pattern analysis. II. Title. III. Series. Q327.N52 1989 006.4–dc20 89-21748

This work is subject to copyright. All rights are reserved, whether the whole or part of the material is concerned,
specifically the rights of translation, reprinting, reuse of illustrations, recitation, broadcasting, reproduction on
microfilms or in other ways, and storage in data banks. Duplication of this publication or parts thereof is only
permitted under the provisions of the German Copyright Law of September 9, 1965, in its version of June 24,
1985, and a copyright fee must always be paid. Violations fall under the prosecution act of the German Copyright
Law.

© Springer-Verlag Berlin Heidelberg 1981 and 1990
Printed in the United States of America

The use of registered names, trademarks, etc. in this publication does not imply, even in the absence of a specific
statement, that such names are exempt from the relevant protective laws and regulations and therefore free for
general use.

2154/3150-543210 – Printed on acid-free paper

Preface to the Second Edition

In this second edition every chapter of the first edition of Pattern Analysis has been updated and expanded. The general view of a system for pattern analysis and understanding has remained unchanged, but many details have been revised. A short account of light and sound has been added to the introduction, some normalization techniques and a basic introduction to morphological operations have been added to the second chapter. Chapter 3 has been expanded significantly by topics like motion, depth, and shape from shading; additional material has also been added to the already existing sections of this chapter. The old sections of Chap. 4 have been reorganized, a general view of the classification problem has been added and material provided to incorporate techniques of word and object recognition and to give a short account of some types of neural nets. Almost no changes have been made in Chap. 5. The part on representation of control structures in Chap. 6 has been shortened, a section on the judgement of results has been added. Chapter 7 has been rewritten almost completely; the section on formal grammars has been reduced, the sections on production systems, semantic networks, and knowledge acquisition have been expanded, and sections on logic and explanation added. The old Chaps. 8 and 9 have been omitted.

In summary, the new edition is a thorough revision and extensive update of the first one taking into account the progress in the field during recent years. The additions to the book are reflected in the modification of the title.

The author is very grateful to Springer-Verlag for the careful preparation of the book, including a rearrangement of the references, and to Dr. A.M. Lahee for her stylistic improvements of the manuscript.

Erlangen, September 1989   *H. Niemann*

Preface to the First Edition

This book is devoted to pattern analysis, that is, the automatic construction of a symbolic description for a complex pattern, like an image or connected speech. Pattern analysis thus tries to simulate certain capabilities which go without saying in any human central nervous system. The increasing interest and growing efforts at solving the problems related to pattern analysis are motivated by the challenge of the problem and the expected applications. Potential applications are numerous and result from the fact that data can be gathered and stored by modern devices in ever increasing extent, thus making the finding of particular interesting facts or events in these hosts of data an ever increasing problem.

It was tried to organize the book around one particular view of pattern analysis: the view that pattern analysis requires an appropriate set of processing modules operating on a common data base which contains intermediate results of processing. Although other views are certainly possible, this one was adopted because the author feels that is is a useful idea, because the size of this book had to be kept within reasonable bounds, and because it facilitated the composition of fairly self-contained chapters.

The book is addressed to the scientists working in the field of pattern analysis and to students with some background in pattern recognition in general and, of course, in mathematics and statistics. The material of the book can be covered in a one semester course, perhaps with varying degrees of emphasis. The first chapter gives a general introduction to the topic. The next two chapters treat low-level processing methods which need little, if any, problem specific knowledge. The fourth chapter gives only a very condensed account of numerical classification algorithms. Chapter 5 is devoted to data structures and data bases, and Chap. 6 to control structures. Representation and utilization of problem-specific knowledge is treated in Chap. 7. The last two short chapters, finally, are concerned with two examples of pattern analysis systems and a guess about future problems.

It is a pleasure to acknowledge the support I have received from many people during the writing of the book. First of all, the editor of the series, Prof. T.S. Huang, now at the University of Illinois, took the labor to read the manuscript, and improved its presentation by numerous suggestions.

Dr. Lotsch from Springer-Verlag provided cooperative and competent support in all editorial problems. Mrs. I. Franken, who typed most of the manuscript, and also Mrs. S. Zett, who typed some parts, did an excellent job. Mr. A. Cieslik completed the original drawings of the figures in an outstanding manner. My family tolerated that for a long time the greater part of my leisure time was absorbed by the book. Considering the first and the last place of an enumeration to be equally prominent, it is acknowledged that the "Bayerisches Staatsministerium für Unterricht und Kultus" (Bavarian Ministry of Education) exempted me from lecturing during the summer semester 1980.

Erlangen, May 1981   *H. Niemann*

Contents

1. **Introduction** 1
   1.1 General Remarks 1
   1.2 Definitions 2
   1.3 Principal Approach 9
   1.4 Scope of the Book 11
   1.5 Applications 14
   1.6 Types of Patterns 16
   1.7 Light and Color 17
       1.7.1 Photometry 17
       1.7.2 Colorimetry 19
   1.8 Sound 23
   1.9 Bibliographical Remarks 24

2. **Preprocessing** 26
   2.1 Coding 27
       2.1.1 Basic Approach 27
       2.1.2 Pulse Code Modulation 27
       2.1.3 Improvements of PCM 30
       2.1.4 Transform Coding 31
       2.1.5 Line Patterns 32
   2.2 Normalization 34
       2.2.1 Aim and Purpose 34
       2.2.2 Size and Time 34
       2.2.3 Intensity or Energy 37
       2.2.4 Gray-Level Scaling 39
       2.2.5 Color 40
       2.2.6 Multispectral Data 41
       2.2.7 Geometric Correction 42
       2.2.8 Image Registration 45
       2.2.9 Pseudocolors 45
       2.2.10 Camera Calibration 46
       2.2.11 Normalized Stereo Images 49
       2.2.12 Speaker Normalization 50
   2.3 Linear Filtering 51
       2.3.1 The Use of Filtering 51

		2.3.2	Linear Systems	52
		2.3.3	Fourier Transform	53
		2.3.4	Computational Aspects	57
		2.3.5	Homomorphic Systems	58
		2.3.6	Filtering of Patterns	59
		2.3.7	Resolution Hierarchies	63
	2.4	Rank Order and Morphological Operations		64
		2.4.1	Rank Order Operations	64
		2.4.2	Morphological Operations on Binary Images	65
		2.4.3	Morphological Operations on Gray Value Images	67
		2.4.4	Morphological Algorithm	68
	2.5	Linear Prediction		69
		2.5.1	Predictor Coefficients	69
		2.5.2	Spectral Modeling	71
	2.6	Bibliographical Remarks		73
3.	Segmentation			74
	3.1	Common Principles		75
	3.2	Thresholding		77
		3.2.1	Obtaining a Binary Image	77
		3.2.2	Operations on Binary Images	79
	3.3	Gray-Level Changes		80
		3.3.1	Differences in Local Neighborhoods	80
		3.3.2	Contour Filters	83
		3.3.3	Zero Crossings	88
		3.3.4	Line and Plane Segments	90
		3.3.5	Statistical and Iterative Methods	92
		3.3.6	Generalization to Several Spectral Channels	94
	3.4	Contour Lines		94
		3.4.1	Computing Curved and Straight Lines	94
		3.4.2	Hough Transform	97
		3.4.3	Linking Contour Points	99
		3.4.4	Characterization of Contours	100
	3.5	Regions		103
		3.5.1	Homogeneity	103
		3.5.2	Merging	105
		3.5.3	Splitting	106
		3.5.4	Split and Merge	107
		3.5.5	Remarks	109
	3.6	Texture		110
		3.6.1	The Essence of Texture	110
		3.6.2	Numerical Characterization	112
		3.6.3	Syntactic Characterization	115
	3.7	Motion		115

	3.7.1	Approaches to Motion Estimation	115
	3.7.2	Optical Flow	116
	3.7.3	Computation of Optical Flow	119
	3.7.4	Matching Corresponding Image Points	120
3.8	Depth		122
	3.8.1	Approaches to Depth Recovery	122
	3.8.2	Stereo Images	123
	3.8.3	Optical Flow	124
	3.8.4	Area Matching	127
	3.8.5	Line Matching	128
3.9	Shape from Shading		129
	3.9.1	Illumination Models	129
	3.9.2	Surface Orientation	131
	3.9.3	Variational Approach Using the Reflectance Map	133
	3.9.4	Local Shading Analysis	134
	3.9.5	Iterative Improvement	135
	3.9.6	Depth Recovery	136
3.10	Segmentation of Speech		138
	3.10.1	Subword Units	138
	3.10.2	Measurements	140
	3.10.3	Segmentation	144
	3.10.4	Vector Quantization	148
3.11	Bibliographical Remarks		149

4. **Classification** 151
   4.1 Statement of the Problem 152
   4.2 Classification of Feature Vectors 154
       4.2.1 Statistical Classification 154
       4.2.2 Distribution-Free Classification 158
       4.2.3 Nonparametric Classification 160
       4.2.4 Template Matching 161
       4.2.5 Relaxation Labeling 163
       4.2.6 Learning 165
       4.2.7 Additional Remarks 166
   4.3 Classification of Symbol Strings 166
   4.4 Recognition of Words 167
       4.4.1 Word Models 167
       4.4.2 Word Recognition Using Hidden Markov Models .. 169
       4.4.3 Basic Computational Algorithm for HMM 171
       4.4.4 Isolated and Connected Words 175
       4.4.5 Continuous Speech 175
       4.4.6 Word Recognition Using Dynamic Programming ... 180
   4.5 Recognition of Objects 182
       4.5.1 Object Models 182

		4.5.2	General Recognition Strategies	187
		4.5.3	Classification of 2D Images Using 2D Models	188
		4.5.4	Classification of 2D Images Using 3D Models	189
		4.5.5	Classification of 3D Images Using 3D Models	193
		4.5.6	Structural Matching	195
		4.5.7	Line Drawings	197
	4.6	Neural Nets		199
		4.6.1	Basic Properties	199
		4.6.2	The Hopfield Net	200
		4.6.3	The Hamming Net	201
		4.6.4	The Multilayer Perceptron	201
		4.6.5	The Feature Map	204
	4.7	Bibliographical Remarks		205
5.	Data			206
	5.1	Data Structures		206
		5.1.1	Fundamental Structures	206
		5.1.2	Advanced Structures	208
		5.1.3	Objects	211
		5.1.4	Remarks	213
	5.2	Data Bases		214
		5.2.1	A General Outline	214
		5.2.2	Hierarchical Data Model	216
		5.2.3	Network Data Model	218
		5.2.4	Relational Data Model	219
		5.2.5	Normal Forms	222
		5.2.6	Data Sublanguages	224
		5.2.7	Semantic Data Model	227
	5.3	Pattern Data		227
		5.3.1	Data Structures for Patterns	227
		5.3.2	Data Bases for Patterns	232
	5.4	Bibliographical Remarks		235
6.	Control			236
	6.1	The Problem		237
	6.2	Some Control Structures and Their Representation		239
		6.2.1	Interaction	239
		6.2.2	Hierarchical and Model-Directed Systems	240
		6.2.3	Heterarchical and Data-Base-Oriented Systems	241
		6.2.4	Network of States	242
		6.2.5	Representation by Abstract Programs	243
		6.2.6	Hierarchical Graphs	244
		6.2.7	Petri Nets	245
	6.3	Designing and Planning		247

			Contents	XIII

- 6.4 Judgements 248
  - 6.4.1 General Aspects 248
  - 6.4.2 Importance 249
  - 6.4.3 Reliability 249
  - 6.4.4 Precision 252
  - 6.4.5 Priority 254
- 6.5 Search Strategies 254
  - 6.5.1 An Algorithm for a State-Space Search 254
  - 6.5.2 An Algorithm for Searching AND/OR Trees 258
  - 6.5.3 Remarks on Pruning 261
  - 6.5.4 Dynamic Programming 264
  - 6.5.5 Heuristic Strategies in Image Analysis 265
  - 6.5.6 Heuristic Strategies in Speech Understanding 268
- 6.6 Bibliographical Remarks 270

7. Knowledge 271
- 7.1 Views of Knowledge 272
  - 7.1.1 Levels and Hierarchies 272
  - 7.1.2 Submodules 275
  - 7.1.3 Aspects 276
  - 7.1.4 Frames 276
- 7.2 Predicate Logic 278
  - 7.2.1 Some Basic Definitions 278
  - 7.2.2 Unification 281
  - 7.2.3 Inferences 282
  - 7.2.4 Remarks 285
- 7.3 Production Systems 286
  - 7.3.1 General Properties 286
  - 7.3.2 Inferences 289
  - 7.3.3 An Example of a Rule Format 291
  - 7.3.4 Examples from Pattern Analysis 293
- 7.4 Semantic Nets 295
  - 7.4.1 Introductory Remarks 295
  - 7.4.2 Components of a Semantic Net 296
  - 7.4.3 Instantiation 302
  - 7.4.4 Control 304
  - 7.4.5 An Application in Image Understanding 308
  - 7.4.6 An Application in Speech Understanding 311
- 7.5 Relational Structures 315
- 7.6 Grammars 316
  - 7.6.1 Augmented Transition Networks 316
  - 7.6.2 Unification-Based Grammar 319
  - 7.6.3 Lexical Organization 327
- 7.7 Acquisition of Knowledge (Learning) 329

		7.7.1	Introductory Remarks	329
		7.7.2	Generalization	330
		7.7.3	Outline of Some Algorithms	334
		7.7.4	Learning Concepts of a Semantic Net	335
	7.8		Explanation	340
	7.9		Bibliographical Remarks	340

References 343

Subject Index 365

1. Introduction

In this introductory chapter we shall give a general view of *pattern analysis and understanding*, relate it to pattern recognition in general, state the general principles underlying all approaches to pattern recognition, and then focus on the view adopted in this book. In addition, some applications and an overview of some aspects of light and sound are discussed.

1.1 General Remarks

A significant aspect of human development is the fact that there is a continuous attempt to support human activities by tools and machines. With these it was possible to do work faster, better, and with less effort. A major breakthrough in this development was the invention of rotating engines, providing power which far surpassed the capabilities of humans or animals. Another breakthrough was the development of devices having *information processing capabilities*; the most important and powerful of these devices is the digital computer.

Digital computers offer for the first time the potential to replace certain intellectual tasks by machines. Intellectual tasks also include realization, evaluation, and interpretation of sensory impressions; we summarize these by *perception*. Any meaningful human activity requires perception. It allows the human to acquire knowledge about the environment, to react to it, and finally to influence it. Although everybody is able to perceive a vast amount of sensory impressions, it is presently impossible to state precisely how this is done. By a precise statement we mean an algorithm which might be implemented on a computer. It is of great scientific interest and opens many practical applications to exploit the mathematical and technical aspects of perception. Some applications are mentioned in Sect. 1.5.

It is not intended and does not seem necessary here to argue whether or not machines can perceive at all, nor whether they can perceive at the present time (the author's opinion is that, at best, they have some rudimentary aspects of perception at present). The important point is that humans demonstrate clearly that perception is possible; they do it excellently with some thousand grams of matter – the central nervous system. There is no reason in principle why this could not be simulated by some other matter, for instance, a digital computer. Of course, simulation is not to be understood in the sense that exactly the same algorithms have to be used, but in the sense that similar results of perception are achieved.

Research activities concerned with the mathematical and technical aspects of perception are the field of *pattern recognition* in the broad sense. More precise definitions will be given in the next section. We emphasize mathematical and technical aspects as distinct from mathematical and biological aspects of perception, which is the field of biocybernetics, physiology, or psychology. The special term *pattern analysis and understanding* is used here to distinguish the subfield of pattern recognition dealing with knowledge-based techniques and fairly complex patterns such as continuous speech, images, and image sequences.

Another view supporting the necessity for pattern recognition in general is provided by Wittgenstein and we quote a few sentences using the numbers introduced in [1.1]

1. The *world* is all that is the case.
2. What is the case – a *fact* – is the existence of states of affairs.
3. A *logical picture* of facts is a thought.
4. A *thought* is a proposition with a sense.
6.21 A sentence of mathematics does not express a thought.

If a machine is to be able to derive its own logical pictures of facts and consequently to derive propositions with a sense, then it must be able to perform pattern analysis and understanding. Without a pattern recognition ability it may be able to manipulate logical pictures of facts (or symbolic structures) provided by somebody else, but it will never be able to judge the adequacy of those externally provided structures. It is certainly an ingredient of *intelligence* not only to manipulate blindly certain symbolic structures, but also to convince oneself that it is meaningful in relation to the real world.

1.2 Definitions

We shall now proceed to define and illustrate by examples some important concepts starting with the *environment* which is the object of perception.

Environment: This is the totality of quantities which can be measured by physical devices. It is represented by a set

$$U = \{b^r(x) | r = 1, 2, \ldots\} \qquad (1.2.1)$$

of functions $b^r(x)$.

The above equation expresses the fact that any object, physical event, fact, or state of affairs may be described by an appropriate number of *functions*. The value of the functions gives a characteristic quantity at each point in space and time. Examples are the density of a body or the electrical field vector of an electromagnetic wave. The dimensionality of b and x is unspecified and may be different for different values of the index r. There is no biological or technical system which is able to record the whole environment. Sensory organs and technical sensors always accept only a section of the environment. A universal system for pattern recognition which is capable of processing the whole environment or even just a large part of it is not feasible at present and always will be uneconomic and inefficient. This suggests that we should limit our interest to a certain *task domain*.

Task domain: A task domain is denoted by Ω and contains only objects or functions belonging to a strictly limited application or subsection of the environment. It is given by a set Ω

$$U \supset \Omega = \{ \boldsymbol{f}^r(\boldsymbol{x}) \,|\, r = 1, 2, \ldots \} \qquad (1.2.2)$$

of functions $\boldsymbol{f}^r(\boldsymbol{x})$ and is a subset of the environment U.

The dimensionality of $\boldsymbol{f}$ and $\boldsymbol{x}$ is left unspecified, but is fixed for all $\boldsymbol{f}^r(\boldsymbol{x})$; it may be different for another task domain. A task domain could be, for example, the classification of electrocardiograms (ECG) or the analysis of remotely sensed images obtained in a certain number of spectral channels, at a certain altitude, and with a certain resolution, in order to ascertain land-use types, their position and area. It is apparent that a particular task domain will require a particular approach and equipment to measure the functions $\boldsymbol{f}^r(\boldsymbol{x})$; on the other hand, a selection of measuring equipment or sensors will imply a certain subset $U \supset \Omega$ of functions which can be measured. This allows us to state what is meant by a *pattern*.

Pattern: The elements of the set Ω, the task domain, are termed patterns. Thus, a pattern is a function

$$\boldsymbol{f}^r(\boldsymbol{x}) = \begin{vmatrix} f_1^r(x_1, x_2, \ldots, x_n) \\ f_2^r(x_1, x_2, \ldots, x_n) \\ \ldots \\ f_m^r(x_1, x_2, \ldots, x_n) \end{vmatrix} . \qquad (1.2.3)$$

There are many different definitions of the term pattern in the literature. A comparison and evaluation might fill pages but would not help very much. So we shall adhere to (1.2.3). As mentioned, in a particular task domain the indices m and n are fixed. For example, an ECG consists of three (or more) time functions $f_l(t)$, $l = 1, 2, 3$ with $m = 3$ and $n = 1$. A color TV picture consists of functions $f_r(x, y, t), f_g(x, y, t), f_b(x, y, t)$ with $m = 3, n = 3$, and continuous speech or isolated words are given by just one function $f(t)$ with $m = n = 1$. It is obvious that any interesting object or fact which might be perceived can be represented by (1.2.3). It is now possible to define *pattern recognition*.

Pattern Recognition: Pattern recognition deals with the mathematical and technical aspects of automatic derivation of logical pictures of facts. At the present state of the art this comprises classification of simple patterns as well as analysis and understanding of complex patterns.

Since in the definition completely general processing of patterns is allowed and the term pattern itself is fairly general, pattern recognition as used here is of a very general scope. Although sometimes the term pattern recognition is associated with and limited to the early approach using feature vectors and statistical decision theory, we stress that it is not useful to adhere to a certain approach and a certain state of the art in a field but to define the general problem. In this book we shall limit the scope as pointed out in Sect. 1.4. The above distinction between simple and complex patterns appeals to the intuitive insight, for example, that a hand-written numeral is simpler than an aerial photograph, but both may be objects of perception. It is not intended to give a quantitative measure and to establish a threshold discriminating between simple and complex. Next the concept of *classification* will be considered in somewhat more detail.

Classification: In classification of (simple) patterns each pattern is considered as one entity and is classified independently of other patterns as belonging to one class Ω_κ out of k classes $\Omega_\lambda, \lambda = 1, \ldots, k$. It is possible to reject a pattern, which means that a $(k+1)$th class Ω_0, the *reject class*, is introduced.

Examples are classification of OCR-A (optical character recognition, alphabet A) characters or of isolated words. When classifying characters, what matters is to put all possible ways of writing or printing a character with a certain meaning, say the numeral "6", into one and the same class. If no unique or reliable classification is possible, it is usually appropriate to reject a pattern, which means, to put it into class Ω_0. Properties of a *class* are the following.

Class (or Pattern Class): Classes Ω_κ are obtained by a partition of the set Ω into k (or $k+1$) subsets $\Omega_\kappa, \kappa = 1, \ldots, k$ or $\kappa = 0, 1, \ldots, k$. It is required that

$$\Omega_\kappa \neq \phi, \kappa = 1, \ldots, k,$$
$$\Omega_\kappa \cap \Omega_\lambda = \phi, \kappa \neq \lambda,$$
$$\bigcup_{\kappa=1}^{k} \Omega_\kappa = \Omega \text{ or } \bigcup_{\kappa=0}^{k} \Omega_\kappa = \Omega.$$

(1.2.4)

The user will be interested in a few, or perhaps in only one of the many possible partitions. This partition of interest is characterized by the fact that patterns of the same class should be *similar* and patterns of different classes should be *dissimilar*. A class contains a subset of patterns of a task domain. If, in connection with isolated word recognition, class Ω_κ has the meaning "house", then all patterns – in this case all functions $f(t)$ – which result from different speakers at different times belong to Ω_κ. The above requirement of disjoint classes is reasonable for many applications. A numeral, for instance, must not be simultaneously classified as being a 1 and a 7. Should a numeral allow both interpretations it may be advisable to reject it. It is possible that in psychological or sociological investigations there are test persons showing criteria of different types. In these cases either the requirement of disjoint classes may be dropped or new, mixed classes, besides the "pure" classes, introduced. With this latter modification, disjoint classes may be used in all cases, however, at the price of an increase in the number of classes. The former approach amounts to attributing more than one class to a certain object or observation; this can be done in a natural way, for example, by assigning several classes with different probabilities. Next we give an idea of a *simple pattern*.

Simple Pattern: A pattern is considered to be simple if the user is interested in a class name only and if it is feasible to classify it as a whole.

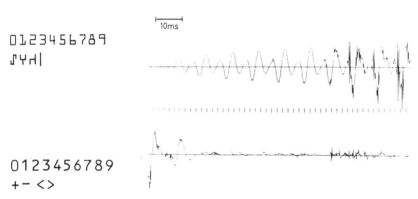

Fig. 1.1. OCR-A characters and an isolated spoken word as examples of simple patterns

The above definition is to be understood as a first approximation, but not as a strict separation between simple and complex. Figure 1.1 gives two examples of patterns which in this sense may reasonably be called simple patterns. Having introduced classification of simple patterns we now turn to *analysis*.

Analysis: When analyzing (complex) patterns each pattern $f^r(x) \in \Omega$ is given an individual symbolic description B^r.

Examples of analysis are understanding of continuous speech, evaluation of multispectral scanner images, determination of circuit elements and electrical connections of a schematic diagram, or interpretation of image sequences, for example, in the medical area. It is obvious, for instance, that in schematic diagrams a class name is not sufficient. Rather, one would like to know the type of circuit elements and their relations (connections). The goal of analysis will vary according to the application and the requirements of the user because this will determine the information required in the description. Thus, B^r may contain the following information:

1. A complete symbolic description of a pattern on an adequate level of abstraction.
2. A list of some interesting events or objects occurring within the pattern.
3. A description of changes which occurred between successive recordings of a pattern.
4. A classification of a complex pattern.

A symbolic description of a pattern contains the most extensive information, whereas information is compressed to its highest degree by classification of a complex pattern. It seems appropriate to point out that classification of simple patterns, like those shown in Fig. 1.1, is different from classification of a complex pattern, like a chest radiograph. A simple pattern may be treated as one entity, but this approach would not be feasible for deciding, for instance, whether a chest radiograph is normal or abnormal. In this case the radiograph will be analyzed in the sense of extracting simpler constituents or segmenting the image into segmentation objects. These are used to classify the image. The term *description* is considered next.

Description: A description of a pattern is its decomposition or segmentation into simpler constituents (or pattern primitives or segmentation objects) and their relations, the identification of objects, events, and situations by symbolic names, and if required, an inference about its implications for the task domain.

In the case of schematic diagrams simpler constituents are, for instance, resistors, transistors, or capacitors. Their relations are electrical connections. We thus may view a class as a partition of the field of problems, and a description as a decomposition and interpretation of a pattern. The complete symbolic description merely translates a pattern to another representation which is better suited for further processing. Usually a description will also be accompanied by a reduction of information and concentration on the important aspects. An example of a description is given below. It shows that a symbolic description may be given on different *levels of abstraction* (or in different conceptual systems) employing an increasing amount of knowledge about the task domain. In the example we distinguish three levels of abstraction; the first and lowest only names *pixels* and *low-level segmentation objects* such as lines or regions; the second identifies *objects and motions by symbolic names*; the third gives an *interpretation* of the meaning of the pattern in relation to a certain *task* which in the example is diagnostic description of image sequences from the beating heart.

Example of a symbolic description:

Level 1 of abstraction: There is a region in the lower right part of all images in the image sequence which is surrounded by a contour C1. The decrease of area of this region is 150 pixels per second from the first to the fifth image.

Level 2 of abstraction: The contraction phase of the left ventricle covers the first half of the heart cycle. During this phase the change of area of the inferioapical segment is 58%.

Level 3 of abstraction: The inferioapical segment of the patient has motional behavior between normal and hypokinetic because the motion is synchronous with the motion of the left ventricle, because the time of stagnation is normal, and because the ejection fraction of 58% is within the interval of inferioapical hypokinesis. The certainty factor of hypokinetic behavior is 0.25 and of normal behaviour 0.1 on a scale from 0 to 1.

We emphasize that the precise content of a symbolic description and also its appropriate level of abstraction depend on the type of the image (or image sequence) and on the requirements of the user. Furthermore, we consider it irrelevant at this point whether a symbolic description is given in natural language or in some formal representation. Similar examples can be given for speech understanding where levels of abstraction might be words, syntactic constituents, sentences, or dialog. Our next point is the idea of a *complex pattern*.

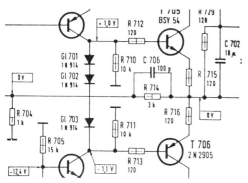 

Fig. 1.2. A schematic diagram and an outdoor scene as examples of complex patterns

Complex Pattern: A pattern is considered to be complex if a class name is not sufficient for the user or if classification of the pattern as a whole is unfeasible.

Figure 1.2 gives two examples of complex patterns. A comparison with Fig. 1.1 shows directly that these patterns are indeed more complex in an intuitive sense. It is clear that simple and complex patterns are not "pure types" but that there is considerable overlap. It may also depend on the particular application, the methods used, and the view of the user whether a pattern is considered to be simple or complex. Nevertheless, we find it useful, in a first approach, to distinguish between these two types.

Our final definition concerns the term *understanding*. It should be noted that the scope here is strictly limited to understanding as used in this book in connection with automatic pattern recognition and that no attempt is made to encompass the psychological, philosophical and so on implications of this term.

Understanding: If a machine is able to map a pattern (which initially is provided by some appropriate sensor like a microphone or a TV camera) to an internal symbolic knowledge structure representing the task domain, then it is said that the machine understands the pattern.

The above definition states that understanding implies the transition from a *numerical representation* of a pattern by an array of integer numbers (which usually is provided at the output of sensors) to a *symbolic representation* within some suitable knowledge representation formalism (which usually is in a format provided by artifical intelligence techniques). Although there is no clear borderline between *recognition* (the statement of a class name), *analysis* (the statement of a symbolic description on an adequate level of abstraction), and *understanding* (the mapping to an internal knowledge structure), the following two examples may give some idea. The first example is taken from speech understanding. It is usually termed "recognition" if a machine is able to identify the words occurring in an utterance, for example, to identify that during a certain time interval the word "Boston" was spoken. It is called "understanding" if the machine can also identify that "Boston" is meant to be, for example, the city where the person speaking would like to go and that this person would like to have information about a convenient flight to Boston. Apparently, what was called recognition requires knowledge about words, pronunciation, and so on; understanding requires detailed knowledge about a specific task domain, in this case knowledge about flight schedules and related utterances. As the second example we may consider the above mentioned example of symbolic descriptions of medical image sequences. What was called in this example "level 2 of abstraction" was more or less a recognition level because objects and motions were identified and described by symbolic names. The "level 3 of abstraction" can be associated with an understanding level because it requires detailed knowledge by the system about medical diagnoses which can be infered from this type of image. Whereas the objects are directly visible, the diagnoses are only implicitly available.

1.3 Principal Approach

In this section the basic assumptions which underlie any approach to pattern recognition are stated in the form of *six postulates*. Although these postulates are usually not given explicitly, without them there would be no pattern recognition. The first postulate is valid for classification as well as analysis.

> **Postulate 1:** In order to gather information about a task domain a representative sample ω
>
> $$\Omega \supset \omega = (\boldsymbol{f}^1(\boldsymbol{x}), \ldots, \boldsymbol{f}^N(\boldsymbol{x})) \tag{1.3.1}$$
>
> of patterns is available.

It is important that the sample contains only patterns belonging to the *task domain* and that it is *representative*. Patterns $\boldsymbol{f}^r(\boldsymbol{x}) \notin \Omega$ present no problem if they are labeled. It is, in general, difficult to decide whether a sample is representative. Some hints may be obtained from consideration of confidence intervals and from testing system performance with patterns not contained in the sample ω. If a representative sample is used to design a system for classification or for analysis of patterns, system performance should be nearly equal when processing patterns which are contained in the sample and patterns which are not. There may be cases where other sources of information than a sample of patterns are available or necessary. These cases occur in particular at high levels of abstraction, for example the "level 3 of abstraction" in the example of the last section. In these cases information about the patterns and the related task domain may be available in the form of printed text books or specific knowledge of human experts. This point concerns knowledge acquisition and is discussed further in Sect. 7.7. For classification of patterns the following two additional postulates are essential.

> **Postulate 2:** A (simple) pattern has features which characterize its membership in a certain class.
>
> **Postulate 3:** Features of patterns of one class occupy a somewhat compact domain of feature space. The domains occupied by features of different classes are separated.

The central and in general unsolved problem is to systematically find or generate features meeting Postulate 3. However, for particular task domains it has been demonstrated empirically that such features do exist. The problem of actually separating the domains, the classification problem, is tractable by general theoretical methods if appropriate assumptions can be made. There are also open problems if general classification problems are treated. This point is discussed in Chap. 4. The general structure of a classification system is given in Fig. 1.3.

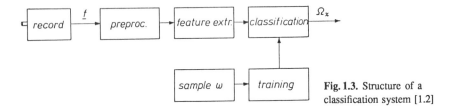

Fig. 1.3. Structure of a classification system [1.2]

A pattern $f^r(x) \in \Omega$ is recorded first. Then the pattern is preprocessed and features are extracted. The features are summarized in the *feature vector* c^r. Finally, the feature vector is classified which amounts to a mapping

$$c^r \to \kappa \in \{1, \ldots, k\} \quad \text{or} \quad \kappa \in \{0, 1, \ldots, k\} \quad (1.3.2)$$

of feature vectors to integers. To do this one has to have information about the domains of classes in feature space. This is acquired from the sample during a phase which is the design, learning, or training phase. It is common practice to first simulate a classification system on a digital computer in order to test its performance. After this one may think of a special hardware realization. Analysis of patterns is based on Postulates 4 and 5, in addition to Postulate 1.

> **Postulate 4:** A (complex) pattern consists of *simpler constituents* or *segmentation objects* which have certain relations to each other. A pattern may be decomposed or segmented into these constituents.
>
> **Postulate 5:** A (complex) pattern belonging to a task domain has a certain *structure*. This implies that not any arrangement of simple constituents will yield a pattern $f^r(x) \in \Omega$ and, furthermore, that many patterns may be represented with relatively few constituents.

The problem of finding appropriate simple constituents or segmentation objects is unsolved in general. Again, it was demonstrated experimentally that, at least in certain applications, they do exist. For instance, phones are appropriate simple constituents for recognition of continuous speech; however, they are not unique since, for example, diphones or demisyllables may be used instead of phones. There are approaches to represent knowledge about structural properties and to carry out analysis. It seems, however, that solutions of this problem are less developed than solutions of classification problems. It is possible to give an outline of the structure of an analysis system following Fig. 1.3. It would consist of recording, preprocessing, extracting simple constituents, and analysis. This diagram is not repeated because a different approach is preferred here. Since this concerns the main topic of this book, discussion is deferred to the next section.

One of the central problems in pattern recognition is the transition from a numerical representation – the pattern $f^r(x)$ – to a symbolic representation – the description B^r. This always requires at certain levels of abstraction the recognition of *similarities* between two representations; one representation may be viewed as representing the original observation, the pattern, or a part of it; the other representation may be viewed as representing some model, prototype, or some general knowledge about the task domain. The notion of similarity is stated in

> **Postulate 6:** Two representations are similar if a suitably defined distance measure is small.

Although this may look self-evident, it is really the basis of most similarity measures. If the representations are vectors of real numbers, similarities may be defined via suitable metrics. If the representations are symbolic structures, transformations may be defined for mapping one representation into the other. The number of transformations and possibly their weight which are required for such a mapping are used to derive a distance measure and a similarity measure. It is emphasized that Postulate 6 only concerns two representations, but not two sets of representations. It is another point to define the similarity of sets which amounts to a definition of the properties of two classes.

1.4 Scope of the Book

To keep this volume within reasonable bounds the scope is limited to some aspects of pattern recognition. First, only *digital methods* of pattern recognition are considered and some comments on this point are made in Sect. 2.1. Second, the emphasis is on *analysis and understanding* of complex patterns, as indicated by the title. Third, the attempt has been made to organize the book around the outline of this section, particularly Fig. 1.4, although a strict separation of concepts is not possible in all cases.

The third point concerns important aspects determining the view adopted here. The system structure of Fig. 1.3 which is termed the *hierarchical structure* is not the only one possible in pattern analysis; other structures are the *model-directed*, the *heterarchical*, or the *blackboard structure*. Other structures are possible, for instance, by hybridizing these four. Which one should be adopted here? The answer is another one, which allows representing all of them. Since pattern analysis systems tend to be fairly involved, care has to be taken to obtain systems which are as easy as possible to design, implement, test, evaluate, and maintain. So adherence to one structure has not been attempted, but a general idea is adopted allowing specialization in order to obtain different system structures which are tailored to specific task domains. This is done by specifying only *components or modules* of the system at this stage.

1. Introduction

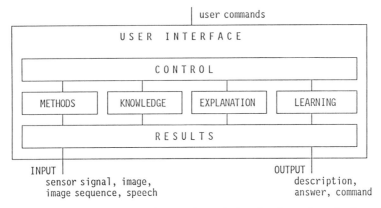

Fig. 1.4. Components of a general system for analysis and understanding of patterns

A system for pattern analysis and understanding may require up to seven basic components:

1. A *data base* which contains intermediate results obtained so far during analysis of a pattern $f^r(x)$.
2. A module which contains *methods* for low level processing and segmentation of patterns.
3. A module which contains *knowledge* about structural properties of patterns and the task domain.
4. A module which executes *control* or determines a *processing strategy* for a pattern.
5. A module for automatic *acquisition or learning* of the knowledge required by the system.
6. A module for the *explanation* of resources, actions, and results of the system to a user.
7. A *user interface* "wrapping" all modules and allowing somebody to use the system without reading lengthy manuals.

This idea is depicted in Fig. 1.4. It may be the case that not all of the above seven components will be required in a certain application, but we consider components 1 to 4 as essential in any knowledge based system for pattern analysis. At the end of analysis, the data base of results will contain a description B^r of the pattern $f^r(x)$, where the information contained in the description depends on the task domain and on the requirements of a user. During analysis intermediate results are obtained. These may contain various possible names for simple constituents, their localization within the pattern, and different levels of abstraction, such as speech segments, phones, syllables, and words. These different levels of abstraction – which were mentioned already in the example description in Sect. 1.2 – can be used to state, in addition to Fig. 1.4, a *general stratified model* of image or speech understanding. This consists of three columns in the horizontal direction representing the explicit knowledge available to the system, the processes using this knowledge, and the intermediate results computed by the processes. In the vertical direction it consists of several rows corresponding to the levels of abstraction, for example, phones, words, syntactic constituents, semantic structures, pragmatic structures, and a dialog level in speech understanding, or line and region segments, lines and regions, objects, object groups, scenes, and the task specifc meaning of scenes in image understanding. The control module embraces the levels and determines the order of activation of processes and the subset of results to be used. This is depicted in Fig. 1.5.

	KNOWLEDGE	PROCESSES	RESULTS	
level n	user requirem.	gen. of reaction	output: high lev. description B^r	
...	...	...		
level j	task	goal oriented	action, advice	
...	...	...		C
				O
level i	objects	matching	symb. names	N
...	...	...		T
				R
level 2	segmentation	segmentation	segmented pattern	O
				L
level 1	distortions	normalization	enhanced pattern	
level 0	pattern form.	recording	input: low level pattern $\underline{f}^r(\underline{x})$	

Fig. 1.5. A stratified model of an image or speech understanding system

There may be a slight difficulty in distinguishing methods and knowledge. *Methods* are operations which work fairly independently of structural properties of patterns and without the explicit representation of task specific knowledge. Examples are linear filtering or Hueckel's operator. Of course, the success and adequacy of these methods does depend on structural properties. On the other hand, *knowledge* contains in an explicit form all facts known in advance about the patterns and the task domain and thus is highly dependent on the particular application. Since we have independent components in Fig. 1.4 and do not want to specify a particular system structure, another component, the control module, is introduced. Its task is to select methods, store and evaluate results, and make use of available knowledge. The control module thus provides a means for determining a suitable processing strategy for every pattern. If an explicit structure is specified in advance as in Fig. 1.3, a control module becomes unnecessary. These four modules will be present in any system for pattern analysis and understanding. In addition it may be useful to have two additional modules. One of them is to acquire the relevant task specific knowledge in order to allow the system to improve its performance. The other one is to outline the reasons for computing a certain description or to point out available system resources. The need for those two additional modules will depend highly on the particular purpose of the system. For example, in a system computing diagnostic descriptions of medical images the explanation capability will be very important because in the end the medical doctor is responsible for the diagnosis. On the other hand, in an industrial vision system picking parts every two or so seconds it would be useless if the system also outputs an explanation every two or so seconds. In this case the explanation facility would be useful only for system design and debugging but not for the operational phase. Finally, the user interface will be desirable in systems allowing or requesting a good deal of user interaction.

1.5 Applications

In this section a short account of some *applications* of pattern recognition will be given without any attempt to be exhaustive. The purpose is to show that there are already domains of applications where analysis methods are in actual use.

The "classical" application is the *reading* of machine- and hand-printed characters. A large area are standardized documents of insurance companies and banks. In these cases the problem is pattern classification with a number k of classes being the ten numerals and a few special signs. Because standardized machine-printed characters are used and the area of printing is prespecified, location and segmentation of characters are simple. The characters are sampled with 100–1000 points and classified on the order of 1000 per second. Another important area is automatic postal address reading. Design philosophies vary considerably according to the postal standards of each country. Here an important step towards analysis is done since independent classification of characters alone would not give satisfactory performance. A further step in reading is the automatic understanding of complete *documents* having an arbitrary format and containing a mixture of text, graphics, and images. These systems will employ general knowledge about documents, they have to perform pattern analysis and understanding, and eventually they will have to exhibit text understanding.

In *speech recognition* several problems are distinguished: speaker identification, speaker verification, isolated word recognition, and continuous speech recognition and speech understanding. The first three problems belong to pattern classification, the last one to pattern analysis and understanding. The time function $f(t)$ which is obtained from speech at the output of a microphone is sampled with 8000–20 000 values per second. In *speaker verification* and *isolated word* recognition the number of classes usually is between 10 and 200; there are experimental systems for isolated word recognition with a vocabulary of up to 20 000 words. Obviously a class name is sufficient in these cases and the classification approach has been proven to be feasible as is demonstrated by several commercial isolated word recognition systems and experimental speaker verification systems. On the other hand, the classification approach has turned out to be infeasible for continuous speech. A class name for the whole speech utterance is not adequate, but one might think of individual classification of single words. This does not work because it is impossible to reliably segment an utterance into single words and to reliably classify words individually. In continuous speech recognition one has to use information about neighboring words, syllables, and phonemes, about the meaning of words, and task-specific information. There has been considerable research and success in this field, and operational systems with vocabularies of the order of 10 000 words have been tested. In this area there are system structures deviating completely from Fig. 1.3.

Medical care offers a rich field of applications comprising time functions $f(t)$, gray-level images $f(x, y)$, time-varying images or time sequences of images $f(x, y, t)$, and volume sequences $f(x, y, z)$. These result from electrocardiograms (ECG), electroencephalograms (EEG), X-ray images, cytologic images, cineangiograms, or tomograms to name a few. In ECG and EEG evaluation the classification predominates. Programs for ECG evaluation are in clinical use. Commercial cell analyzers are now available which classify blood cells in a population and count cell types. Their main task is classification of individual cells, but localization of cells, determination of boundaries, or extraction of textural information is much more involved than, for instance, in the character classification case. Radiographic images and image sequences require methods of analysis, particularly segmentation into simpler constituents as is demonstrated by corresponding research projects. For instance, diagnosis of a chest radiograph may require determination of rib contours first. This is complicated since dorsal and ventral rib contours overlap in the image. It requires methods for contour detection as well as knowledge about possible appearance of contours. Therefore, this single step of processing already requires involved methods of analysis. In order to obtain a diagnostic description it is necessary that the system has explicit knowledge about the medical aspects of a certain type of image.

Remotely sensed images are available in ever increasing number such that automatic processing is imperative. There are various projects, such as the American LANDSAT and SEASAT, or the German FMP. Images are obtained by multispectral scanners, for instance, with four spectral channels in LANDSAT and eleven channels in FMP, by synthetic aperture radar, for instance, in SEASAT, or by several other devices. They are carried by satellites or airplanes and have varying resolution, for instance, 80×80 m resolution in a 185×185 km image field in LANDSAT. From these images information can be obtained about agriculture and forestry, geology, hydrology, geography, environment, cartography, and meteorology. Processing requires image normalization, extraction of lines, regions, and textures, classification of regions or individual picture elements, clustering, and extraction of objects, so that this is a problem of pattern analysis. Because of the inherent difficulties interactive systems are widely in use. These allow combination of human insight and knowledge with data processing and storage capabilities of digital computers. A human operator selects processing steps on the basis of desired results and intermediate results obtained from previous processing. If a successful sequence of processing steps is found, it may be placed in a working file and executed in a batch mode.

Automation of *industrial production* is another field requiring techniques of pattern recognition for advanced systems. They are required for checking quality of products and automating assembly processes. Checking quality may be done by classification of sounds from engines or by inspection of critical areas of products. The last problem requires methods of image analysis since defects have to be detected and located in an image. To automate assembly processes it is necessary to locate machine parts, to distinguish a particular part, and to locate important areas, such as screw holes. A class name for the machine part is not sufficient, but must be supported by additional descriptive details. There are some experimental systems available, for instance, for mounting wheels to cars or for assembling small electromotors. There also are some systems in actual use, for example, for inspecting printed circuit boards and for bonding integrated circuits. Automation by using pattern recognition methods will be very important in the factory of the future. In the industrial area, but also in related areas, the *autonomous vehicle* and the *autonomous moving robot* are under development. They have to move in an environment which is unknown or not completely known and need vision capabilities to perform specified tasks.

There are other areas of application for pattern recognition which are not discussed here, partly because they may be considered as special cases of the above areas. For instance, speaker identification is interesting for criminalistics and remotely sensed image processing for military use.

1.6 Types of Patterns

Several basic types of *images* – as the elements of visual perception – may be distinguished:

1. *Reflection images*, that is images which result from the reflection of appropriate illuminating waves from the surface of an object. The recorded images $f(x, y)$ depend on the *surface* of the object (the notation f indicates that there may be several spectral channels), the properties and position of the illumination source, and the properties and position of the recording device. A standard configuration in computer vision is an object illuminated by visible light and recorded by a (color) TV camera. The basic relations for transforming a three-dimensional object point to a two-dimensional image point are given in Sect. 2.2.10. By recording images at subsequent time instances one obtains a time sequence of images $f(x, y, t)$ showing the motional properties of the object.

2. *Transmission images*, that is images resulting from the transmission of appropriate radiation through the object. The resulting image $f(x, y)$ depends on the properties of the *volume* of the object and shows a projection of this volume on the image plane. Standard examples are X-ray images and time sequences of X-ray images.

3. *Tomographic images* which show the *cross sections* of an object. The image $f(x, y)$ depends on the properties of the *volume* and the technique used for tomography. Standard examples are X-ray and MR tomograms. From several cross sections one obtains a volume image sequence $f(x, y, z)$ showing the volume of the object; doing this at subsequent times gives a time sequence showing the time-varying volume $f(x, y, z, t)$ of an object.

4. Whereas in the above cases a real world object or scene is recorded, there are practically important cases where the image has a standardized *man-made* format representing certain facts or situations. Examples are schematic diagrams, weather charts, maps, or documents.

The variability of *speech* – as the most important element of auditory perception – is much more limited than that of images. Speech is restricted to a frequency interval of about 20–20 000 Hz, and about 300–3 400 Hz is sufficient for intelligibility of speech. It is common practice to distinguish the following types of speech:

1. *Isolated words*, that is, single words or a few words of 2–4 s duration spoken with a certain minimum time separation between them.
2. *Connected words*, that is, a short but continuously spoken sequence of words, for example, a complete telephone number.
3. *Continuous speech*, that is, in the limit, unrestricted speech as used in everyday human communication.

1.7 Light and Color

Since important examples of patterns result from visual stimuli, it is useful to summarize some of their basic physical properties in relation to characteristics of the human visual system. It is not intended here to discuss the visual system.

1.7.1 Photometry

Let $F'(\lambda)d\lambda$ be the radiation energy emitted by a light source in the wavelength interval $(\lambda, \lambda+d\lambda)$, measured in Watts. In photometry this is related to subjective visual perception by the relative sensitivity function $K(\lambda)$ of the human eye. The *luminous flux*, measured in *Lumens* (lm), is defined by

$$F = \frac{\int_0^\infty K(\lambda)F'(\lambda)d\lambda}{\int_0^\infty K(\lambda)d\lambda} = K_\lambda \int_0^\infty K(\lambda)F'(\lambda)d\lambda \quad . \tag{1.7.1}$$

It gives the perceptual brightness of a light source perceived by an average observer. The weighting function $K(\lambda)$ is a C.I.E. (Commission Internationale de l'Eclairage) standard. It is shown in Fig. 1.6 and denoted there by $\overline{y}(\lambda)$. The constant is $K_\lambda = 683\,\text{lm/W}$.

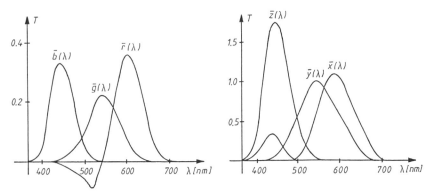

Fig. 1.6. The color matching functions of the C.I.E. standard

The luminous flux per 1 Steradian, the unit of solid angle Θ, is the *luminous intensity*

$$I = \frac{F}{\Theta} \quad \text{or} \quad I = \frac{\partial F}{\partial \Theta} \quad . \tag{1.7.2}$$

Its unit is 1 *Candela* (cd), so that 1 Lumen = 1 Candela × 1 Steradian. The candela is a basic unit of physics (like the meter or the second). The unit of solid angle is defined as the solid angle of a cone cutting an area of 1 m² out of a unit sphere having its center at the apex of the cone.

The luminous intensity per projected area of the sender and solid angle is the *luminance*

$$B = \frac{F}{A\Theta \cos \varphi} \quad \text{or} \quad B = \frac{\partial F}{\partial A \partial \Theta \cos \varphi} \quad . \tag{1.7.3}$$

The unit of luminance is $1\,Stilb = 10^4$ Candela/m². Luminous flux, luminous intensity, and luminance are properties of the sender.

For the receiver the *illumination* is the relevant quantity. It is defined by

$$E = \frac{F}{S} \quad \text{or} \quad E = \frac{\partial F}{\partial S} \quad , \tag{1.7.4}$$

where S is receiver surface. The unit is 1 *Lux* = 1 Candela × 1 Steradian/1 m².

Assume an object having surface S and being illuminated by a light source emitting illuminance $E(\lambda)$, and an observer seeing luminance $B(\lambda)$. If the surface orientation is assumed to have no influence on reflectivity, the *reflectivity* of the surface is defined by

$$\varrho(\lambda) = \frac{B(\lambda)}{E(\lambda)} \quad . \tag{1.7.5}$$

In pattern analysis one usually is interested in ϱ since it is an invariant property of the object, whereas B and E may be highly variable. The problem is to recover ϱ from B when E and S are unknown. Apparently this is only possible with simplifying assumptions. This gives a simple distinction between image processing and computer vision: The former deals with E (or a quantity related to it), the latter tries to recover ϱ and S.

1.7.2 Colorimetry

A radiation $F'(\lambda)$ having nonzero values in the range of visible light, that is approximately $380\,\text{nm} \leq \lambda \leq 780\,\text{nm}$, is perceived as *color* or *colored light*. It has been verified experimentally that most colors may be represented by a linear superposition of three suitably chosen *primary colors* or *primaries*. In general, different systems of primary colors can be used and are in practical use. Some of them are mentioned below. The only restriction is that the primary colors are independent. This means that no primary color can be represented by a linear superposition of the remaining primaries. Usually colors in the range of red, green, and blue are chosen as primary colors which are denoted here by $C_r(\lambda), C_g(\lambda)$, and $C_b(\lambda)$, respectively.

Representation of a color by linear superposition means that it is possible to match a certain $F(\lambda)$ by a weighted sum of primary colors such that they are perceived as identical colors. Depending on the primaries used there may be some colors requiring a negative proportion of one or two primaries for matching. A match in such cases means that the sum of the positive primaries is perceptually equal to the sum of the color and the negative primary (or primaries). A color match is performed by a perception experiment and is the basis for the definition of the *tristimulus values*. First, select three primary colors as above, define a reference white light $w(\lambda)$, and determine weights w_r, w_g, and w_b such that

$$w(\lambda) =\text{p}\ w_r C_r(\lambda) + w_g C_g(\lambda) + w_b C_b(\lambda) \quad, \tag{1.7.6}$$

where "=p" stands for "perceptually equivalent". Now choose an arbitrary color denoted by $C(\lambda)$ and determine weights a_r, a_g, a_b such that

$$C(\lambda) =\text{p}\ a_r C_r(\lambda) + a_g C_g(\lambda) + a_b C_b(\lambda) \quad. \tag{1.7.7}$$

The tristimulus values are then defined by

$$T_r = \frac{a_r}{w_r}\ , \quad T_g = \frac{a_g}{w_g}\ , \quad T_b = \frac{a_b}{w_b} \quad. \tag{1.7.8}$$

Defining normalized primaries

$$P_j = w_j C_j\ , \quad j \in \{r,g,b\} \tag{1.7.9}$$

gives with (1.7.7)

$$C(\lambda) =\text{p}\ T_r P_r + T_g P_g + T_b P_b \quad. \tag{1.7.10}$$

One may now measure the tristimulus values of the primaries which are necessary to perceptually match a certain monochromatic (or single wavelength) color C_φ having unit energy of 1 Watt or energy distribution $\delta(\lambda - \varphi)$ by the steps outlined above. Doing this for every value of φ in the range of visible light gives the tristimulus values $T_{\mathrm{p}j}(\varphi)$ of the primary colors for every spectral color C_φ. These special tristimulus values are called *color matching functions*. They have to be measured and averaged for several observers in order to obtain the color matching functions of a *standard observer*.

If the tristimulus values of the primary colors are known, it is possible to compute the tristimulus values of an arbitrary color having spectral distribution $C(\lambda)$. Using some weighting functions $k(\lambda)$ defined similar to (1.7.1), the energy of some color $C(\lambda)$ is

$$E_k(C) = \int k(\lambda)C(\lambda)d\lambda = \sum \int k(\lambda)w_j T_j C_j(\lambda) d\lambda \quad . \tag{1.7.11}$$

In the above equation and those to follow the integral extends from 0 to ∞ and the sum is over the three values of j denoting the primary colors, that is $j \in \{r, g, b\}$. From the definition of the tristimulus values of the primary colors we get for a unit energy spectral (or monochromatic) color C_φ

$$C_\varphi = \sum w_j T_{\mathrm{p}j}(\varphi) C_j(\lambda) \quad . \tag{1.7.12}$$

This spectral color has the energy E_k given by

$$E_k(C_\varphi) = \int k(\lambda)C_\varphi d\lambda = \sum \int k(\lambda) w_j T_{\mathrm{p}j}(\varphi) C_j(\lambda) d\lambda \quad . \tag{1.7.13}$$

If we consider a fixed wavelength $\lambda = \varphi$ of the color $C(\lambda)$, we get in the interval $(\varphi, \varphi + d\varphi)$ the contribution

$$dE_k(C(\varphi)) = \sum \int k(\lambda) w_j T_{\mathrm{p}j}(\varphi) C_j(\lambda) d\lambda\, d\varphi \quad . \tag{1.7.14}$$

Integrating with respect to φ gives a second equation for E_k,

$$E_k(C(\lambda)) = \sum \int \left[\int T_{\mathrm{P}j}(\varphi) C(\varphi) d\varphi \right] k(\lambda) w_j C_j(\lambda) d\lambda \quad . \tag{1.7.15}$$

Since the expressions for E_k in (1.7.11) and (1.7.15) must be equal, we get

$$T_j = \int T_{\mathrm{p}j}(\varphi) C(\varphi) d\varphi \quad . \tag{1.7.16}$$

If in a color camera three color channels are provided with filters $T_{\mathrm{p}j}$, the tristimulus values of a color are directly recorded.

It follows from the above discussion that two colors $C_1(\lambda)$ and $C_2(\lambda)$ are perceived as identical if their tristimulus values are identical, that is if

$$\begin{aligned} T_{1j} &= \int T_{\mathrm{p}j}(\lambda) C_1(\lambda) d\lambda = T_{2j} \\ &= \int T_{\mathrm{p}j}(\lambda) C_2(\lambda) d\lambda\,, \quad j \in \{r, g, b\} \quad . \end{aligned} \tag{1.7.17}$$

Two colors meeting the criterion (1.7.17) are called *metameric colors*.

In 1931 the C.I.E. adopted as a standard the three primary colors red at $\lambda = 700$ nm, green at $\lambda = 546.1$ nm, and blue at $\lambda = 435.8$ nm. These primaries are *monochromatic colors* or *spectral colors*. The reference white mentioned above in (1.7.6) is chosen in the C.I.E. standards as having constant energy in the visible interval. Tristimulus curves (or color matching functions) normalized by this standard white and measured for the C.I.E. standard primaries are usually denoted by $\bar{r}(\lambda), \bar{g}(\lambda), \bar{b}(\lambda)$; they are shown in Fig. 1.6. The tristimulus values resulting for a color $C(\lambda)$ are denoted by R, G, B. As can be seen from (1.7.8), an uncolored or white light has tristimulus values $R = G = B$ and reference white has $R = G = B = 1$. In TV standards other types of standard white have been defined since a constant energy white is hard to realize. Denoting the C.I.E. primaries by $\boldsymbol{R}, \boldsymbol{G}, \boldsymbol{B}$ yields from (1.7.10)

$$C(\lambda) = R\boldsymbol{R} + G\boldsymbol{G} + B\boldsymbol{B} \quad . \tag{1.7.18}$$

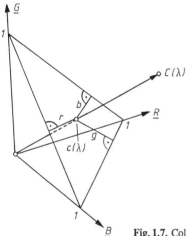

Fig. 1.7. Color space and chromaticity diagram

Thus a color may be represented as a vector in a space spanned by $\boldsymbol{R}, \boldsymbol{G}$, and $\boldsymbol{B}$ which usually are chosen as orthogonal axes. This is shown in Fig. 1.7. It is convenient to depict color independent of intensity in a two-dimensional space. This can be done by using the point $c(\lambda)$ where the vector $C(\lambda)$ intersects the plane $R + G + B = 1$, as a measure of color. The special coordinates r, g, b of $c(\lambda)$ indicated in Fig. 1.7 are called the *chromaticity coordinates* and are given by

$$r = \frac{R}{L}, \quad g = \frac{G}{L}, \quad b = \frac{B}{L}, \quad L = R + G + B \quad . \tag{1.7.19}$$

Since $r + g + b = 1$, two coordinates are sufficient to represent chromaticities, e.g. r and g. These two coordinates are conveniently drawn in an orthogonal (r, g)-*chromaticity diagram*.

It was mentioned above that different systems of primaries may be used. The tristimulus values R, G, B measured for some color with respect to primaries R, G, B are related to the tristimulus values R', G', B' of the same color with respect to some other primaries R', G', B' by a linear transformation

$$R' = a_{11}R + a_{12}G + a_{13}B ,$$
$$G' = a_{21}R + a_{22}G + a_{23}B , \quad (1.7.20)$$
$$B' = a_{31}R + a_{32}G + a_{33}B ,$$

where the a_{ij} are suitable transformation coefficients. From (1.7.19, 20) it follows that chromaticities are related by the projective transformation

$$r' = \frac{a_{11}r + a_{12}g + a_{13}b}{D} ,$$

$$D = (a_{11} + a_{21} + a_{31})r + (a_{12} + a_{22} + a_{32})g + (a_{13} + a_{23} + a_{33})b . \quad (1.7.21)$$

Similar relations hold for g' and b'. Finally, coefficients a_{ij} are also used to transform the tristimulus curves $\bar{r}(\lambda), \bar{g}(\lambda), \bar{b}(\lambda)$ to the new curves $\bar{r}'(\lambda)$ and so on.

A problem with the C.I.E. primaries is that the tristimulus curves $\bar{r}, \bar{g}$ and $\bar{b}$ also have negative values. Therefore, a virtual or normalized color system X, Y, Z has been defined having tristimulus curves $\bar{x}(\lambda), \bar{y}(\lambda), \bar{z}(\lambda)$ with only positive values as shown in Fig. 1.6. The linear transformation between the two systems is

$$X = 0.490R + 0.310G + 0.200B$$
$$Y = 0.177R + 0.813G + 0.011B \quad (1.7.22)$$
$$Z = 0.000R + 0.010G + 0.990B .$$

In addition the color matching function $\bar{y}$ has the desirable property that $\bar{y}(\lambda) = K(\lambda)$, where $K(\lambda)$ is the C.I.E. standard introduced in (1.7.1).

The C.I.E. primaries are spectral colors. Commercial color displays have three phosphors, red, green, and blue. However, they do not emit monochromatic lines but have a certain continuous distribution of energy. Thus it is not useful to display some color C, represented by C.I.E. tristimulus values, on such a display. Rather one should represent the color by tristimulus values R', G', B' obtained for the display's red, green, and blue as primaries. For these purposes two slightly different standards were defined – the 625 line EBU standard and the 525 line FCC standard – allowing the representation of a subset of the C.I.E. colors. The linear transformation of the FCC standard is

$$R' = 0.842R + 0.156G + 0.091B$$
$$G' = -0.129R + 1.320G - 0.203B \quad (1.7.23)$$
$$B' = 0.008R - 0.069G + 0.897B ,$$

$$X = 0.607R' + 0.174G' + 0.201B'$$
$$Y = 0.299R' + 0.586G' + 0.115B' \qquad (1.7.24)$$
$$Z = 0.000R' + 0.066G' + 1.117B'.$$

The R', G', B' values may be obtained directly from the output of a color TV camera if appropriate filters are used. In the later chapters we will denote pixels of an image f in the red, green, and blue color channel by f_r, f_g, and f_b, respectively. For the NTSC color TV standard the (Y, I, Q)-system was introduced. It contains the *luminance* signal Y and the so called *chrominance* signals I and Q. The transformation from R', G', B' to Y, I, Q is

$$Y = 0.299R' + 0.587G' + 0.114B'$$
$$I = 0.596R' - 0.274G' - 0.322B' \qquad (1.7.25)$$
$$Q = 0.211R' - 0.523G' + 0.312B'.$$

It should be noted that in TV a so-called γ-correction is usually performed to account for nonlinearities of the color tube; we assumed $\gamma = 1$ above. Otherwise one would have to replace R' by R'^{γ} and so on. Some other transformations are introduced in Sect. 2.2.5.

1.8 Sound

Another important type of pattern results from auditory stimuli. Therefore, a short account of some basic principles is given here. Acoustic waves in the frequency range of about 20 to 20 000 Hz are perceived as sounds. Similar to the electromagnetic waves above, acoustic waves are characterized by the physical quantities of frequency and energy. For human perception it is important to know how the physical quantities are related to subjective sensory impressions.

A subjective measure of *pitch* may be derived as follows. A person is given a reference tone of frequency ν_1 and a test tone of frequency ν_2. The person has to judge whether the test tone sounds lower- or higher-pitched than half the reference tone. The result obtained from experiments with different persons is shown in Fig. 1.8.

A second measure may be derived from frequency groups at the threshold of audibility. A narrowband noise having adjustable bandwidth or also several closely spaced sinusoidal tones are presented to a test person and the sound level of the threshold of audibility is determined. The experiments show that up to a certain number of sinusoidal tones (or also up to a certain noise bandwidth) the number of tones may be doubled and the amplitude may be halved, that is, the ear integrates over a certain frequency interval. The interval of integration is called a *frequency group*. Outside the interval it is not possible to reduce the amplitude by adding more tones. The frequency group depends on the center frequency of the tones. If frequency groups are arranged *non-overlapping* in the

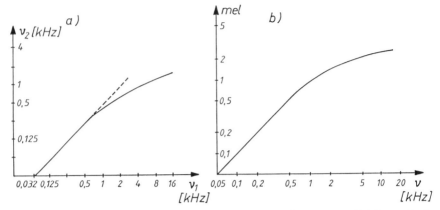

Fig. 1.8. (a) The result of subjective tests to find the pitch ν_2 of a test tone which is perceived to have half the pitch of a reference tone ν_1; for example, a test tone of about 0.75 kHz is perceived to have half the pitch of a reference tone of 2 kHz. (b) Defining a tone of 131 Hz to have subjective pitch of 131 mel (melodic pitch) gives from (a) the relation shown here; for example, a tone of 2 kHz has melodic pitch of 1500 mel

interval 20–16 000 Hz, there are 24 frequency groups. A person can, of course, distinguish more than 24 tones since frequency groups may be formed at any part of the audible frequency range. This is often used in speech recognition to divide the frequency interval into 24 groups.

The energy of a sound is measured by the relative quantity

$$L = 20 \log \left(\frac{p_s}{p_a} \right) \quad , \tag{1.8.1}$$

where p_s is the pressure of the sound to be measured and p_a is the pressure of a 1 kHz tone at the threshold of audibility. Sometimes the sound energy is called the *loudness*. A subjective measure of loudness is defined as follows. The 1 kHz tone is defined to have the loudness L given in (1.8.1). For another tone or sound a test person has to determine the sound energy corresponding to a subjective impression of equal loudness. Figure 1.9 shows the lines of constant loudness for a standard observer.

1.9 Bibliographical Remarks

There are numerous books on various aspects of pattern recognition. The classification problem using the feature vector approach is treated, for example, in [1.3–7], speech processing and understanding is covered in [1.8–11], and image processing and understanding in [1.12–19]. General techniques for knowledge representation as developed in artificial intelligence are treated in [1.20–25]. Examples of knowledge based systems for speech or image understanding are

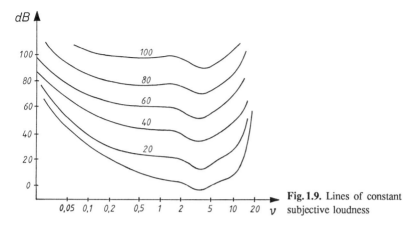

Fig. 1.9. Lines of constant subjective loudness

described in [1.13, 1.26–35]. A comprehensive treatment of color standards can be found in [1.36]. Properties of the auditory system are discussed in [1.37].

2. Preprocessing

In *preprocessing* a pattern is transformed to some other pattern which is expected to be more suited for further processing and which should yield improved results of classification or analysis. It is, in general, very difficult to judge the success of preprocessing by objective criteria. An obvious but laborious method would be to evaluate the performance of analysis with and without a particular preprocessing algorithm. Although this method can clearly identify the contribution of preprocessing to the success of the analysis, it is usually not adopted because of the cost involved. Rather, preprocessing is judged on a subjective and intuitive basis. This may be done by looking at a preprocessed image or listening to a preprocessed sound and judging subjectively whether its quality has improved. Furthermore, it is often possible to use intuitive reasoning. For instance, if a pattern is corrupted by noise it seems resonable to try to reduce noise; or if patterns have different energy it may be reasonable to normalize energy. It is thus possible to select methods for preprocessing without implementing a complete analysis system. However, it is not known whether the effort of preprocessing is balanced by subsequent gains which may consist of less effort in further processing or more successful analysis. Experience gathered so far indicates that careful preprocessing is very valuable in many task domains.

We consider the following methods of preprocessing:

1. Coding – methods to efficiently store and represent data (patterns).
2. Normalization – methods to adjust some parameters of the patterns to prespecified values.
3. Linear filtering – methods to operate on spectral properties of patterns.
4. Rank order and morphological operations – nonlinear operations using a local neighborhood.
5. Linear prediction – modeling the characteristics of speech or other waveforms.

Our discussion will be brief, leaving many details, for example, proofs of theorems, to the references.

2.1 Coding

2.1.1 Basic Approach

The discussion of this section is limited to some general results concerning the transition from the pattern $f(x)$ in (1.2.3), which is continuous with respect to the values of f (the amplitudes) and with respect to x, to a digital version which is discrete with respect to amplitude and independent variable. This is one of the goals of *coding*; other goals not treated here may be, for example, the detection or correction of errors in a coded pattern or the minimization of the number of bits required in a digital version of a pattern. The success of coding may always be judged independently of further processing by evaluation of the number of bits necessary to achieve a specified quality of pattern representation and by evaluation of the effort for coding and decoding. A general block diagram for pattern coding is given in Fig. 2.1.

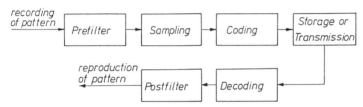

Fig. 2.1. Block diagram for recording and reproduction of patterns

The prefilter is to assure band limitation of the pattern. Sampling is assumed here to be equally spaced and in orthogonal directions in the case of images or three-dimensional scenes. Storage and transmission may involve modulation, error correction and error detection coding, and their inverse operations. This is not considered because it is of little interest to pattern analysis. It can be shown that the pre- and postfilters may be chosen as ideal low-pass filters with cutoff frequencies chosen to avoid aliasing effects. These filters will minimize mean-square error between recorded and reconstructed patterns for a fixed number of samples per pattern.

2.1.2 Pulse Code Modulation

A basic method to obtain a digital version of a pattern $f(x)$ is *pulse code modulation* (PCM), which is illustrated for a pattern $f(t)$ of one variable. The first step is quantization of the independent variable t to a finite number of discrete *sample values* (or *pixels* in the case that f is an image). This yields

$$f(t) \to f_i = f(t_0 + i\Delta t) = f(i) \quad \text{if} \quad t_0 = 0 \quad \text{and} \quad \Delta t = 1; \; i = 0, \ldots, M-1$$

$$f(x, y) \to f_{ij} = f(x_0 + i\Delta x, y_0 + j\Delta y) = f(i, j)$$

$$\text{if} \quad x_0 = y_0 = 0 \quad \text{and} \quad \Delta x = \Delta y = 1 \quad . \tag{2.1.1}$$

Since the amplitudes taken at discrete sample points in general take continuous values f_i, the second step is to quantize them to discrete amplitude values f'_i represented by $L = 2^B$ bits per sample point for digital processing, storage, and transmission. This idea, which works for patterns $\boldsymbol{f}(\boldsymbol{x})$ with any number m of functions and any number n of variables, is shown in Fig. 2.2. It can be seen that the continuous range $f_{\min} \le f_i \le f_{\max}$ is mapped to $L = 2^B$ integers $0, 1, \ldots, 2^B - 1$; these integers may conveniently be stored with B bits. To apply PCM, three choices have to be made:

1. The *sampling rate* (or number M of sample points per coordinate).
2. The *number B* of bits per sample point.
3. The *quantization characteristic* which maps a continuous amplitude interval to a discrete amplitude value.

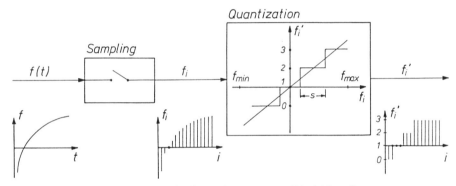

Fig. 2.2. Pulse code modulation (PCM) of a continuous pattern $f(t)$ yielding f'_i

The sampling rate is determined by the *sampling theorem*:

Theorem 2.1.1. Let $f(t)$ be a function with *Fourier transform*

$$F(\omega) = \int_{-\infty}^{\infty} f(t) \exp(-i 2\pi \omega t) dt \quad , \tag{2.1.2}$$

such that

$$F(\omega) = 0 \quad \text{if} \quad |\omega| \ge 2 f_g \quad \text{(band-limitation)} \quad . \tag{2.1.3}$$

In this case $f(t)$ is completely determined by the sample values

$$f_i = f(t_0 + i \Delta t) \, , \quad i = 0 \, , \, \pm 1, \pm 2, \ldots$$

defined in (2.1.1) if the *sampling period* is $\Delta t \le 1/(2 f_g)$. The continuous function $f(t)$ may be reconstructed by the *interpolation formula*

$$f(t) = \sum_{i=-\infty}^{\infty} \frac{f_i \sin\left[2\pi f_g(t - i\Delta t)\right]}{2\pi f_g(t - i\Delta t)} \quad . \tag{2.1.4}$$

This theorem states that a band-limited function may be represented by discrete sample values. It has an obvious generalization to n variables.

Proof. See, for example, [2.1], pp. 206–210, or [2.2], pp. 65–76. □

If the sampling period is fixed at $\Delta t = 1/(2f_g)$, the bit rate will be $2f_g B$ bits/s or a total of $M \times B$ bits. Reduction of bit rate is one object of different coding methods. Since the sampling rate is determined by the sampling theorem, it is usually tried to reduce B. It is obvious from Fig. 2.2 that B determines the accuracy with which f'_i approximates f_i. In the following, two results are presented which give some insight into the dependence of accuracy upon the number B and the sampling characteristic.

The accuracy of a PCM representation f'_i of a sample point f may be based on *quantization noise*

$$n_i = f_i - f'_i \quad . \tag{2.1.5}$$

A measure of the accuracy of a PCM representation is

$$r' = \frac{E\{f_i^2\}}{E\{n_i^2\}} \quad , \tag{2.1.6}$$

which is the ratio of signal power to quantization noise power. In (2.1.6) the expected value is denoted by $E\{.\}$. A relation between B and the variable

$$r = 10 \log_{10} r' \tag{2.1.7}$$

is given by the following theorem.

Theorem 2.1.2. If $E\{n\} = E\{f\} = 0$, if the quantization is fairly fine (e.g. $B > 6$), and if the quantizer is not saturated, then the relation

$$r = 6B - 7.2 \tag{2.1.8}$$

holds.

Proof. See, for example, [2.3] pp. 611–632. □

From (2.1.8) it is evident that for speech recognition $B = 11$ bits is sufficient for about 60 dB *signal-to-noise ratio* and that a saving of 1 bit means a loss of 6 dB. According to experience about $B = 8$ bits per spectral channel are sufficient to represent images with satisfactory (subjective) quality. In speech recognition between 8 and 12 bits are used.

In Fig. 2.2 *linear quantization* was assumed. It may be desirable to employ a quantization scheme which is optimal in some sense. Using the notation of Fig. 2.3 the mean-square quantization error

30   2. Preprocessing

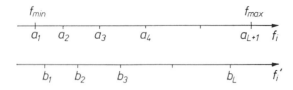

Fig. 2.3. The range (f_{min}, f_{max}) of sample values f_i is divided into $L = 2^B$ intervals $(a_\nu, a_{\nu+1})$, by input threshold levels $a_\nu, \nu = 1, \ldots, L$. The intervals are represented by L discrete values b_ν which are the possible output levels used to encode f_i

$$\varepsilon = \sum_{\nu=1}^{L} \int_{a_\nu}^{a_{\nu+1}} (f - b_\nu)^2 p(f) df \tag{2.1.9}$$

is defined as a criterion for measuring the quality of the quantization characteristic. An optimal characteristic uses values a_ν, b_ν yielding minimal error ε.

Theorem 2.1.3. The optimal values of a_ν, b_ν are given by

$$a_\nu = (b_{\nu-1} + b_\nu)/2, \quad \nu = 2, 3, \ldots, L = 2^B \tag{2.1.10}$$

if $p(f = a_\nu) \neq 0$ and

$$b_\nu = \frac{\int_{a_\nu}^{a_{\nu+1}} f p(f) df}{\int_{a_\nu}^{a_{\nu+1}} p(f) df}, \quad \nu = 1, \ldots, L \quad . \tag{2.1.11}$$

Proof. Taking partial derivatives of ε with respect to a_ν, b_ν, setting them equal to zero and solving for a_ν, b_ν gives (2.1.10), (2.1.11). For details see [2.4]. □

The above theorem shows that *uniform* or *linear quantization*, as used in Fig. 2.2, is optimal only for a uniform distribution of sample values in (f_{min}, f_{max}). Otherwise, nonuniform quantization is advantageous. This can be done by companding the signal with a nonlinear device and applying it to a linear quantizer. When reconstructing the coded samples they have to be expanded before reconstruction. Logarithmic devices are frequently used in speech and image processing. Usually, a PCM version of a pattern is the starting point for further processing since it is obtained conveniently by analog-to-digital converters combined with companding amplifiers and sampling units. In the following it is assumed that $f_i = f_i'$ within the specified accuracy so that no distinction is made between them, if not otherwise necessary.

2.1.3 Improvements of PCM

There are various improvements of PCM which allow a reduction of the bit rate without loss of accuracy. Among these are *adaptive quantization, differential pulse code modulation* (DPCM), *delta modulation* (DM), and a combination of adaptive quantization with DPCM and DM. Adaptive quantization makes the step size s dependent on previous samples. DPCM makes use of the correlation between successive samples f_i, f_{i+1}. If the correlation is greater than 0.5, it is useful to code the difference between f_{i+1} and f_i instead of f_{i+1}. By oversampling a pattern, i.e., using $\Delta t < 1/(2f_g)$, it is possible to increase the correlation. This is used in DM where in its basic form a staircase approximation of the pattern is obtained. Speech coding with a logarithmic quantizer requires $B = 7$ bits to obtain 60 dB signal-to-noise ratio and with adequate oversampling $B = 1$ may used in DM.

2.1.4 Transform Coding

The idea of *transform coding* is not to store f, but a *linear transform*

$$c = \Phi f_s \quad , \tag{2.1.12}$$

where Φ is a suitable transformation matrix. The first step is to choose such a matrix and compute c, the second is to quantize components c_ν of the vector c. If f is a pattern vector or matrix with very few elements (say less than 100) it is possible to transform the whole pattern. In this case $f_s = f$ in (2.1.12); otherwise f_s is a subpattern of f. This is particularly useful in image coding since images may be sampled by $M^2 = 512^2$ or more points.

Since in any case c is a linear combination of elements of f_s, it is possible to treat f_s as a column vector, regardless of whether it is a one- or two-dimensional subpattern of a one- or two-dimensional pattern. Elements of a pattern matrix may be rearranged to elements of a vector. A useful transformation matrix in (2.1.12) is one which yields uncorrelated components c_ν of c. In this case good compression of information is achieved. The transformation matrix is given by the following result.

Theorem 2.1.4. If the rows of the matrix Φ in (2.1.12) are the eigenvectors φ_ν, defined by

$$\lambda_\nu \varphi_\nu = K_s \varphi_\nu \tag{2.1.13}$$

of the *covariance matrix*

$$K_s = E\{(f_s - E\{f_s\})(f_s - E\{f_s\})_t\} \quad , \tag{2.1.14}$$

then the components c_ν of c in (2.1.12) are uncorrelated. (The subscript t denotes transposition of a vector.)

Proof. See, for example, [2.1] pp. 383 for the continuous case. □

The transformation resulting from Theorem 2.1.4 is known as the discrete *Karhunen-Loeve transformation*, eigenvector transformation, Hotelling transformation, or *principal axis transformation*. If f is an image with M^2 sample points and with no division into subpatterns, K_s would be a matrix with $M^2 \times M^2$ elements. So even for moderate M it is necessary to use subpatterns in order to reduce the size of K_s. The inverse transformation of (2.1.12) is

$$f_s = \Phi_t c \quad . \tag{2.1.15}$$

Let the subpattern be arranged into a vector f_s with M_s components. It is often unnecessary to use all M_s eigenvectors $\varphi_\nu, \nu = 1, \ldots, M_s$ in (2.1.13). If only the $n_s < M_s$ eigenvectors φ_ν belonging to the largest eigenvalues λ_ν are used as rows of Φ, reconstruction of f_s by (2.1.15) will be imperfect but with minimum mean-square error. Since a rearrangement of rows and columns of K_s leaves the eigenvalues λ_ν unchanged, it is evident that a two-dimensional subpattern may be arranged to a vector in an arbitrary, but fixed, manner. A rowwise, columnwise, or any other arrangement of the subpattern to a vector will result only in a different arrangement of rows and columns of the matrix K_s. Efficient algorithms to compute eigenvalues and eigenvectors are available. Of course, other transformation matrices may be used in (2.1.12), for example, the discrete Fourier or Hadamard transformation.

Having transformed a number of subpatterns which cover the whole pattern the next step is to quantize the coefficients c_ν of the subpatterns. It is useful to normalize c_ν by its variance and to quantize the normalized coefficients. This may be done by the methods of Sect. 2.1.2.

The best transformation in the sense of minimum mean-square error and also of subjective quality is the discrete Karhunen-Loeve transformation, but the Fourier and Hadamard transformations perform nearly as well. In order to obtain good reproductions of patterns about 7 bits per coefficient c_ν are necessary and about $n_s = 0.5 M_s$ coefficients should be stored for a subpattern with M_s sample values. When images are coded the subpatterns should be square arrays with $M_s \geq 4 \times 4 = 16$ sample values. For one-dimensional patterns one should use $M_s = 16$. It is possible to use the same matrix Φ in (2.1.12) to code different subpatterns in many cases. This reduces the effort in computation of transformation matrices and in bookkeeping. The number n_s of coefficients may be chosen according to local structure, for instance, by taking the smallest value of n_s such that

$$\frac{\sum_{\nu=1}^{n_s} |c_\nu|^2}{\sum_{\nu=1}^{M_s} |c_\nu|^2} > 0.99 \quad . \tag{2.1.16}$$

The above remarks apply to "average images". The numbers given may vary according to particular pattern properties.

2.1.5 Line Patterns

An important class of patterns are images containing only black (or white) lines on white (or black) background. These line patterns may be diagrams or contours obtained from gray-level images. Each sample value may be coded by 1 bit. Since there is considerable correlation in line patterns and often the number of pixels on a line is much less than the number of background pixels, a more efficient code than representation of M^2 sample values is possible.

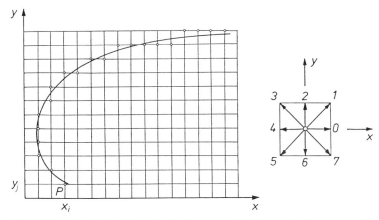

Fig. 2.4. Example of the chain code of a line pattern. In this case the line starts at $P(x_i, y_j)$, and the code is 332221210101000001000

One possibility is *chain code*. The line pattern is superimposed by a grid which is fine enough to yield the desired accuracy. Coding of a line starts at a point P whose coordinates are stored. From P the next point on the line is searched for and the direction from P to this point is recorded. The direction is one of eight possible directions as shown in Fig. 2.4. In this way the whole line pattern is coded by storing the direction to the next point on the line, starting at P, as shown in Fig. 2.4. If a point on a line does not lie on a grid point the nearest grid point is taken. For some regular patterns it is possible to obtain the chain code from the generation function and to perform operations, like enlargement by a factor n, directly on the chain code. It is also possible to compute some parameters of a pattern, like length of the line or area between the line and the x-axis from the chain code. If there are only a few lines, considerably less than M^2 bits are necessary to code them. Some additional code words are necessary to encode special conditions like crossing or branching of lines. A further reduction is achieved by using four instead of eight possible directions.

Other possibilities are *run-length coding* and the use of grammars. In run-length coding, which may also be used to code gray-level images, the image is scanned line by line and for each line the number of zeros and ones encountered during scanning is stored. To encode a gray-level image a line is coded by a set of pairs (b_i, l_i), where b_i is the gray value and l_i the number of successive pixels having value b_i. Apparently, if the number of possible gray values increases, the efficiency of run-length coding decreases. The use of grammars is possible if a grammar generating the line pattern is known. The code is obtained by the rule that in generating the line the right-most symbol is rewritten and the number of the alternative used for rewriting is stored in the code.

2.2 Normalization

2.2.1 Aim and Purpose

Often a pattern contains parameters which may vary within a certain range, but the result of analysis (or classification) should be independent of these variations. Rather than trying to develop an analysis algorithm which tolerates these variations, it seems reasonable to try to eliminate them by proper *normalization* prior to analysis. Thus it is expected that analysis will become easier if parameter variations are eliminated in advance.

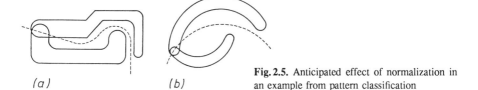

Fig. 2.5. Anticipated effect of normalization in an example from pattern classification

(a) (b)

Some examples will illustrate the possible variations which are to be eliminated. The meaning of a letter is independent of its size; equivalently the meaning of a spoken word is independent of its duration. Therefore, size or time should be normalized in advance. The result of analyzing an image should be independent of illumination or geometric distortions. Pitch or energy of speech should have no effect on a recognition procedure. Of course, one may think of a "robust" analysis (or classification) method which is independent of parameter variations. The anticipated advantage of normalization is shown in Fig. 2.5 with an example from classification. Let there be two classes and two features. Without normalization, the conditional densities may be nonzero in the indicated areas of Fig. 2.5a, whereas with normalization the much simpler areas of Fig. 2.5b result. To obtain minimum error rate a much more complicated decision surface is necessary in (a). On the other hand, with given effort – for example, a linear or quadratic decision function – a lower error rate is achieved in (b).

To summarize, the purpose of normalization is to achieve a certain performance with less effort. Experience in many areas of pattern recognition indicates that normalization is of extreme importance.

2.2.2 Size and Time

Often normalization of size or time of a pattern with n variables, $n = 1, 2, 3, \ldots$, which occupies an n-dimensional interval in R^n, may be done by linear mapping of this interval to a standard interval. In this way the duration of words ($n = 1$) or the size of characters ($n = 2$) has been normalized and several algorithms and hardware for linear mapping of digitized patterns have been developed. Assume that $f = [f_i], i = 0, 1, \ldots, M - 1$ is a pattern according to (2.1.1) having M

sample values in the interval $t_0 = 0 \leq t \leq M - 1 = t_1$. Linear normalization of the discrete $\boldsymbol{f}$ can be viewed in two (equivalent) ways:

1. Reconstruct the continuous function $f(t)$ from the discrete $\boldsymbol{f}$, map $f(t)$ to a standard interval $0 \leq t \leq t'_1$, and sample the normalized function at Δt with M' normalized values $h_k, k = 0, \ldots, M' - 1$. In actual normalization one avoids the reconstruction of $f(t)$ and only maps the sample points $i = 0, \ldots, M - 1$ to new values $x'(i)$.
2. Reconstruct the continuous $f(t)$, resample it at $\Delta t'$ to obtain M' sample points of the normalized function h_k.

The result of both approaches is that every pattern $\boldsymbol{f}^r$ is represented by the same number M' of sample points. This type of normalization may require a new prefiltering in order to obey the conditions of the sampling theorem. This follows from step 2 above because it requires a sampling with a new step size $\Delta t'$. If we assume the conditions of (2.1.1), the ith sample value is mapped to the new coordinate

$$x'(i) = \frac{i(M' - 1)}{(M - 1)} \quad . \tag{2.2.1}$$

Since x' in general will not be a point on the sampling grid of the normalized pattern, the amplitude at x' has to be determined by interpolation from the non-normalized pattern. In order to reduce computing time this is usually done by linear interpolation between two neighboring values and not by (2.1.4). Let k be a grid point of the normalized pattern h_k, let $x'(i-1)$ be the largest value according to (2.2.1) with $x'(i-1) \leq k$, and $x'(i)$ the smallest value with $k \leq x'(i)$, then linear interpolation between f_{i-1} and f_i yields

$$h_k = f_{i-1} + \frac{(f_i - f_{i-1})(k - x'(i-1))}{x'(i) - x'(i-1)} \quad . \tag{2.2.2}$$

This approach can be extended to two- or three-dimensional patterns. In this case (2.2.1) is evaluated per coordinate. However, interpolation of the amplitude at the normalized pattern becomes somewhat more complicated. For example, in the two-dimensional case a normalized value h_{kl} may be interpolated from the four surrounding values $f_{i-1,j-1}, f_{i,j-1}, f_{i-1,j}, f_{i,j}$ as indicated in Fig. 2.6. The simplest method is to give h_{kl} the value of the nearest point, which is $f_{i-1,j-1}$ in the case of Fig. 2.6. A better approach is to compute h_{kl} from *bilinear interpolation*. This is done by computing the two intermediate values p_{kj} and $p_{k,j-1}$ from linear interpolation between $f_{i-1,j}, f_{i,j}$ and $f_{i-1,j-1}, f_{i,j-1}$, respectively. The h_{kl} is computed from another linear interpolation between p_{kj} and $p_{k,j-1}$. It may be shown that the same result is obtained if the two intermediate values $q_{i-1,l}$ and $q_{i,l}$ to the left and right of h_{kl} are computed.

Linear normalization is a simple and in many cases an efficient preprocessing step. On the other hand, it became evident in word recognition that nonlinear mapping may be required. The reason is that, for example, the duration of a vowel may vary within larger limits than the duration of a consonant.

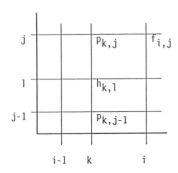

Fig. 2.6. Bilinear interpolation of the normalized function value h_{kl} from four surrounding values of f

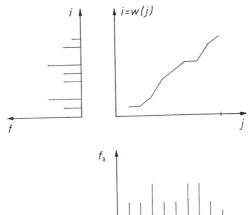

Fig. 2.7. Example of nonlinear mapping by the warping function $i = w(j)$

Determination of a *nonlinear mapping* is illustrated in Fig. 2.7 with a test pattern $\boldsymbol{f} = [f_i], i = 0, 1, \ldots, M - 1$ and a reference pattern $\boldsymbol{f}_\lambda = [f_{\lambda j}], j = 0, 1, \ldots, M_\lambda - 1$. The mapping is defined by the warping function

$$i = w(j)\, ;\ \ 0 = w(0),\ M - 1 = w(M_\lambda - 1)\ . \tag{2.2.3}$$

Assume that there are M_w tuples $(j, i = w(j))$ of the warping function which are ordered by an index l starting at the lower left of Fig. 2.7 and ending at the upper right. The warping function then is given by a sequence of indices $(j(l), i(l)), l = 1, 2, \ldots, M_w$ defining that the value $f_{i(l)}$ of the test pattern is to be compared with the value $f_{\lambda j(l)}$ of the reference pattern. If $d(f_{\lambda j}, f_i)$ is a measure of the distance between reference and test pattern, the total distance is

$$D_\lambda = \sum_{l=1}^{M_w} d\big(f_{\lambda j(l)}, f_{i(l)}\big) = \sum_j d\big(f_{\lambda j}, f_{i=w(j)}\big)\ . \tag{2.2.4}$$

The *optimal warping function* w^* is defined by

$$D_\lambda^* = \min_w D_\lambda\ . \tag{2.2.5}$$

Usually it is possible to restrict the number of warping functions by constraining the slope of w, for example, according to

$$w(j+1) - w(j) = \begin{cases} 0, 1, 2 & \text{if } w(j) \neq w(j-1) \\ 1, 2 & \text{if } w(j) = w(j-1) \end{cases} \quad (2.2.6)$$

Another possibility is to restrict the set of index tuples which may be predecessors of a tuple $(j(l), i(l))$, for example, according to

$$(j(l), i(l)) = (j(l-1), i(l-1)+1) \mid (j(l-1)+1, i(l-1)+1) \mid$$
$$(j(l-1)+1, i(l-1)) \quad . \quad (2.2.7)$$

The optimal warping function w^* can be computed by *dynamic programming* using the following algorithm.

FOR $j = 0, 1, \ldots, M_\lambda - 1$, that is for all columns in Fig. 2.7:
    Determine lower limit $L(j)$ and upper limit $U(j)$ of index i, for example, by using (2.2.6) or (2.2.7); if no constraints are available, use $L = 0$, $U = M - 1$
    FOR $i = L(j), \ldots, U(j)$, that is for all rows in column j:
        Compute $D^*(j, i) = d(f_{\lambda j}, f_i) + \{$ distance of that predecessor of (j, i) having minimal distance $D^* \}$; use $D^*(j, i) = 0$ for predecessors of $(0, 0)$.
        Store back pointer to predecessor with minimal D^*.
    END FOR
END FOR

The minimal distance is $D^*(M_\lambda - 1, M - 1)$, the optimal warping function can be traced from the back pointers.

If there are K reference patterns, K warping functions have to be computed. Therefore, nonlinear mapping becomes fairly complex since it determines the mapping with respect to a reference pattern and does not compute a single mapping to a reference interval as is the case with linear mapping. Actually, this method of nonlinear normalization is an example of a classification algorithm which tolerates parameter variations. Computing D_λ^* in (2.2.5) for K reference patterns $f_\lambda, \lambda = 1, \ldots, K$ allows selection of that reference pattern f_κ which is most similar to the test pattern f.

2.2.3 Intensity or Energy

Variations in mean intensity or energy of patterns are usually of no importance for classification and analysis. They result, for example, from changes of illumination or loudness. Multispectral images are treated in Sect. 2.2.6. Energy of speech is defined by

$$E_\nu = \sum_{n=0}^{m-1} w_n f_{\nu-n}^2 \quad \text{or} \quad E_\nu = \sum_{n=0}^{m-1} |w_n f_{\nu-n}| \quad . \quad (2.2.8)$$

In (2.2.8) w_n is a suitable weighting sequence or *window function*, for example,

$$w_n = 1, \; n = 0, 1, \ldots, m - 1, \qquad \text{– rectangular window,}$$
$$w_n = 0.54 - 0.46 \cos(2\pi n/(m-1)) \qquad \text{– Hamming window,} \qquad (2.2.9)$$
$$w_n = 0.5 - 0.5 \cos(2\pi n/(m-1)) \qquad \text{– Hanning window} \;\; .$$

A simple normalization results if the energy of the whole pattern is normalized to a standard value, e.g. $E_{n\nu}$. The summation in (2.2.8) then has to be taken over the full length of the pattern f_j. Normalization is achieved by replacing every value f_j by

$$h_j = f_j \sqrt{E_{n\nu}/E_\nu} \quad \text{or} \quad h_j = f_j E_{n\nu}/E_\nu \;\; . \qquad (2.2.10)$$

Usually one will not use the sequence f_j or $h_j, j = 0, 1, \ldots, M-1$, for further processing, but some parameters or features derived from those values. In this case it may be advantageous to normalize not f_j but the relevant parameters. The normalization according to (2.2.10) may not be adequate for speech processing. For example, values of f belonging to background noise during a speech pause should be omitted and very loud portions should not have a significant effect. This can be accomplished by normalizing energy to the interval $0 \le E_n \le E'$. The whole utterance is divided into frames of 20 ms or so duration and energy E_ν is computed per frame; so the summation in (2.2.8) now is taken over all values in a frame. A noise threshold $E_{\min}$ and a maximal energy $E_{\max}$ are defined and energy is normalized by

$$E_{n\nu} = \begin{cases} 0 & \text{if } E_\nu < E_{\min} \;, \\ E'(E_\nu - E_{\min})/(E_{\max} - E_{\min}) & \text{if } E_{\min} \le E_\nu \le E_{\max} \;, \\ E' & \text{if } E_\nu > E_{\max} \;. \end{cases} \qquad (2.2.11)$$

If E_ν is the energy of a frame according to (2.2.8), the normalized value h_j is computed by (2.2.10). A slightly different normalization procedure for speech is obtained by computing energy only from sample values which are larger than a noise (or pause) threshold $E_{\min}$. In an interval of length m the sample values f_j are thresholded by

$$f'_j = f_j \text{ if } f_j > E_{\min} \;, \text{ and } f'_j = 0 \text{ otherwise} \;\; . \qquad (2.2.12)$$

Using f'_j the energy E_U of the whole utterance and the energy E_I of an interval within the utterance are computed by (2.2.8). A suitable length of an interval is about 400 ms which is approximately one word. The normalized sequence h_j is computed from (2.2.10) by replacing $E_{n\nu}$ with E_U and E_ν with E_I. This normalizes the energy of every interval to the energy of the whole utterance.

Other possibilities for normalization are to reduce the pattern f to mean value 0 and variance 1 according to

$$h_j = (f_j - \mu)/\sigma \;,$$
$$\mu = \sum_j f_j/M \;, \quad \sigma = \left(\sum_j f_j^2/M\right) - \mu^2 \;, \qquad (2.2.13)$$

or to normalize the amplitudes of f to a standard interval, say $[0, a]$, by

$$h_j = \frac{a(f_j - f_{\min})}{(f_{\max} - f_{\min})},\qquad(2.2.14)$$
$$f_{\min} = \min\{f_j\},\quad f_{\max} = \max\{f_j\}.$$

2.2.4 Gray-Level Scaling

The gray level of an image f may deviate from the original object due to noise, scanner characteristic, or film exposure. Suitable alteration of gray levels may improve the objective and subjective quality of an image. Therefore, methods for normalization of gray levels are very important.

Invariance to linear scaling and translation of gray levels results from PCM (see Sect. 2.1.2 and Fig. 2.2) if the interval $(f_{\min}, f_{\max})$ is linearly mapped to L discrete levels. Gray level may be measured by film transmittance $\tau(x,y)$ or photographic density $D(x,y)$. If $I_i(x,y)$ is the light intensity incident at point (x,y) and $I_t(x,y)$ is the intensity transmitted at (x,y), the transmittance is defined by

$$\tau(x,y) = \text{local average}\left\{\frac{I_t(x,y)}{I_i(x,y)}\right\}\qquad(2.2.15)$$

and the density is

$$D(x,y) = \log[1/\tau(x,y)]\ .\qquad(2.2.16)$$

The local average in (2.2.15) should be evaluated over an area which is large with respect to film grain and small with respect to image detail. Usually this is accomplished by the scanning device since it measures intensity not at a point but in a small area. The logarithmic conversion of (2.2.16) was found to be appropriate, for instance, for radiographic images.

In any case, the discrete values f_{ij} of an image matrix f will occur with a certain relative frequency which is obtained from the *gray-level histogram*. Let l_i be the ith quantization level, $l_i = 0, 1, \ldots, L-1$. Then the relative frequency p_i of level l_i is

$$p_i = M(l_i)/M^2\ ,\qquad(2.2.17)$$

where $M(l_i)$ is the number of sample points f_{ij} with gray level l_i, and M^2 is the total number of sample points as in (2.1.1). A common technique is to change the gray-level histogram to obtain a normalized histogram p_{ni}. The special case of constant p_{ni}, that is uniform distribution, is well known from statistics. If ξ is a random variable with continuous distribution $P(x) = P(\xi \leq x)$, then

$$\eta = P(\xi)\qquad(2.2.18)$$

is a random variable with uniform distribution. A straightforward application of this result to the discrete case of (2.2.17) is shown in Fig. 2.8. It works well if $L_n \ll L$, but there are problems if $L_n \approx L$.

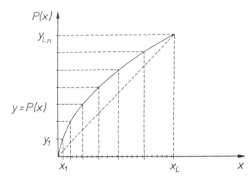

Fig. 2.8. An almost uniform distribution is obtained by dividing the y axis into L_n equal intervals and mapping them back to the x axis via the distribution $y = P(x)$. The number of intervals of the x axis mapped to one interval of the y axis thus depends on $P(x)$

2.2.5 Color

A short account of perceptually weighted measurement and matching of colors has been given in Sect. 1.7. Therefore, it is sufficient here to consider only particular transformations useful in image analysis. The chromaticity coordinates introduced in (1.7.19) may be considered as normalized colors since they are independent of the intensity; also the luminance, chrominance signals in (1.7.25) separate intensity and color information. Another approach to normalization is to represent colors in a space such that points of identical perceived color difference have the same distance in this space. Such a space, also called a *uniform color space*, can be obtained only approximately because the perceived difference of colors depends to a large degree on the viewing conditions. Such conditions can be controlled in precise experimental environments but not in actual situations of image processing.

In 1976 the C.I.E. defined an approximately uniform color space by the three coordinates

$$L^* = 116(Y/Y_n)^{1/3} - 16 ,$$
$$u^* = 13L^*(u' - u'_n) , \qquad (2.2.19)$$
$$v^* = 13L^*(v' - v'_n) ,$$

$$u' = 4X/(X + 15Y + 3Z) , \qquad v' = 9Y/(X + 15Y + 3Z) ,$$
$$u'_n = 4X_n/(X_n + 15Y_n + 3Z_n) , \qquad v'_n = 9Y_n/(X_n + 15Y_n + 3Z_n) . \qquad (2.2.20)$$

These equations are valid in many experimental conditions and if $Y/Y_n > 0.01$. If $Y/Y_n \leq 0.01$, L^* is obtained from $L^* = 903Y/Y_n$. The values X, Y, Z are those defined in (1.7.22), and X_n, Y_n, Z_n are the tristimulus values of the standard white object. The color difference between two colors is

$$D_{uv} = \sqrt{(\Delta L^*)^2 + (\Delta u^*)^2 + (\Delta v^*)^2} \quad . \tag{2.2.21}$$

Another approximately uniform color space is given by the coordinates (L^*, a^*, b^*), where L^* is as above and the other terms are

$$\begin{aligned} a^* &= 500 \left[(X/X_n)^{1/3} - (Y/Y_n)^{1/3} \right] , \\ b^* &= 200 \left[(Y/Y_n)^{1/3} - (Z/Z_n)^{1/3} \right] . \end{aligned} \tag{2.2.22}$$

The color difference D_{ab} is computed by an equation analogous to (2.2.21). Finally, color may be expressed in terms of *lightness, chroma, saturation* and *hue*. The term L^* is a measure of lightness. Chroma is measured by either of

$$C_{uv} = \sqrt{u^{*2} + v^{*2}} , \quad C_{ab} = \sqrt{a^{*2} + b^{*2}} . \tag{2.2.23}$$

A measure of *saturation* is

$$S = C_{uv}/L^* = 13\sqrt{(u' - u'_n)^2 + (v' - v'_n)^2} \quad . \tag{2.2.24}$$

A measure of the *hue* angle is either of

$$h_{uv} = \arctan(v^*/u^*) \quad , \tag{2.2.25}$$

$$h_{ab} = \arctan(b^*/a^*) \quad . \tag{2.2.26}$$

Hue is thus represented by an angle in degrees; the hue difference is

$$\Delta h_{uv} = \sqrt{D_{uv}^2 - (\Delta L^*)^2 - (\Delta C_{uv})^2} \quad , \tag{2.2.27}$$

and similar for Δh_{ab}. This allows one to describe a total color difference D in terms of difference in lightness, chroma, and hue.

A first prerequisite for the application of the above equations is to determine what is the output of a color scanner (primary colors, reference white, and so on as discussed in Sect. 1.7). From (1.7.5) it is seen that the observed luminance is the product of surface reflectivity and source illuminance. The statement of Helmholtz that "the observer tends to see the true color of an object independent of the illumination" cannot be realized by simple equations of the type given above. So far it seems that this capability of the human observer is still a research goal.

2.2.6 Multispectral Data

In multispectral scanner data the recorded intensity f_ν of the νth spectral channel is

$$f_\nu = \alpha_\nu \varrho_\nu I_\nu + \beta_\nu \quad , \tag{2.2.28}$$

where I_ν is the intensity of the source, ϱ_ν is the reflectivity of the object, α_ν is the atmospheric transmissivity, and β_ν is the background intensity. Neglecting β_ν one obtains for two adjacent channels ν and $\nu + 1$ the ratio

$$\frac{\varrho_\nu}{\varrho_{\nu+1}} = \frac{f_\nu}{f_{\nu+1}} \frac{\alpha_{\nu+1} I_{\nu+1}}{\alpha_\nu I_\nu} \quad . \tag{2.2.29}$$

If $\alpha_{\nu+1} I_{\nu+1} = \alpha_\nu I_\nu$, this reduces to

$$\frac{\varrho_\nu}{\varrho_{\nu+1}} = \frac{f_\nu}{f_{\nu+1}} \quad . \tag{2.2.30}$$

In this way an m-channel image

$$\boldsymbol{f}(x,y) = [f_1(x,y),\ f_2(x,y),\ \ldots,\ f_m(x,y)]_t \tag{2.2.31}$$

is normalized to an $(m-1)$-channel image

$$\boldsymbol{f}_n(x,y) = \left[\frac{f_1(x,y)}{f_2(x,y)},\ \ldots,\ \frac{f_{m-1}(x,y)}{f_m(x,y)}\right]_t \quad . \tag{2.2.32}$$

If it can be assumed that the terms $\alpha_\nu I_\nu$ of any three adjacent channels are equal and so are β_ν, another normalization is

$$\frac{\varrho_{\nu-1} - \varrho_\nu}{\varrho_\nu - \varrho_{\nu+1}} = \frac{f_{\nu-1} - f_\nu}{f_\nu - f_{\nu+1}} \quad . \tag{2.2.33}$$

Similar to (2.2.32) the m-channel image (2.2.31) is thus normalized to an $(m-2)$-channel image.

We mention in passing that inhomogeneities of film, display screens, or sensor targets may be corrected, provided they are time invariant and known in advance, perhaps from a calibration measurement. Normalization is achieved by multiplying with a matrix of correction coefficients.

2.2.7 Geometric Correction

An image may have geometrical distortions which result from imperfect recording devices and other reasons. For instance, in remotely sensed images distortions may result from earth curvature and rotation, from instabilities of the aircraft, or from the variable sweep velocity of oscillating mirror scanners. It is also often necessary to compare two images recorded under different viewing conditions. In any case, there is an ideal image as indicated in Fig. 2.9. A distorted image is available, and the problem is to correct geometric distortions.

Let $s(x,y)$ be the ideal (undistorted image) and $f(u,v)$ the distorted image. Geometric correction is accomplished by mapping the distorted corrdinates (u,v) back to the original. This can be done if the *distortion mapping*

$$u = \Phi_1(x,y), \quad v = \Phi_2(x,y) \tag{2.2.34}$$

is known. In this case the corrected image is

$$s(x,y) = f[\Phi_1(x,y), \Phi_2(x,y)] \quad . \tag{2.2.35}$$

The operation in (2.2.35) may be considered as spatial warping (see Sect. 2.2.2 for time warping). It remains to determine suitable distortion mappings and to

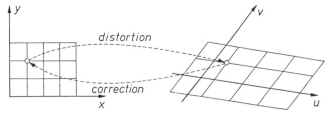

Fig. 2.9. An image in the original (x, y) system is distorted to an image in the (u, v) system. By a geometric correction it is mapped back to the (x, y) system

adapt (2.2.35) to discrete images. For instance, if $s(x, y)$ is sampled by a regular lattice of points, a lattice point in (x, y) mapped to (u, v) by (2.2.34) will not, in general, correspond to a lattice point of the (u, v) system; therefore, image intensity is not directly available. This was pointed out already in Sect. 2.2.2. A mapping as in (2.2.34) is also a fairly general procedure for normalization of characters.

Application of geometric correction to sampled images involves three basic steps.

1. Selection of *control points* CP in order to specify the distortion mapping (2.2.34).
2. Selection of a particular mapping which contains some parameters. The parameters follow from step 1.
3. Computation of (u, v) coordinates for the chosen (x, y) lattice and *resampling* of $f(u, v)$.

Control points are points of the image with known (x, y) coordinates. In the case of remotely sensed images these may be obtained from ground information. For instance, the coordinates of an airport in a geographic reference system, such as universal transverse mercator, may be known. In the image the airport may be located by edge correlation.

Commonly used mappings are projective and polynomial transformations. The general *projective transformation* in homogeneous coordinates is linear, but since the last homogeneous coordinate is unknown, the equations

$$u = \frac{a_0 + a_1 x + a_2 y}{c_0 + c_1 x + c_2 y}, \quad v = \frac{b_0 + b_1 x + b_2 y}{c_0 + c_1 x + c_2 y}, \qquad (2.2.36)$$

result. These are nonlinear in the coefficients. Equation (2.2.36) is the most general transformation which maps straight lines into straight lines. A special case is the *affine transformation*

$$u = a_0 + a_1 x + a_2 y, \quad v = b_0 + b_1 x + b_2 y, \qquad (2.2.37)$$

which maps parallel lines into parallel lines. This would be adequate for the correction in Fig. 2.9. A *polynomial transformation*

$$u = \sum_{i=0}^{r}\sum_{j=0}^{r-i} a_{ij}x^i y^j \ , \quad v = \sum_{i=0}^{r}\sum_{j=0}^{r-i} b_{ij}x^i y^j \ , \tag{2.2.38}$$

is a generalization of (2.2.37) which is also linear in the coefficients. It maps straight lines into curved lines. The degree r of the polynomials is limited by the number of control points available to determine the unknown coefficients a_{ij}, b_{ij}. Let (u_ν, v_ν) be a control point in the distorted image whose true coordinates (x_ν, y_ν) in the undistorted image can be determined. Then the equation

$$u_\nu = a_{00} + a_{10}x_\nu + a_{01}y_\nu + a_{20}x_\nu^2 + a_{11}x_\nu y_\nu + a_{02}y_\nu^2 + \ldots \tag{2.2.39}$$

must hold. A polynomial of degree r with two variables x, y has

$$R = \binom{2+r}{r} \tag{2.2.40}$$

coefficients a_{ij}. If there are $N \geq R$ control points (x_ν, y_ν), (2.2.39) yields a system of N linear equations with R unknowns. If $\boldsymbol{u}$ is a column vector containing N coordinates of control points, $\boldsymbol{a}$ is a column vector with R unknown coefficients a_{ij}, and $\boldsymbol{Z}$ is a NR matrix of terms $x_\nu^i y_\nu^j$ as exemplified for $r = 2, R = 6$ in (2.2.39), a compact notation is

$$\boldsymbol{u} = \boldsymbol{Z}\boldsymbol{a} \ , \tag{2.2.41}$$

and a similar equation holds for the coefficients b_{ij} of (2.2.38). The solution of (2.2.41) for $\boldsymbol{a}$ is well known from the following theorem.

Theorem 2.2.1. If $(\boldsymbol{Z}_t \boldsymbol{Z})^{-1}$ exists, the solution of (2.2.41) minimizing the mean-square error

$$\varepsilon = (\boldsymbol{u} - \boldsymbol{Z}\boldsymbol{a})_t (\boldsymbol{u} - \boldsymbol{Z}\boldsymbol{a}) \tag{2.2.42}$$

is given by

$$\boldsymbol{a} = (\boldsymbol{Z}_t \boldsymbol{Z})^{-1} \boldsymbol{Z}_t \boldsymbol{u} \ , \tag{2.2.43}$$

where $\boldsymbol{a}$ is the unique vector of minimum norm.

Proof. See, for instance, [2.5] pp. 15–23. □

From geometric considerations it is evident that the vector $(\boldsymbol{u} - \boldsymbol{Z}\boldsymbol{a})$ has to be orthogonal to the columns $\boldsymbol{z}_i$ of $\boldsymbol{Z}$, yielding

$$\boldsymbol{Z}_t(\boldsymbol{u} - \boldsymbol{Z}\boldsymbol{a}) = 0 \ , \quad \boldsymbol{Z}_t \boldsymbol{u} = \boldsymbol{Z}_t \boldsymbol{Z}\boldsymbol{a} \ . \tag{2.2.44}$$

This last equation is fulfilled if $\boldsymbol{a}$ is chosen as in (2.2.43). This is the *orthogonality principle* for the minimization of a quadratic error criterion.

The final step after determination of a mapping is to obtain (u, v) coordinates and resample the image $f(u, v)$. This is done by choosing a lattice of sampling points $(x_i, y_j), i, j = 0, 1, \ldots, M - 1$ in the (x, y) system; usually this will be a regular lattice. A point (x_i, y_j) is mapped by (2.2.37), (2.2.38) or in general (2.2.34) to the (u, v) system yielding a point (u_i, v_j). Then, by (2.2.35),

$$s(x_i, y_j) = f(u_i, v_j) \tag{2.2.45}$$

gives the intensity of the corrected image. But since (u_i, v_j) in general will not be a lattice point in the (u, v) system, $f(u_i, v_j)$ has to be obtained by interpolation. The simplest technique is to set $f(u_i, v_j)$ equal to the intensity of the nearest lattice point; it is also possible to use bilinear interpolation according to Sect. 2.2.2.

2.2.8 Image Registration

The problem of comparing two images recorded under different viewing conditions is called *image registration*. Often a method based on the correlation of subimages is used. Fast algorithms for translational registration are *sequential-similarity detection algorithms*. If f is a M^2 reference image and s is a $M_s^2, M_s < M$, search image to be matched with f, the distance criterion

$$d(r, i, j) = \sum_{\varrho=1}^{r} |f_{i+k(\varrho), j+l(\varrho)} - s_{k(\varrho), l(\varrho)}|^p \tag{2.2.46}$$

is used. In (2.2.46) $(k(\varrho), l(\varrho))$ is a random nonrepeating sequence of coordinates and $p \geq 1$. The distance $d(r, i, j)$ for fixed (i, j) increases with r, $r \leq M_s^2$. If d increases too fast, the images are considered to be dissimilar and another (i, j) may be tried in order to find out whether a match occurs at another position.

2.2.9 Pseudocolors

It is difficult to get an intuitive impression of a multispectral image $f(x, y)$ if $m > 3$. Furthermore, if there are many adjacent spectral channels, their output is highly correlated. Thus it is reasonable to try to compress information to $m_0 \leq 3$ channels in such a way that the new channels are uncorrelated and contain most of the information. According to the results of Sect. 2.1.4 this can be achieved by a discrete Karhunen-Loeve transformation. If f_{ij} is the vector containing samples of m spectral channels at point (x_i, y_j), the transformed vector is

$$h_{ij} = \Phi f_{ij} \quad . \tag{2.2.47}$$

The rows of matrix Φ are eigenvectors of

$$K_f = \sum_{i,j} \frac{(f_{ij} - m)(f_{ij} - m)_t}{M^2} \quad ; \quad m = \sum \frac{f_{ij}}{M^2} \quad . \tag{2.2.48}$$

Usually a subset of the M^2 samples will be sufficient to estimate K_f. Using the m_0 eigenvectors belonging to the largest eigenvalues gives a h_{ij} which allows reproduction of f_{ij} with minimum mean-square error. If $m_0 = 2$ or $m_0 = 3$ is used, each spectral channel of h_{ij} can be assigned arbitrarily to one of three primary colors of a color display. In this way the points h_{ij}, $i, j = 0, 1, \ldots, M - 1$, can be viewed with *pseudocolors* and processing is possible with less expenditure. Often the term pseudocolors is also used if gray values of a gray-level image are coded by colors in order to obtain better contrast or also "nice" pictures. This type of color can be assigned by look-up tables.

2.2.10 Camera Calibration

Two-dimensional images of three-dimensional objects are an important type of pattern as mentioned in Sect. 1.6. They are obtained according to the imaging properties of a lens. As shown in Fig. 2.10 a (thin) lens provides an image point P' of an object point P. If P is at distance z from the center of the lens, P' at distance z', and if f_l is the *focal length* of the lens, the relation

$$\frac{1}{z} + \frac{1}{z'} = \frac{1}{f_l} \qquad (2.2.49)$$

holds.

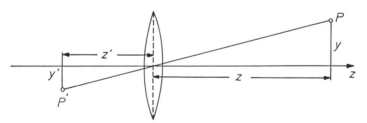

Fig. 2.10. An image point P' recorded by a thin lens from an object point P

If in (2.2.49) z becomes sufficiently large, $z' \approx f_l$. This leads to the simple *pinhole camera model* depicted in Fig. 2.11. An object point P having coordinates (x_c, y_c, z_c) in a camera coordinate system is projected to an image point P' having image coordinates (x_p, y_p). If the distance of P from the lens is sufficiently large, the image P' will be observed in the distance f_l behind the lens center. Obviously, it gives the same geometric relations if P' is assumed to occur at the distance f_l before the center. From geometric relations it is observed that

$$\frac{y_p}{f_l} = \frac{y_c}{z_c} \quad \text{or} \quad y_p = \frac{f_l y_c}{z_c} \quad . \qquad (2.2.50)$$

Camera calibration is the process of determining the unknown parameters of a certain camera model by means of a suitable *calibration image*; this is an image

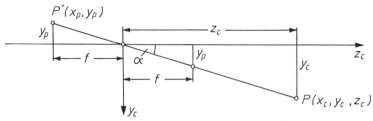

Fig. 2.11. The pinhole camera model showing the image point P' of an object point P, where f_l is the focal length of the camera lens

having a simple but precisely measured set of geometric properties, for example, a set of equally spaced rectangles. In the following we distinguish between two three-dimensional coordinate systems, the *world coordinates* $\boldsymbol{x}_w = (x_w, y_w, z_w)_t$ and the *camera coordinates* $\boldsymbol{x}_c = (x_c, y_c, z_c)_t$, where the subscript t denotes vector transposition. The world coordinates may be chosen arbitrarily, for example, with respect to a reference point and reference direction of a room or with respect to the calibration image. To use a particular example, we will assume in the following that the calibration image is of rectangular shape and fixed on the floor of a room. The world coordinates have their origin in a corner of this rectangle, their x- and y-axes parallel to the sides of the rectangle, and the z-axes chosen such that a right handed orthogonal system results. The camera coordinates have their origin in the center of the camera, the x- and y-axes parallel to the sides of the image plane, and the z-axis pointing in the direction of the optical axis. A third coordinate system are the *object coordinates* (x_o, y_o, z_o) which are defined with respect to some reference point of a real world object. However, we will not use this system here.

The imaging geometry is shown in Fig. 2.12. If $\boldsymbol{R}$ is a rotation matrix and $\boldsymbol{t}$ a translation vector, the relation between world coordinates $\boldsymbol{x}_w$ and camera coordinates $\boldsymbol{x}_c$ is

$$\boldsymbol{x}_c = \boldsymbol{R}\boldsymbol{x}_w + \boldsymbol{t} \quad . \tag{2.2.51}$$

The camera, modeled by the pinhole camera, projects an object point P having camera coordinates $\boldsymbol{x}_c$ to an image point having two-dimensional image or picture coordinates $\boldsymbol{x}_p = (x_p, y_p)_t$. If f_l is the focal length of the camera, the relation

$$x_p = \frac{f_l x_c}{z_c}, \quad y_p = \frac{f_l y_c}{z_c} \tag{2.2.52}$$

follows from (2.2.50).

A lens may cause certain *radial distortions* of an image which are described by parameters d_1, d_2. Between distorted (and measured) image coordinates $\boldsymbol{x}_d$ and undistorted picture coordinates $\boldsymbol{x}_p$ the relation

$$x_d(1 + d_1 r^2 + d_2 r^4) = x_p, \quad y_d(1 + d_1 r^2 + d_2 r^4) = y_p \tag{2.2.53}$$

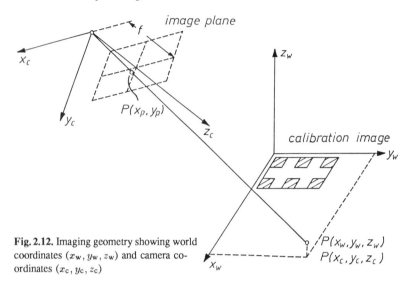

Fig. 2.12. Imaging geometry showing world coordinates (x_w, y_w, z_w) and camera coordinates (x_c, y_c, z_c)

holds where $r = \sqrt{x_d^2 + y_d^2}$. Although higher order terms could be used in the above equations, this usually is sufficient for standard lenses. The result of a radial distortion by (2.2.53) is that a straight object line is observed as a curved image line. Combining equations gives the relation between other world coordinates of an object point and recorded (distorted) image coordinates. For x_d, y_d this equation is

$$x_d(1 + d_1 r^2 + d_2 r^4) = f_l \frac{r_1 x_w + r_2 y_w + r_3 z_w + t_x}{r_7 x_w + r_8 y_w + r_9 z_w + t_z} \quad , \qquad (2.2.54)$$

$$y_d(1 + d_1 r^2 + d_2 r^4) = f_l \frac{r_4 x_w + r_5 y_w + r_6 z_w + t_y}{r_7 x_w + r_8 y_w + r_9 z_w + t_z} \quad . \qquad (2.2.55)$$

In these equations r_i and t_j are the elements of $\boldsymbol{R}$ and $\boldsymbol{t}$, respectively. The parameters to be determined by camera calibration are the elements of $\boldsymbol{R}$ and $\boldsymbol{t}$, the distortion parameters d_1 and d_2, and the focal length f_l. If the calibration image contains a sufficient number of points with known world coordinates and if the corresponding image coordinates can be determined, then it is possible to compute the unknown parameters by solving equations (2.2.54, 55). So these parameters can be computed from one calibration image.

An additional parameter results if the image coordinates are to be transformed into the common integer pixel coordinates (i, j). From (2.1.1) we observe $(x_d, y_d) = (i \Delta x_d, j \Delta y_d)$ if $(x_0, y_0) = (0, 0)$. Since the optical axis will meet the image plane at some point $(i_0 \Delta x_d, j_0 \Delta y_d)$, the distorted picture coordinate y_d is given by

$$y_d = (j - j_0) \Delta y_d \quad . \qquad (2.2.56)$$

To cope with some irregularities in the horizontal scanning an additional *timing factor* s_x is introduced to yield

$$s_x x_\mathrm{d} = (i - i_0) \Delta x_\mathrm{d} \quad . \tag{2.2.57}$$

The step sizes Δx_d and Δy_d can be obtained from the data sheet of the camera. The values of i_0 and j_0 may be adjusted to be in the center of the image. So the only unknown parameter is s_x which again can be obtained by calibration as outlined below.

Without going into too much detail, the solution of the relevant equations for the unknown parameters will now be outlined. Every calibration point of the calibration image having world coordinates $(x_\mathrm{w}, y_\mathrm{w}, z_\mathrm{w})$ gives a pixel (i, j). From each calibration point two equations result by inserting (2.2.56, 57) into (2.2.54, 55); these two equations contain 18 unknown parameters. Therefore, the minimum number of calibration points is 9, but in order to get good results the number of calibration points should be significantly larger than 9. The first step is to divide (2.2.54) by (2.2.55) after insertion of (2.2.56, 57) yielding

$$\frac{(i-i_0)\Delta x_\mathrm{d}}{s_x(j-j_0)\Delta y_\mathrm{d}} = \frac{r_1 x_\mathrm{w} + r_2 y_\mathrm{w} + r_3 z_\mathrm{w} + t_x}{r_4 x_\mathrm{w} + r_5 y_\mathrm{w} + r_6 z_\mathrm{w} + t_y} \quad . \tag{2.2.58}$$

This equation contains only 11 unknown parameters. The equation may be converted to a system of linear equations and solved for the unknown parameters. In the second step the remaining parameters are f, t_z, d_1, and d_2. These may be computed by solving a nonlinear problem or by conversion to a linear system as shown in detail in the references.

All unknown parameters, except s_x, may be computed from the so-called *single plane method*, which requires a two-dimensional calibration image; s_x may be obtained from the *multi-plane method*, which requires a three-dimensional calibration image and the procedure outlined above. In the single plane method s_x is not an unknown and z_w may be set to zero. This results in further simplification.

2.2.11 Normalized Stereo Images

In processing of *stereo images* measurements of pixel positions are used to determine the displacement of corresponding image points and depth of object points. Therefore, accurate geometric information is necessary. It is useful to transform the actual stereo images into ideal or *normalized stereo images* (we assume *binocular stereo* for simplicity) meeting the following criteria:

1. The two stereo images are taken by an ideal pinhole camera.
2. The optical axes of the two imaging cameras are parallel.
3. The two images are in the same plane which is parallel to the stereo basis.
4. The vertical position and angular orientation of the images are the same, assuming a horizontal stereo basis.
5. The focal lengths of the cameras are the same.

Due to imperfections of the cameras these conditions will not be met initially. From the actual images the normalized images can be computed as outlined below where we omit the details of the several transformation steps.

1. Determine the relative positions, focal lengths, and distortion coefficients of the cameras by *camera calibration* as discussed in the last section.
2. Determine the relative positions of the cameras with respect to the same *world coordinates*.
3. Determine the *stereo basis* (the line connecting the optical centers of the two cameras).
4. Determine the orientation of the cameras with respect to the stereo basis.
5. Determine a common focal length as the mean value of the two focal lengths.
6. Rotate the images using the orientation parameters determined in step 4.
7. Resample the images.

Sometimes a stereo geometry with non-parallel axes may be convenient. In this case the axis of one ideal image is rotated by an angle α towards the axis of the other image. Again, normalized images are useful as a starting point.

If non-normalized stereo images are used, the two corresponding image points of an object point lie on an *epipolar line* in each of the images. The epipolar lines are obtained as intersections of the epipolar plane with the two image planes. The *epipolar plane* is the plane defined by the object point and the optical centers of the two cameras. If normalized stereo images are used, the epipolar lines coincide with a (horizontal) scan line in the images.

2.2.12 Speaker Normalization

One of the severe problems of speech recognition is the fact that a system which has been trained with speech samples of a certain speaker usually shows significantly inferior performance when recognizing speech from another speaker. It is an obvious idea to "normalize" the speech samples of the new speaker in order to reduce or eliminate his speaker-dependent characteristics. However, experimental evidence shows that this is hard to achieve. Another idea is to collect a representative sample of speech (in the sense of Sect. 1.3) covering all speaker characteristics and giving (almost) the same performance for all speakers. A problem with this approach is that, in general, performance tends to decrease as the number of speakers in the sample is increased. Yet another idea is to find some speaker-independent features giving the same performance for all speakers. So far no such features have been found. In this section we shall give one example of the first approach, that is, normalization of features.

Let us assume that the first step of speech recognition is to obtain some parametric representation of this speech signal. Examples are the Fourier, LPC, or cepstral coefficients in Sects. 2.3.3, 2.5, or 3.10, respectively. In any case one obtains a vector c_d from the design sample every few ms. The design sample is spoken by some reference speaker (or speakers). Similarly in a new speech sample one obtains a vector c_s every few ms. The mean vector and covariance matrix of parameter vectors in the design and the new speech sample are m_d, K_d and m_s and K_s, respectively. For simplicity only the diagonal elements of the covariance matrices are considered further and are denoted by $D_d = \mathrm{diag}\,(d_{d1}, d_{d2}, \ldots, d_{dn})$

and D_s, respectively. One approach to normalization is to find a linear transformation

$$c_a = Ac_s + b \tag{2.2.59}$$

giving parameter vectors c_a having the same mean vector and diagonal elements of the covariance matrix as c_d. This can be achieved by a diagonal transformation matrix A having elements a_i and a vector b having elements b_i defined by

$$a_i = \sqrt{d_{di}/d_{si}}\,, \quad b_i = m_{di} - a_i m_{si}\ . \tag{2.2.60}$$

An obvious extension of this approach is not only to consider a diagonal matrix, but either the full matrix or a band matrix having one or more nonzero bands of elements in addition to the main diagonal. Experimental studies have shown that even the simple approach (2.2.60) gives an improvement of performance; additional improvements can be achieved using additional elements in the transformation matrix A. The normalization can be applied to all parameter vectors regardless of their class, or it can be applied using special transformations for certain groups of vectors. For example, if in advance a sufficiently reliable distinction between vowels and consonants is possible, different transformations for these two groups can be used.

2.3 Linear Filtering

2.3.1 The Use of Filtering

A powerful method of preprocessing is to take a system T having as its input a pattern $f(x)$ and as output a transformed pattern

$$h(x) = T\{f(x)\}\ . \tag{2.3.1}$$

If T is properly chosen, $h(x)$ will be better suited for further processing than $f(x)$. In order to make the problem mathematically tractable certain restrictions and assumptions about T are necessary as discussed later, but even with these restrictions useful operations are obtained.

It is possible to achieve one of the following results or a combination of them, provided some conditions, which will be mentioned later, are met:

1. Noise may be reduced.
2. Certain properties of a pattern can be enhanced.
3. Linear distortions can be corrected.
4. The uncorrupted pattern can be restored by computing a linear estimate.

A combination would be reduction of distortions in a pattern which also contains noise. In this case, not geometric distortions as in Sect. 2.2.7 are meant, but distortions of the values taken by the function $f(x)$ – that is, a blurring of the pattern. Enhancement of a pattern was also the objective of Sect. 2.2.4. Here different techniques are discussed. They encompass systems or transformations which are known as *filters*. A most important class are *linear filters*, which are introduced next.

2.3.2 Linear Systems

For simplicity of notation a pattern $f(x)$ or its sampled version $\boldsymbol{f}$ is considered in the following. The case $f(x,y)$ is a straightforward generalization and will be mentioned sometimes.

A system is called a *linear system* if for two arbitrary and real constants a_1, a_2 and two arbitrary functions $f_1(x), f_2(x)$ the superposition property

$$T\{a_1 f_1(x) + a_2 f_2(x)\} = a_1 T\{f_1(x)\} + a_2 T\{f_2(x)\} \tag{2.3.2}$$

holds. The same is valid for sampled functions $\boldsymbol{f}_1, \boldsymbol{f}_2$.

When dealing with linear systems, it is extremely useful to have available the idealized function $\delta(x)$. The δ-*function* is defined by the limit

$$\delta(x) = \lim_{\sigma \to 0} (2\pi\sigma^2)^{-1/2} \exp\left[-x^2/(2\sigma^2)\right] . \tag{2.3.3}$$

The most important properties of this function are summarized by the following theorem.

Theorem 2.3.1. For the function $\delta(x)$ the relations

$$\int_{-\infty}^{\infty} \delta(x) dx = 1 ,$$

$$\int_{-\infty}^{\infty} f(x)\delta(x - x_0) = f(x_0) , \tag{2.3.4}$$

$$\int_{-\infty}^{\infty} \exp(i\xi x) d\xi = 2\pi\delta(x) ,$$

hold, where $f(x)$ is a function continuous at $x = x_0$.

Proof. See, for instance, [2.6]. □

The function may be viewed as a *unit impulse* centered at $x = 0$, taking the value 0 if $x \neq 0$, and being infinite at $x = 0$. The discrete version is the unit sample

$$\delta_j = \begin{cases} 1 & \text{if } j = 0 \\ 0 & \text{if } j \neq 0 \end{cases} . \tag{2.3.5}$$

If a unit impulse $\delta(x - x_0)$ at point $x = x_0$ is given as input to a linear system, the system response

$$g(x, x_0) = T\{\delta(x - x_0)\} \tag{2.3.6}$$

is termed the *impulse response* $g(x, x_0)$. If in addition the system has the property that

$$g(x, x_0) = g(x - x_0) \;, \tag{2.3.7}$$

it is said to be *shift invariant*. In this case x_0 may be chosen to be zero without loss of generality. In the discrete shift-invariant case we have

$$g_{j-k} = T\{\delta_{j-k}\} \tag{2.3.8}$$

as equivalent to (2.3.7). The impulse response is very useful because it allows one to compute the output $h(x)$ of a linear, shift-invariant system for any input $f(x)$.

Theorem 2.3.2. If $g(x)$ is the impulse response of a linear, shift-invariant system, the output $h(x)$ is given by the *convolution integral*

$$h(x) = \int_{-\infty}^{\infty} f(u)g(x - u)du = \int_{-\infty}^{\infty} f(x - u)g(u)du = f(x) \otimes g(x) \;. \tag{2.3.9}$$

In the discrete case the convolution integral is replaced by the *discrete convolution*

$$h_j = \sum_{\nu=-\infty}^{\infty} g_{j-\nu} f_\nu = \sum_{\nu=-\infty}^{\infty} g_\nu f_{j-\nu} = f_j \otimes g_j \;. \tag{2.3.10}$$

Proof. Since $f(u)$ at place $u = \nu\Delta u$ may be approximated arbitrarily closely by an impulse of height $f(\nu\Delta u)$ and width Δu, the whole function may be represented by a sum of impulses. Then (2.3.9) follows as the limit $\Delta u \to 0$ with (2.3.2, 6, and 7). Details may be obtained from [2.7] for the continuous case and from [2.8] for the discrete case. □

2.3.3 Fourier Transform

The result of Theorem 2.3.2 is one reason for the importance of linear, shift-invariant sytems. It gives a fairly simple closed-form solution. Particularly, (2.3.10) allows direct implementation on a digital computer. Evaluation of (2.3.10) may, however, require extensive computations, depending on the number of points with nonzero values of g_j. If this number is large, it may be advantageous to use the *Fourier transform* (FT). For a function of two variables the following relations hold for the Fourier transform FT and its inverse FT^{-1}:

Theorem 2.3.3. If the Fourier transform of a continuous function $f(x, y)$ is defined by

$$F(\xi,\eta) = \int\int f(x,y)\exp[-i(\xi x+\eta y)]dx\,dy = \text{FT}\{f(x,y)\} \quad , \qquad (2.3.11)$$

which also is called the *spectrum* of $f(x,y)$, then the inverse transformation is

$$f(x,y) = \frac{1}{4\pi^2}\int\int F(\xi,\eta)\exp[i(\xi x+\eta y)]d\xi\,d\eta$$

$$= \text{FT}^{-1}\{F\xi,\eta)\} \quad , \qquad (2.3.12)$$

where the double integrals extend from $-\infty$ to ∞. For a function of n variables the factor $(4\pi^2)$ is replaced by $(2\pi)^n$ and the double integrals by analogous n-fold integrals.

Proof. An identity results if (2.3.11) is substituted into (2.3.12) and Theorem 2.3.1 is used. A proof without use of δ-functions is presented, for instance, in [2.9]. □

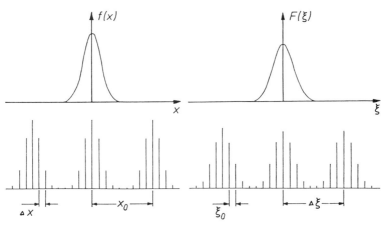

Fig. 2.13. A continuous function $f(x)$ has a continuous spectrum $F(\xi)$ and a periodic discrete function has a periodic discrete spectrum

In digital processing of patterns a continuous function $f(x,y)$ is sampled at points spaced $\Delta x, \Delta y$, as discussed in Sect. 2.1 and shown in Fig. 2.13 for a function of one variable only. The resulting discrete sequence f is supposed to be repeated with period x_0, y_0. The sequence is uniquely determined by the sampling interval, the period, and the sampling values f_{jk}. The following result states that the spectrum of such a sequence is also a periodic sequence of coefficients $F_{\mu\nu}$. It is evident from Fig. 2.13 that *aliasing* occurs if periodic repetitions overlap. Obeying the conditions of the sampling Theorem 2.1.1 guarantees that aliasing does not occur.

Theorem 2.3.4. If f_{jk}, $j, k = 0, 1, \ldots, M-1$ is a periodic sequence of sample values spaced at $\Delta x, \Delta y$ and having period x_0, y_0, then the spectrum is a periodic sequence of values $F_{\mu\nu}$, spaced at ξ_0, η_0 having period $\Delta\xi, \Delta\eta$. In the following let $\Delta x = \Delta y = 1$ and $x_0 = y_0 = M$. The coefficients $F_{\mu\nu}$ or f_{jk} may be computed from the *discrete Fourier transform* (DFT) or the *inverse discrete Fourier transform*, respectively, defined by

$$F_{\mu\nu} = \sum_{j=0}^{M-1} \sum_{k=0}^{M-1} f_{jk} \exp[-i2\pi(\mu j/M + \nu k/M)] = \mathrm{DFT}\{f_{jk}\}, \quad (2.3.13)$$

$$f_{jk} = \sum_{\mu=0}^{M-1} \sum_{\nu=0}^{M-1} F_{\mu\nu} \exp[i2\pi(\mu j/M + \nu k/M)]/M^2 = \mathrm{DFT}^{-1}\{F_{\mu\nu}\}, (2.3.14)$$

$$\Delta\xi\Delta x = \Delta\eta\Delta y = \xi_0 x_0 = \eta_0 y_0 = 2\pi. \quad (2.3.15)$$

Proof. The above relations may be verified directly by substituting (2.3.13) into (2.3.14). Using the closed form for the sum of geometric series an identity results. Unnecessary complication is avoided when doing this for the one-dimensional case. Details are presented, for example, in [2.8]. □

The above equations may be evaluated by the *fast Fourier transform* (FFT). The importance of FT, DFT, and FFT for linear systems rests on the fact that system output may be computed efficiently via FFT by use of the following result which is given for a function of one variable.

Theorem 2.3.5. If $F(\xi), G(\xi), H(\xi)$ are the FT of an input $f(x)$, impulse response $g(x)$, output $h(x)$ of a linear, shift-invariant system, respectively, then

$$H(\xi) = F(\xi)G(\xi). \quad (2.3.16)$$

Since multiplication is associative, this result shows that also convolution is associative. Similarly in the discrete case,

$$H_\nu = F_\nu G_\nu. \quad (2.3.17)$$

In the discrete case the periodic continuation has to be chosen such that $M' \geq M + m - 1$, where M is the number of sample values of f_j, m is the number of sample values of g_j, and M' is the common period of both functions. A proper M' can always be achieved by filling in zeros at the end of f and g.

Proof. See, for instance, [2.7, 8]. □

Since there are many detailed treatments of the FFT in the literature, we only show briefly that FFT can be formulated via factorization of a suitably chosen transform matrix. From (2.3.13) and the well-known relation $\exp(a+b) = \exp(a)\exp(b)$ it is obvious that in general an n-dimensional DFT may be computed via n one-dimensional DFTs. Therefore, it suffices to consider the FFT of a 1-dimensional discrete function f. One formulation of the FFT can then be given in the following steps:

1. Consider the one-dimensional DFT

$$F_\mu = \sum_{j=0}^{M-1} f_j \exp[-i2\pi(\mu j/M)] = \text{DFT}\{f_j\} \quad , \tag{2.3.18}$$

and define the number $W_M = \exp(-i2\pi/M)$ and matrix W_M'' by

$$W_M'' = [W_M^{mj}], \quad m, j = 0, 1, \ldots, M-1 \quad . \tag{2.3.19}$$

2. Reduce elements of W_M'' modulo M to obtain the matrix W_M'. The reduction modulo M does not change the value of the elements since the function exp (.) is periodic.

3. Reorder the rows of W_M' by bit reversal to obtain the matrix W_M. This is the transformation matrix used for the FFT. Bit reversal means that the number of a row in W_M', represented as a binary number, is read from right to left to yield the new row number in W_M.

4. The next theorem states that this matrix can be factored into a product of three very simple matrices.

Theorem 2.3.6.

$$W_M = \begin{vmatrix} W_{M/2} & O_{M/2} \\ O_{M/2} & W_{M/2} \end{vmatrix} \begin{vmatrix} I_{M/2} & O_{M/2} \\ O_{M/2} & K_{M/2} \end{vmatrix} \begin{vmatrix} I_{M/2} & I_{M/2} \\ I_{M/2} & -I_{M/2} \end{vmatrix} \tag{2.3.20}$$

where $O_{M/2}$ and $I_{M/2}$ are the zero- and unit-matrix, respectively, of size $(M/2) \times (M/2)$,

$$\begin{aligned} W_{M/2} &= [W_{M/2}^{mj}], \quad m, j = 0, 1, \ldots, (M/2) - 1 \\ K_{M/2} &= \text{diag}(W_M^m, m = 0, 1, \ldots, (M/2) - 1) \end{aligned} \tag{2.3.21}$$

Proof. See, for example, [2.10, 11]. $\square$

5. Iterate the process of matrix factorization until it stops with matrix

$$W_2 = \begin{vmatrix} 1 & 1 \\ 1 & -1 \end{vmatrix} = W_{M/(M/2)} = [W_M^{Mmj/2}] \quad m, j = 0, 1 \quad . \tag{2.3.22}$$

6. Observe that $F'' = W_m'' f$ requires M^2 complex multiplications if f has M elements and that $F = $ [factored matrices]f requires only $MldM$ complex multiplications.

Computation of the system response by (2.3.17) is advantageous if the impulse response g_j is nonzero in a sufficiently large number of points. This short account of linear systems is to give only the main ideas. Many additional results are contained in the references. A system acting in the way described above on an input signal is also called a *linear filter* or, for short, a *filter*. It should be mentioned that besides Fourier transforms, transforms over a finite field may also be used. These transforms may also be used for computation of FFT. For a treatment of recursive and nonrecursive, causal and noncausal filters as well as filter design the reader is referred to the literature, where also a detailed discussion of practical considerations for computation of discrete convolutions is given.

2.3.4 Computational Aspects

The above results show that a (discrete or continuous) convolution can be computed in either of two ways:

1. Computation in the space domain using (2.3.9) or (2.3.10). In the discrete case the computational effort depends on Mm, if f and g have M and m nonzero values, respectively.
2. Computation in the frequency domain using (2.3.11, 16, 12) or (2.3.13, 17, 14), where the DFT is computed via an FFT algorithm such as (2.3.20–22). The required effort depends on $MldM$ assuming $M \gg m$.

In the two-dimensional case the discrete convolution is

$$h_{jk} = \sum_{\mu=0}^{M-1} \sum_{\nu=0}^{M-1} g_{j-\mu,k-\nu} f_{\mu\nu}$$

$$= \sum_{\mu=0}^{m-1} \sum_{\nu=0}^{m-1} f_{j-\mu,k-\nu} g_{\mu\nu} , \quad j,k = 0,1,,,,M+m-2 \qquad (2.3.23)$$

if f and g have M^2 and m^2 nonzero values, respectively. In the space domain the computational effort depends on $M^2 m^2$. If the impulse response g is *separable*, that is, if

$$g_{jk} = g_{x,j} g_{y,k} \quad , \qquad (2.3.24)$$

the two-dimensional discrete convolution reduces to

$$h_{jk} = \sum_{\mu} \sum_{\nu} f_{j-\mu,k-\nu} g_{x,\mu} g_{y,\nu} = \sum_{\mu} \left(\sum_{\nu} f_{j-\mu,k-\nu} g_{y,\nu} \right) g_{x,\mu} \quad . \qquad (2.3.25)$$

The computational effort in the space domain now depends on $Mm + Mm$ and is drastically reduced.

If a function $f(x)$ has FT $F(\xi)$, then its derivative $df(x)/dx$ has FT $i\xi F(\xi)$ as can be deduced easily from (2.3.11, 12). Using (2.3.16) it follows that

$$\left[\frac{df(x)}{dx} \right] \otimes g(x) = \frac{d[f(x) \otimes g(x)]}{dx} \quad , \qquad (2.3.26)$$

that is, differentiation and convolution are associative. This result can sometimes be used to simplify computations.

2.3.5 Homomorphic Systems

A generalization of the definition of a linear system in (2.3.2) is obtained by *homomorphic systems*. They are a class of *nonlinear systems* which may be treated such that results from linear systems are applicable. Assume that o is an operation combining functions $f_1(x)$ and $f_2(x)$ and $\triangle$ is an operation combining a real constant α and a function. Then a homomorphic system T has the property

$$T\{f_1 \circ f_2\} = T\{f_1\} \circ T\{f_2\} ,$$
$$T\{\alpha \triangle f_1\} = \alpha \triangle T\{f_1\} . \tag{2.3.27}$$

It can be seen that (2.3.2) is a special case of (2.3.27). The operations o and $\triangle$ have to obey the rules of vector addition and of multiplying a vector and a scalar. If it is possible to find a *characteristic system* T_c and its inverse T_c^{-1} such that

$$T_c\{f_1 \circ f_2\} = T_c\{f_1\} + T_c\{f_2\} , \quad T_c\{\alpha \triangle f_1\} = \alpha T_c\{f_1\} , \tag{2.3.28}$$

$$T_c^{-1}\{f_1 + f_2\} = T_c^{-1}\{f_1\} \circ T_c^{-1}\{f_2\} , \quad T_c^{-1}\{\alpha f_1\} = \alpha \triangle T_c^{-1}\{f_1\} , \tag{2.3.29}$$

then it is possible to represent a homomorphic system as shown in Fig. 2.14. In this figure T_L is a linear system. As an example, the multiplicative homomorphic system is also shown. In this case the operation o is multiplication, $\triangle$ is exponentiation, and

$$T_c\{f\} = \log\{f\} , \quad T_c^{-1}\{f\} = \exp\{f\} . \tag{2.3.30}$$

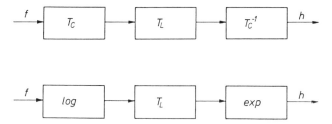

Fig. 2.14. General homomorphic system and multiplicative homomorphic system as a special example

If the patterns obey $f > 0$, no problems arise. Otherwise the complex log and exp have to be used. So the operation of the multiplicative homomorphic system is

$$h = T\{f\} = \exp\{T_L\{\log\{f\}\}\} . \tag{2.3.31}$$

Another class of homomorphic systems results if o is convolution. The characteristic system has to provide

$$T_c\left\{\int f_1(u)f_2(x-u)du\right\} = T_c\{f_1\} + T_c\{f_2\} \quad . \tag{2.3.32}$$

Using (2.3.9,16) this may be done by a system

$$T_c\{f\} = \text{FT}^{-1}\{\log\{\text{FT}\{f\}\}\} \quad , \tag{2.3.33}$$

where FT denotes the Fourier transform and FT^{-1} its inverse. The inverse characteristic system is

$$T_c^{-1}\{f\} = \text{FT}^{-1}\{\exp\{\text{FT}\{f\}\}\} \quad . \tag{2.3.34}$$

The operation of a convolutional homomorphic system is, therefore,

$$h = T\{f\} = T_c^{-1}\{T_L\{T_c\{f\}\}\} \quad , \tag{2.3.35}$$

with T_c and T_c^{-1} as in (2.3.33, 34) and T_L a linear system. The output of T_c in (2.3.33) is termed the *complex cepstrum* of f. If $f > 0$, the complex log is not required and the output of T_c is referred to as the *cepstrum*. Mathematical details are discussed in the references. In any case the characteristic system turns the operation o to addition and thus allows processing by a linear system.

2.3.6 Filtering of Patterns

Most applications of filters to *preprocessing* of patterns follow a particular model of the observed pattern $f(x)$ or $f(x,y)$. This is obtained from the assumption that an ideal signal $s(x)$ is distorted by a linear shift-invariant system with impulse response $g(x)$ and that the result is corrupted by additive noise $n(x)$. By "noise" we denote here any undesired component, be it stochastic or not. Therefore, the observed pattern is

$$f(x) = \int s(u)g(x-u)du + n(x) \quad . \tag{2.3.36}$$

Using this model we consider the four cases noted in Sect. 2.3.1.

The first case is *noise reduction* which results if $g(x) \approx \delta(x)$. In this case (2.3.36) reduces to

$$f(x) = s(x) + n(x) \quad , \tag{2.3.37}$$

that is, the pattern consists of a signal corrupted by additive noise. It is desired to find a filter with impulse response $g_0(x)$ such that filtering of f by g_0 will reduce n as much as possible. By (2.3.11) the spectrum of $f(x)$ is

$$F(\xi) = S(\xi) + N(\xi) \quad , \tag{2.3.38}$$

where $S(\xi)$ and $N(\xi)$ are the spectra of signal and noise, respectively. The fundamental requirement is that S and N are adequately separated on the ξ axis; for instance, as shown in Fig. 2.15. In this case a filter G_0 may be used, and $H = FG_0 \approx S$. This means that the output of the filter is approximately the signal. It is often the case that the spectrum of the noise is located at higher frequencies, so that the noise is attenuated by a *low-pass filter* as shown in Fig. 2.15. This results in a smoothing of the pattern since abrupt changes have high frequency content. Therefore, care has to be taken that no important details of the pattern are blurred.

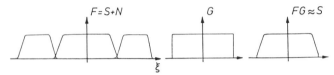

Fig. 2.15. Suppression of noise by a filter if the spectrum of the signal is separated from that of the noise

An example of noise reduction is band limiting of speech. This reduces noise as well as aliasing effects when sampling the speech signal. Another example is local averaging of patterns – for instance, pictures – to reduce irregularities. A different approach to noise reduction is possible if N observations $f^r(x)$, containing the same signal and independent noise functions $n^r(x)$, are available. In this case a cleaned version is

$$f(x) = N^{-1} \sum_{r=1}^{N} f^r(x) = s(x) + N^{-1} \sum n^r(x) \approx s(x) \quad , \qquad (2.3.39)$$

if $E\{n\} = 0$. This is used in processing of periodic signals such as those recorded from heart motion or rotating machines. More involved examples of preprocessing result if signal and noise are related nonadditively. For multiplicative and convolutional relations homomorphic techniques may be used.

A multiplicative relation results, for instance, in image processing if an observed intensity $f(x,y)$ consists of an illumination component $f_i(x,y)$ and a reflectance component $f_r(x,y)$. In this case

$$f(x,y) = f_i(x,y) \cdot f_r(x,y) \quad . \qquad (2.3.40)$$

Usually the signal $s(x,y)$ corresponds to the reflectance f_r, and the illumination f_i corresponds to "noise". Since $f, f_i, f_r > 0$, the logarithm of (2.3.40) yields

$$\log f = \log f_i + \log f_r \quad . \qquad (2.3.41)$$

This corresponds to the output of the characteristic system (2.3.30). Now a linear system is chosen by making use of the fact that f_i varys slowly and f_r rapidly in space. The result of this processing is an enhancement of the image f. Another example is satellite images disturbed by clouds. Ground reflectance and cloud transmission are assumed to be $f_r(x,y)$ and $f_t(x,y)$, respectively. Cloud reflection plus transmission of sunlight are assumed to equal one. Sun illumination is the constant L, and illumination on earth is approximated by a constant aL. This results in a scanner image

$$f(x,y) = aLf_r(x,y)f_t(x,y) + L(1 - f_t(x,y)) \quad . \tag{2.3.42}$$

A multiplicative relation between the desired f_r and the undesired f_t is obtained from

$$L - f(x,y) = f_t(x,y)(L - aLf_r(x,y)) \quad . \tag{2.3.43}$$

Again, the characteristic system (2.3.30) may be used.

A convolutional relation results, for instance, in the processing of speech and seismic patterns. This may also be treated as a special case of (2.3.36) with $n(x) = 0$. Equivalently, as done here, one may set $g(x) = \delta(x)$, generalize addition of $n(x)$ in (2.3.36) to convolution, and use homomorphic processing. Echo effects in speech and seismic waves may be treated this way. Separation of pitch period and vocal tract properties in voiced speech is another example. If the glottal pulses are denoted by s and the impulse response of the vocal tract by g, the speech signal is

$$f = s \otimes g \quad . \tag{2.3.44}$$

The characteristics of the vocal tract are slowly varying in time in comparison to s. Therefore, influencing the complex cepstrum of f by a low-pass filter (and passing the result through T_c^{-1}) yields g, whereas a high-pass filter yields s.

Another special case of (2.3.36) is obtained if $n(x) \approx 0$ in addition to $g(x) \approx \delta(x)$. This is the case of *pattern enhancement*. Examples are known from speech and image processing. Usually this enhancement is done by emphasizing high frequencies. For example, high-frequency emphasis is used to reduce the effect of the glottal source and to enhance high-frequency formants in order to separate peaks of adjacent formants. To enhance radiographic images a filter

$$G_0(\omega) = \begin{cases} 1 & \omega \leq \omega_c \\ \alpha\left[1 - (\omega_T - \omega)^2/(\omega_T - \omega_c)^2\right] + 1 & \omega_c < \omega < \omega_T \\ 1 + \alpha & \omega_T \leq \omega \end{cases} \tag{2.3.45}$$

$$\omega = \sqrt{\xi^2 + \eta^2}$$

was suggested. Another useful filter is

$$G_0(\xi, \eta) = 1 + \alpha(\xi^2 + \eta^2) \quad . \tag{2.3.46}$$

This corresponds in the space domain to the substraction of the *Laplacian operator* yielding

$$h(x,y) = f(x,y) - \alpha \frac{\partial^2 f}{\partial x^2} + \frac{\partial^2 f}{\partial y^2} \quad . \tag{2.3.47}$$

The above operation tends to enhance image contrast.

A more complicated case of (2.3.36) arises if *linear distortions* occur. In this case $n(x) \approx 0$ and $f = s \otimes g$. In the context of (2.3.44) it was assumed that the complex cepstra of s and g were separated sufficiently. Here we assume that the distorting system g is known in advance. This allows one, in principle, to solve the integral equation

$$f(x) = \int s(u)g(x-u)du \tag{2.3.48}$$

for the signal $s(x)$. Using Fourier transformation of (2.3.48) the simple solution

$$S(\xi) = \frac{F(\xi)}{G(\xi)} = F(\xi)G_\mathrm{i}(\xi) \tag{2.3.49}$$

is obtained where G_i is the inverse filter. When processing sampled patterns the equivalent of solving an integral equation is solving a system of linear equations, as is evident from (2.3.10). This may be written in compact form as

$$\boldsymbol{f} = \boldsymbol{s}\boldsymbol{\Gamma} \quad , \tag{2.3.50}$$

where $\boldsymbol{\Gamma}$ is the matrix obtained from the discrete impulse response. Unfortunately, (2.3.50) may be ill conditioned, and there are problems with (2.3.49) because $G(\xi)$ may have zeros. The general case of *pattern restoration* results if $n(x) \neq 0$ and $\delta(x) \neq 0$, yielding

$$f(x) = s(x) \otimes g(x) + n(x) \quad , \tag{2.3.36}$$

$$F(\xi) = S(\xi) \cdot G(\xi) + N(\xi) \quad . \tag{2.3.51}$$

Substituting (2.3.49) into (2.3.51) gives as an estimate $\hat{S}$ of the signal spectrum

$$\hat{S} = \frac{F}{G} = S + \frac{N}{G} \quad , \tag{2.3.52}$$

which amplifies noise if G takes small values. One approach is to look for an estimate $\hat{\boldsymbol{s}}$ of $\boldsymbol{s}$ which minimizes mean-square error. Considering only the discrete case

$$\boldsymbol{f} = \boldsymbol{\Gamma}\boldsymbol{s} + \boldsymbol{n} \tag{2.3.53}$$

the estimate $\hat{\boldsymbol{s}}$ of $\boldsymbol{s}$ should satisfy

$$\varepsilon = E\{(\boldsymbol{s} - \hat{\boldsymbol{s}})_\mathrm{t}(\boldsymbol{s} - \hat{\boldsymbol{s}})\} = \min_{\{\hat{\boldsymbol{s}}\}} E \quad . \tag{2.3.54}$$

If the estimate is constrained to be linear, that is

$$\hat{\boldsymbol{s}} = \boldsymbol{\Gamma}_0 \boldsymbol{f} \quad , \tag{2.3.55}$$

the matrix $\boldsymbol{\Gamma}_0$ may be derived from (2.3.54) and is given by the following theorem.

Theorem 2.3.7. The linear estimate $\hat{s}$ in (2.3.55) minimizing (2.3.54) is determined by the matrix

$$\Gamma_0 = \left(\Gamma_t K_n^{-1} \Gamma + K_s^{-1}\right)^{-1} \Gamma_t K_n^{-1} \quad , \tag{2.3.56}$$

where matrices K_n, K_s are defined by

$$K_n = E\{nn_t\} \; , \; K_s = E\{ss_t\} \quad . \tag{2.3.57}$$

Proof. See, for instance, [1.18, 2.12]. □

It would be beyond the scope of this volume to treat restoration techniques in detail. The reader is referred to the references which give comprehensive overviews, treat numerical problems in finding solutions, evaluate various other approaches to restoration, consider the problem of determining impulse response $g(x)$ in (2.3.36), and obtain generalizations to shift-variant systems. Although the above examples are not intended to be exhaustive, they indicate that a wealth of useful preprocessing operations may be realized by filters.

2.3.7 Resolution Hierarchies

A common computational approach used in many preprocessing and segmentation techniques is to employ a *resolution hierarchy* or an *image pyramid* representing the pixel data in different resolutions. Global image properties can be obtained with less computational effort from the coarse resolution; from this it is often possible to guide further processing and restrict processing at the fine image resolution to areas of interest. The full resolution image has $M \times M$ pixels and it is assumed that $M = 2^p$. The image at level l of the resolution hierarchy is $f^{(l)}$ and has $2^{p-l} \times 2^{p-l}$ pixels, the image at the lowest level $l = 0$ is $f^{(0)}$ having the full resolution. Given an image at a certain level l it is then necessary to compute the image at the next reduced resolution level $l + 1$.

The reduction of resolution is achieved by a *generating kernel* g_{jk}, $j, k = 0, 1, \ldots, m-1$. It is useful to have a kernel whose values sum to unity (normalized kernel), and which is symmetric, unimodal, and separable in the sense of (2.3.24). The optimal kernel would be the ideal low-pass filter having frequency response unity in the interval $0 \leq \xi \leq \xi_B$ and response zero outside. Since this filter has an impulse response of infinite extent, it is not a practically useful filter. Two simple generating kernels are the 2×2 and 4×4 unweighted averaging kernels

$$g = \frac{1}{4} \begin{vmatrix} 1 & 1 \\ 1 & 1 \end{vmatrix}, \quad g = \frac{1}{16} \begin{vmatrix} 1 & 1 & 1 & 1 \\ 1 & 1 & 1 & 1 \\ 1 & 1 & 1 & 1 \\ 1 & 1 & 1 & 1 \end{vmatrix} \quad . \tag{2.3.58}$$

These kernels are separable, for example, the 2×2 kernel separates into

$$g_{x,0} = g_{x,1} = g_{y,0} = g_{y,1} = \frac{1}{2} \quad . \tag{2.3.59}$$

A reduced resolution image at level $l+1$ is obtained from an image at level l by convolution of image and kernel

$$f_{jk}^{l+1} = \sum_{\mu=0}^{m-1} \left[\sum_{\nu=0}^{m-1} f_{2j-\mu+m-1,2k-\nu+m-1}^{(l)} g_{y,\nu} \right] g_{x,\mu} ,$$

$$j, k = 0, 1, \ldots, 2^{p-l-1} - m/2 . \tag{2.3.60}$$

However, the spectral properties of the kernels in (2.3.58) are not very good. Better kernels are obtained from an equiripple filter, an example of which is

$$\begin{aligned}
g_{x,1} &= g_{x,11} = 0.051 , \quad g_{x,3} = g_{x,9} = -0.087 \\
g_{x,5} &= g_{x,7} = 0.298 , \quad g_{x,6} = 0.475 , \\
g_{x,j} &= 0.0 , \quad j = 0, 2, 4, 8, 10, 12 , \\
g_{y,j} &= g_{x,j} , \quad j = 0, 1, \ldots, 12 .
\end{aligned} \tag{2.3.61}$$

Since equiripple filters can be obtained with different parameters, there are other possible solutions. The filter (2.3.61) gives better results than (2.3.58).

2.4 Rank Order and Morphological Operations

Several *nonlinear operations*, other than homomorphic systems, have been developed. In the following we discuss the *rank order operations* and *morphological operations* and we will see that for certain operations the former are just a special case of the latter. As a special case we consider a pattern f which is a two-dimensional array of sample values, that is a gray level or a binary image, taking gray levels from a finite set of discrete values as introduced in Sect. 2.1. However, as will be evident from the discussion, the operations generalize to any number of dimensions.

2.4.1 Rank Order Operations

Let $S = \{(\mu, \nu)\}$ be a certain finite set of discrete coordinate values which may or may not contain the origin (0,0). Define a neighborhood V_{ij} of an image point f_{ij} by

$$V_{ij} = \{f_{i+\mu,j+\nu} | \mu\nu \in S\} . \tag{2.4.1}$$

If there are M tuples (μ, ν) in S, then there are M values f_{kl} in V_{ij} if we neglect boundary points of f. For example, if $|\mu| \leq 1$, $|\nu| \leq 1$, one may visualize S as a 3×3 *mask* or *structuring element* with the origin at its center. This defines a 3×3 neighborhood for any pixel f_{ij}. In general, it is common to use sets S defining rectangular or circular masks. In general, the origin may be at the center of the mask, at the lower left corner, or at some other arbitrary point. Define the *rank order* of the gray values in the neighborhood of f_{ij} by

$$R_{ij} = \{r_1 \leq r_2 \leq \ldots \leq r_k \leq \ldots \leq r_M | r_k \in V_{ij}\} \quad . \tag{2.4.2}$$

A *rank order operation* is defined as a function

$$h_{ij} = \varphi(R_{ij}) \quad ; \tag{2.4.3}$$

where h_{ij} is the new pixel value computed from the neighborhood of f_{ij}. Some standard operations are

$$h_{ij} = r_1 , \quad \text{(erosion)} \tag{2.4.4}$$

$$h_{ij} = r_M , \quad \text{(dilation)} \tag{2.4.5}$$

$$h_{ij} = r_M - r_1 , \quad \text{(contour detection)} \tag{2.4.6}$$

$$h_{ij} = r_{(M+1)/2} , \quad \text{(median)} \tag{2.4.7}$$

$$h_{ij} = \begin{cases} r_1 & \text{if } (f_{ij} - r_1 < r_M - f_{ij}) \\ r_M & \text{otherwise} \end{cases} \quad \text{(extremum sharpening)}. \tag{2.4.8}$$

The operation in (2.4.6) can be used to detect contour changes. The median and extremum sharpening operations often are used iteratively until no more changes occur. It was mentioned in the preceding section that the use of lowpass filtering for noise reduction also tends to blur details in the pattern. This is avoided if median filtering is used. It is based on the definition of the median of a distribution function $P(x)$. The median x_m of $P(x)$ is given by

$$P(x_m) = 0.5 \quad . \tag{2.4.9}$$

The desirable properties of this operation are shown in Fig. 2.16. In contrast to a mean-value operation of size 3, a median operation of size 3 does not smear abrupt changes if these are large enough; this is evident from Fig. 2.17. An example of median filtering is shown in Fig. 2.18.

2.4.2 Morphological Operations on Binary Images

Morphological operations usually are represented in set theoretic terms. Details of the set theoretic notation may be obtained from the references; the principle is to define a discrete pattern f by a set of tuples $(x1, x2, \ldots, xn; f1, f2, \ldots, fm)$, where the first n components are coordinate values and the last m components are function values, see (1.2.3). However, in order to keep to one notation we will continue to view a pattern as introduced in (1.2.3) and (2.1.1). First, the special case that f is a binary image is considered. A binary image f only has values $f_{ij} \in \{0, 1\}$. Usually one considers sets of "ones" as object points and sets of "zeros" as background points. Let S be a structuring element as introduced in the preceding section. Two elementary morphological operations are

$$\boldsymbol{h} = \boldsymbol{f} \ominus \boldsymbol{s} , \quad \text{(erosion)}$$
$$h_{ij} = 1 \text{ if } [f_{i+\mu, j+\nu} = 1 \text{ for all } (\mu, \nu) \in S] , \quad \text{else } h_{ij} = 0 ; \tag{2.4.10}$$

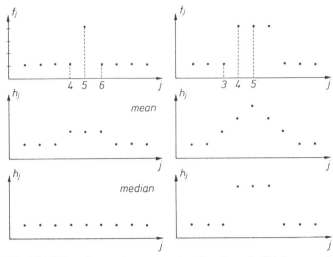

Fig. 2.16. Effect of a running mean and median filter of width 3 on two functions f_j (influence of beginning and end of f_j is neglected). The small discontinuity is completely erased by the median operation, the broad discontinuity is completely preserved

Fig. 2.17a,b. In (a) the median is shown for $f_j, j = 4, 5, 6$ of the upper left of Fig. 2.16; (b) shows the same for $f_j, j = 3, 4, 5$ of the upper right

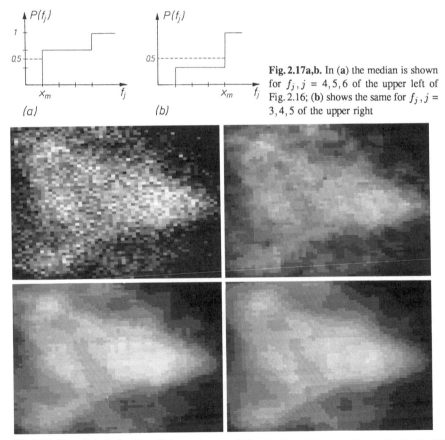

Fig. 2.18. The original of a diastolic image in a gated blood pool study (*upper left*), result after 3×3 median filtering (*upper right*), 5×5 median filtering (*lower left*), and 7×7 median filtering (*lower right*)

$$h = f \oplus s, \quad \text{(dilation)} \tag{2.4.11}$$

$h_{ij} = 1$ if $[f_{i+\mu,j+\nu} = 1$ for at least one $(\mu, \nu) \in S]$, else $h_{ij} = 0$.

If one visualizes S as containing a "one" at every position $(\mu, \nu) \in S$ and a "zero" in the other positions, then a binary mask s results and binary erosion and dilation may be computed from s. The origin of the mask s is centered on an image point f_{ij}. If every one-element in s is covered by a one-element in f, the pixel h_{ij} in the eroded image has value one. If at least one one-element in s is covered by a one-element in f, the pixel h_{ij} in the dilated image has value one. Two additional elementary operations are defined by

$$h = f \circ s = (f \oplus s) \ominus s, \quad \text{(closing)} \tag{2.4.12}$$

$$h = f \bigcirc s = (f \ominus s) \oplus s, \quad \text{(opening)}. \tag{2.4.13}$$

The above-defined operations are often applied iteratively. The erosion and dilation operations tend to remove image details or to smooth the image. In addition, erosion causes a shrinking, dilation an expansion of the image. The opening of an image eliminates small object structures like narrow bridges between larger structures and sharp peaks, the closing tends to fill in small background areas like small holes or gaps. Figure 2.19 illustrates these operations with a simple example. Several other operations are available, but the above ones should suffice to demonstrate the basic principles.

2.4.3 Morphological Operations on Gray Value Images

Morphological operations may be generalized to gray level images by using the max and min operators and allowing a set S having tuples $(\mu, \nu, s_{\mu\nu})$. Now one may visualize S as a structuring element s having gray values $s_{\mu\nu}$ defined at certain coordinates (μ, ν). We assume $s_{\mu\nu} \in \{0, 1, \ldots, L-1\}$, that is, there may be L gray levels in s; similarly, there may be L gray levels in f. The elementary morphological operations are defined by

$$h = f \ominus s, \quad \text{[(gray level) erosion]}$$
$$h_{ij} = \min \{f_{i+\mu,j+\nu} - s_{\mu\nu}, \text{ for all } (\mu, \nu) \in S\}, \tag{2.4.14}$$

note that h_{ij} is undefined, if $f_{i+\mu,j+\nu}$ is undefined for some μ, ν;

$$h = f \oplus s, \quad \text{[(gray level) dilation]}$$
$$h_{ij} = \max \{f_{i+\mu,j+\nu} + s_{\mu\nu}, \text{ for all } (\mu, \nu) \in S\}, \tag{2.4.15}$$

note that for definition of h_{ij} it is sufficient if $f_{i+\mu,j+\nu}$ is defined for some μ, ν.

The gray level closing and opening are defined in direct analogy to (2.4.12,13). When comparing the morphological operations in (2.4.14, 15) to the convolution operation in (2.3.10), it becomes evident that the role of the impulse response g is in analogy to the role of the structuring element s. When comparing (2.4.4, 5) to (2.4.14, 15), it becomes evident that the former are a special case of the latter with $s_{\mu\nu} = 0$.

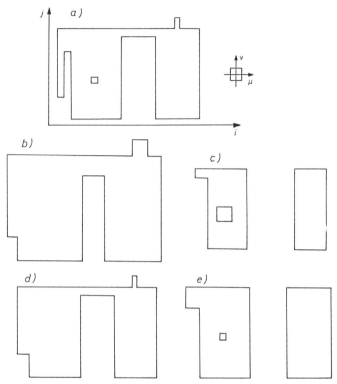

Fig. 2.19a–e. An example of morphological operations showing: (**a**) the original binary image; (**b**) the dilated image; (**c**) the eroded image; (**d**) the result of closing the image; (**e**) the result of opening the image

2.4.4 Morphological Algorithm

A generalization of a morphological operation may be defined in various ways, for example, by allowing input images f_1, f_2, f_3, structuring elements s_1, s_2, the m-fold iteration of an elementary morphological operation (*emo*), and the combination of partial results by arithmetic or logical operations (*alo*). An example of a *generalized morphological operation* is

$$h = (f_1 \text{ emo } s_1^m) \text{ alo } (f_2 \text{ emo } s_1 \text{ emo } s_2) \text{ alo } f_3 \quad . \tag{2.4.16}$$

A *morphological algorithm* is a sequence of elementary or generalized morphological operations. As with many synthesis problems, there is in general no systematic way to synthesize a morphological algorithm given an input image (e.g. a gray level image taken from a certain task domain) and a desired output image (e.g. containing the contours of objects). A simple morphological algorithm is (2.4.6) which has the form $(f \text{ emo}_1 s) \text{ alo } (f \text{ emo}_2 s)$. In comparison to the linear filters discussed above the morphological operations have the advantage of requiring no or only very simple arithmetic operations. Several special hardware structures have been designed to realize morphological operations.

2.5 Linear Prediction

2.5.1 Predictor Coefficients

Although the *linear prediction method* is closely related to the spectrum, that is the Fourier transform of a one-dimensional pattern, it is treated in a separate section because of its paramount importance for speech processing and processing of waveforms in general. Only a brief account is given here treating the basic properties.

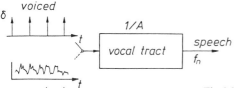

Fig. 2.20. A model for speech production

Generation of speech may be modeled as indicated in Fig. 2.20. A linear system $1/A$ is used to approximate the *human vocal tract* for *speech production*. Its input $e(t)$ is a train of impulses $\delta(t)$ spaced by $l\triangle T$, with l integer, for voiced speech and white noise for unvoiced speech. Its output is the speech wave $f(t)$. Sampling $e(t)$ at time $t = n \triangle T$ and denoting the z transform of the discrete signals by $E(z)$ and $F(z)$ one obtains

$$F(z) = \frac{E(z)}{A(z)} \tag{2.5.1}$$

where A is referred to as the *inverse system* or *inverse filter*. It is approximated by

$$A(z) = \sum_{\mu=0}^{m} a_\mu z^{-\mu}, \quad a_0 = 1 \quad . \tag{2.5.2}$$

Multiplying (2.5.1) by $A(z)$ and going back to the discrete time domain results in

$$\begin{aligned} e_n &= \sum_{\mu=0}^{m} a_\mu f_{n-\mu} \quad , \\ &= f_n + \sum_{\mu=1}^{m} a_\mu f_{n-\mu} \quad , \\ f_n &= e_n - \sum_{\mu=1}^{m} a_\mu f_{n-\mu} \quad . \end{aligned} \tag{2.5.3}$$

This may be interpreted in the form that the nth output sample is determined by the nth input sample and the m preceding output samples – strictly speaking, this is an approximation of f_n since $A(z)$ in (2.5.2) is only an approximation of the vocal tract. Since e_n is not available to the observer, one may try to predict f_n only by observation of the m preceding samples, which results in the linear predictor equation

$$\hat{f}_n = -\sum_{\mu=1}^{m} a_\mu f_{n-\mu} \quad . \tag{2.5.4}$$

It remains to determine the m *predictor coefficients* a_μ. This is done by minimization of the sum of squared errors

$$\varepsilon = \sum_{n=n_0}^{n_1} (f_n - \hat{f}_n)^2 \quad . \tag{2.5.5}$$

Taking partial derivatives of ε with respect to $a_\nu, \nu = 1, \ldots, m$ and equating them to zero results in a system of m linear equations for a_μ

$$\sum_\mu a_\mu \sum_n f_{n-\mu} f_{n-\nu} = -\sum_n f_n f_{n-\nu}, \quad \nu = 1, \ldots, m \quad . \tag{2.5.6}$$

In the *autocorrelation approach* it is assumed that $n_0 = -\infty, n_1 = \infty$ and $f_n = 0$ for $n < 0$ and $n \geq N$. With the notation

$$r_{|\nu-\mu|} = \sum_{n=0}^{N-1-|\nu-\mu|} f_n f_{n+|\nu-\mu|} \tag{2.5.7}$$

(2.5.6) may be written as

$$\sum_{\mu=1}^{m} a_\mu r_{|\nu-\mu|} = -r_\nu, \quad \nu = 1, \ldots, m \quad , \tag{2.5.8}$$

where $r_{|\nu-\mu|}$ is the *short-time autocorrelation*. In this case the system of linear equations can be solved very efficiently. For the so-called *covariance method* the reader is referred to the references. The matrix of the linear system of equations in (2.5.8) has a special form since the elements are $r_{\nu,\mu} = r_{|\nu-\mu|}$; it is called a *Toeplitz matrix*. The resulting system of equations can be solved iteratively, for example. The iteration starts with a linear prediction equation of order $j = 0$, and proceeds from j to $j+1$ in the steps

$$a_0^j = 0 \quad , \tag{2.5.9}$$

$$a_\mu^j = a_\mu^{j-1} + k_j a_{j-\mu}^{j-1}, \quad \mu = 1, \ldots, m-1 \tag{2.5.10}$$

$$a_j^j = k_j , \tag{2.5.11}$$

$$k_j = -\frac{\sum_{\mu=0}^{j-1} a_\mu^{j-1} r_{|j-\mu|}}{\sum_{\mu=0}^{j-1} a_\mu^{j-1} r_\mu} \quad . \tag{2.5.12}$$

The above equations are solved starting with $j = 1$ and continuing until the final number m of predictor coefficients is computed.

It can be seen from the above discussion that the predictor coefficients may be obtained in a simple manner from samples of speech. This determines also the inverse filter in (2.5.2). In general, f_n may be any sample value, not necessarily taken from speech; other areas of application for linear prediction methods are, for instance, seismics and EEG processing.

2.5.2 Spectral Modeling

An important technique in speech processing is the characterization of the spectral properties of speech. This may be done by a short-time Fourier transform, sampling the output of a bank of band-pass filters, analysis by synthesis, and also by linear prediction. Some advantages of the last approach are that a smoothed speech spectrum is obtained, that resonances or *formants* of voiced speech are accurately represented, and that a small number m of predictor coefficients suffices. From (2.5.1) it is evident that a filter

$$G(z) = \frac{\sigma}{A(z)} \tag{2.5.13}$$

should be used to model the vocal tract. The constant σ is a gain factor which matches energy of the data to unit sample response. It can be shown that

$$\sigma^2 = \sum_{\mu=0}^{m} a_\mu r_\mu \quad . \tag{2.5.14}$$

The model for the spectrum of the speech data then is

$$|G(z)|^2\big|_{z=\exp(i\alpha)} = \frac{\sigma^2}{|A(z)|^2}\big|_{z=\exp(i\alpha)} \quad , \tag{2.5.15}$$

that is, $1/|A|^2$ evaluated on the unit circle, and α is angular frequency in cycles/rad.

Numerical computation of the *model spectrum* requires specification of some parameters. Since a set of patterns $\{f^r(t)\}$ representing speech is much more restricted than a set $\{\boldsymbol{f}^r(\boldsymbol{x})\}$ representing images (which may be gray-level, color, or multispectral images or image sequences from objects having sizes in the range from μm to km), some remarks on the choice of parameters can be made based on experimental evaluations. A standard sampling frequency for speech is between 8 and 20 kHz; in particular accurate representation of some fricatives may require up to 20 kHz. To avoid aliasing as mentioned in Sect. 2.3.3 speech has to be low-pass filtered at half the sampling frequency. If f_s is the sampling frequency in kHz, from experimental results the number m of predictor coefficients should be $m = f_s + 4$ or $m = f_s + 5$. The number N of samples in (2.5.7) is based on an analysis interval (or *window* or *frame*) of 15–35 ms, which for 20 ms and 10 kHz results in $N = 200$ sample values. Computation of a_μ is repeated about every 10 ms. Since only N sample values $f_0, f_1, \ldots, f_{N-1}$ are used, this amounts to a rectangular windowing of the data. But in order to avoid discontinuities, which may cause distortions of the spectrum, it is useful to apply a nonrectangular window. Examples are given in (2.2.9). With weights w_n a windowed sequence

$$f'_n = w_n f_n, \quad n = 0, 1, \ldots, N - 1 \tag{2.5.16}$$

is computed and used in (2.5.7). The useful effect of high-frequency emphasis was already mentioned in Sect. 2.3.6. A simple way of doing this is to use instead of (2.5.16) the differenced signal

$$f'_n = w_n (f_{n+1} - f_n), \quad n = 0, 1, \ldots, N - 1 \quad . \tag{2.5.17}$$

The spectrum usually is graphically represented as the logarithm of the absolute value of the Fourier transform and denoted here by $\text{LS}^{(x)}$ as the *logarithmic spectrum* of sample values f_n. If F_ν are the discrete Fourier coefficients of f_n from the one-dimensional version of (2.3.13) (where M should be replaced by N), then the *logarithmic data spectrum* is

$$\text{LS}^{(f)} = 10 \log_{10} |F_\nu|^2 \quad . \tag{2.5.18}$$

The *logarithmic model spectrum* from (2.5.15) is computed efficiently as follows. The discrete Fourier transform of

$$a_t = (1, a_1, a_2, \ldots, a_m, 0, \ldots, 0) \tag{2.5.19}$$

is computed from (2.3.13) yielding discrete Fourier coefficients α_ν. The number M of components of a depends on the desired frequency resolution f_r and sampling frequency f_s by

$$M > \frac{f_s}{f_r} \quad . \tag{2.5.20}$$

With $f_s = 10$ kHz, $f_r = 0.04$ kHz one gets $M > 250$ such that $M = 256$ would be adequate for a FFT computation. The logarithmic model spectrum then is

$$\text{LS}^{(\hat{f})}_\nu = 10 \log_{10} \sigma^2 - 10 \log_{10} |\alpha_\nu|^2 ,$$

$$= 10 \log_{10} \frac{\sigma^2}{|A \exp(i 2\pi \nu / M)|^2} , \quad \nu = 0, 1, \ldots, M - 1 . \tag{2.5.21}$$

Because of symmetry only half of the coefficients have to be computed. The appearance of a model spectrum of voiced speech is shown in Fig. 2.21. It is mentioned that *pole enhancement* is possible if a_μ in (2.5.19) is replaced by $a_\mu d^\mu, d > 1$, and that modeling may be restricted to a certain frequency range by *selective linear prediction*. The predictor coefficients a_μ, the model spectrum

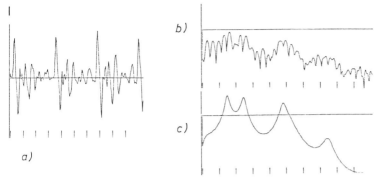

Fig. 2.21. (a) A 20 ms window of the vowel "a" in the German word "Faß" – this is pronounced approximately as the vowel "u" in the English word "but"; (b) the DFT spectrum corresponding to 10 kHz sampling frequency; (c) the logarithmic model spectrum using $m = 13$ predictor coefficients

$LS^{(f)}$ or other parameters derived from a_μ are used to parametrically represent a short window of speech data.

2.6 Bibliographical Remarks

A comprehensive treatment of coding techniques is given in [2.3, 13, 14]. Additional material on transform coding and coding of line patterns may be found in [2.15, 16] and [2.17–19], respectively. The effects of a finite sampling function are investigated in [2.20]. Recent results in image coding are given in [2.21].

Linear and nonlinear normalization of size and time are treated in [2.22–24]. Examples of papers on gray level scaling are [2.25–29]. Various color metrics are given in [2.20, 30]. Estimation of surface reflectivity from color images is treated in [2.31]. Examples of geometric corrections and registration of images are given in [2.32–35]. Additional material on camera calibration can be found in [2.36–39].

There is a lot of material on linear systems and filters, textbooks are [2.8, 40]. Discrete and fast Fourier transform are covered in detail in [2.10, 41]. Homomorphic systems were introduced in [2.42] and are also treated in [2.8, 43]. Various applications are discussed in [2.40, 44–48]. Image restoration is covered in detail in [2.12, 49]. Linear filters can also be realized by optical methods [2.50–52]. The computation of resolution hierarchies was investigated in [2.53].

The linear prediction method is presented with full details in [2.54–56]. Applications in speech processing, EEG analysis, and seismic waveform processing are treated in [2.23, 57, 58].

Additional material on morphological operations can be found, for example, in [2.59–65]. Filtering using order statistics is treated in [2.66].

The approach to speaker normalization outlined in Sect. 2.2.12 has been studied in detail in [2.67] where a comparison with other approaches is also given, additional investigations are given in [2.68].

3. Segmentation

In this chapter it is assumed that a pattern is available which was preprocessed in the best possible way. Referring back to Sect. 1.2, where "analysis" and "description" were defined, it is necessary to decompose or to segment a pattern into simpler constituents or *segmentation objects*. Since the most important examples of complex patterns are images and connected speech, these will be treated in the following with emphasis on images.

Decomposition or *segmentation* of a pattern in such a way that meaningful simpler constituents are obtained is a very complicated task. Usually this task is structured into several steps which are related to each other. Obtaining a meaningful segmentation requires a good deal of information about the task domain. As outlined in Sect. 1.4, the approach to pattern analysis advocated here is to modularize the problem of analysis. One of the modules is *"methods"* as indicated in Fig. 1.4. Therefore, this chapter is limited to methods of segmentation which use little, if any, knowledge about structural properties of patterns. Such an initial segmentation is necessary in a flexible and general system for pattern analysis. Incorporation of knowledge is another module and will be discussed later. Judging the success of segmentation involves problems similar to those mentioned at the beginning of Chap. 2.

The methods of segmentation discussed here are the following:

1. *Common principles* – some remarks which may help to unify the abundance of different methods.
2. *Thresholding* – cutting slices out of a pattern.
3. *Gray level changes* – looking for changing properties in an image.
4. *Lines* – computing lines given a sequence of points.
5. *Regions* – looking for homogeneous properties.
6. *Texture* – describing complicated surface properties.
7. *Motion* – making use of time as an additional variable.
8. *Depth* – recovering three-dimensional information.
9. *Shape from shading* – recovering the 3D shape from gray values.
10. *Segmentation of speech* – finding subword units of (connected) speech.

It is beyond the scope of this chapter to treat all these methods in detail.

3.1 Common Principles

It seems reasonable to look for some basic ideas which underlie many different methods of segmentation. At a fairly high level of generalization such ideas may be found. We state two such ideas. The first is that the result of segmentation is a set of *segmentation objects* where each object has certain *attributes*; the second is that segmentation requires some changes or fluctuations of suitable pattern properties.

In image segmentation important types of segmentation objects are the junction (an image point where two or more lines meet), the line segment, the line, the region, and the volume. Possible types of attributes are the location in two-dimensional image coordinates or in three-dimensional world coordinates, gray level and color, texture, motion, depth, surface normal, shape, and reliability or certainty. Apparently, not every type of attribute is adequate for every type of segmentation object or every type of image (see Sect. 1.6). Alternative or competing results of segmentation are represented, for example, by two segmentation objects of type "region" having overlapping positions in the image. There is evidence that in human perception of objects an image is segmented such that a set of volumetric objects is obtained. Such volumetric objects are blocks, cylinders, wedges, or cones. A small set (≈ 40) of segmentation objects is derived from attributes and relations like curvature, collinearity, symmetry, parallelism, and cotermination of lines; these are fairly invariant under different viewing angles.

In speech segmentation important types of segmentation objects are the phone component, the phone, the diphone, the demisyllable, and the syllable. A short discussion of these units is given in Sect. 3.10. Examples of types of attributes are the class name, the begin and end time, the reliability, and some acoustic properties like pitch or loudness. The number of different types of segmentation objects is fairly small (≈ 60), if phones are selected.

Results of segmentation may be represented by a structure

I: (name of segmentation object SO: type T
(attribute A: type of attribute TA, value is a real number R or an element from a finite set of names of attribute values AV)*,
reliability CF: real number R).

This is abbreviated by

$$\text{I: } (\text{SO: T; } (\text{A: TA, } (R \cup AV))^*, \text{ CF : R}) \; . \qquad (3.1.1)$$

Often it will be desirable to combine the simple results to larger entities already at the stage of segmentation. For example, one may wish to group a set of (nearly) collinear line segments to a straight line. To represent this result requires a structure

I: (name of segmentation object SO: type T
(attribute A: type of attribute TA, value is a real number R or an element from a finite set of names of attribute values AV)*,

(part P: structure I)*,
(structural relation between parts and/ or attributes S(A, P): real number measuring the fulfilment of the relation)*,
certainty CF: real number R),

$$I:(SO: T,(A:(TA,R \cup AV))^*, (P:I)^*, (S(P):R)^*, CF:R) . \quad (3.1.2)$$

In the above example the "straight line" is a structure I having as parts P a set of line segments which must meet the structural relation S(P) that they are collinear. If we allow in (3.1.2) the special case that there are no parts and no structural relations, the simpler structure (3.1.1) results. The structure I in (3.1.2) is sufficiently general to combine all results of segmenting a pattern $f^r(x)$ in an *initial description* I^r

$$f^r \to I^r \ . \qquad (3.1.3)$$

The second idea was that segmentation requires some changes or fluctuations in the pattern. Segmentation of a pattern represented by a function $f(x)$ is obviously impossible if this function is just a constant. To a human a constant function may invoke some sensory impressions, but is evidently without structure. Therefore, an idea common to all techniques of segmentation of patterns is that simple constituents are related to changes in $f(x)$. This is obvious for images and speech. Changes in $f(x)$ give hints for possible borders between segmentation objects. This idea is independent of any information about structural properties or the task domain.

On the other hand, it is obvious that changes may be neither necessary nor sufficient to make sure that a border exists. An example from speech demonstrates that changes are not necessary. Assume that simple constituents are words. The complex pattern "givemesomemilk" then should be segmented into words as "give-me-some-milk". But the pronunciation of the last two words is "s o m m i l k" with no change in pronunciation at the immediate "meaningful" border. In this case the segmentation requires consideration of changes in the neighborhood of the border and information about "meaningful" words (the segmentation "give-me-some-ilk" would not make sense). The same example also clarifies that changes are not sufficient because there are, for instance, also changes inside the word "some", but these do not separate it because this, again, would not make sense.

This discussion shows that changes are necessary to allow segmentation of patterns, but that they alone, in general, are not adequte to find the exact location of the border between simple constituents. Rather, they give a hint as to where such borders might be expected. Sometimes the task is facilitated by the fact that there are task domains where a change is indeed necessary for a border between constituents. An example would be a figure on a background. Always the problem is complicated by the fact that noise makes detection of changes prone to errors. There seem to be two basically different approaches to detection of changes.

The first approach tries to directly find changes. To do this a suitable set of parameters is chosen and evaluated for different values of x on $f(x)$, that is, at different locations of space and/or time. If the values of the parameters change significantly, this is taken as an indication of a possible transition from one simple constituent to another one. A simple example is to choose the gray level as parameter and to compute the difference in gray level of adjacent points as a measure of change.

The second approach tries to trace the homogeneous part of a pattern. Again a set of parameters is chosen and evaluated for different values of x. If the values of the parameters do not change significantly, this is taken as evidence that no transition to another segmentation object occurred. A simple example might be to use color as a parameter and to investigate whether points in the vicinity of a point have similar color or not.

Despite these common aspects there is an abundance of different methods available. For instance, they differ in the set of parameters, the attempts to overcome errors caused by noise, the judgement of significant changes, and so on. The extreme importance of adequate pattern recording becomes evident here. A parameter which is excluded by the recording device cannot contribute to the detection of changes or homogeneity.

3.2 Thresholding

3.2.1 Obtaining a Binary Image

A *threshold operation* transforms a sampled image f into another one h having only binary values. The basic operation is

$$f \to h,$$
$$h_{jk} = \begin{cases} 1 & \text{if } f_{jk} > \theta \\ 0 & \text{otherwise.} \end{cases} \quad (3.2.1)$$

Proper selection of θ is essential for the success of thresholding. Three types of thresholds are the global, local, and dynamic threshold.

A *global threshold* is

$$\theta = \theta(f) \quad , \quad (3.2.2)$$

which depends on the gray values of the image. This kind of threshold is used extensively in cell analysis to separate objects from the background. For instance, blood cells may be scanned in two spectral channels (yellow and green). The histogram of the yellow channel has the trimodal shape of Fig. 3.1 with the modes belonging to background, red cells, and nucleus (from left to right). On the basis of this histogram thresholds may be selected to separate various objects from the background; an example is Θ in Fig. 3.1. A modification of this technique is to first apply a Laplacian operator to the image. Then only points where the value

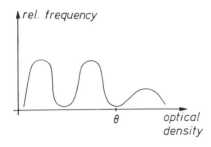

Fig. 3.1. Histogram of gray levels of blood cells which exhibits three modes with the leftmost belonging to background

of the Laplacian is above a certain limit are considered. Thus, most points of f are ignored except those in the vicinity of a border between objects or object and background. Because of the symmetry of the Laplacian, points on the object and background occur nearly equally frequently. Therefore, this *filtered histogram* has two peaks of nearly equal height. This technique is useful if the (unfiltered) histogram has a broad valley and two modes of very different size.

Another approach to threshold selection assumes that the gray level histogram is a linear superposition of two nearly normal distributions, one belonging to the object, the other to the background. Two pairs of parameters (m_i, σ_i), $i = 1, 2$ are determined such that the sum of the two distributions $N(m_i, \sigma_i)$ gives the best approximation of the actual histogram. The parameters may be computed by selecting an arbitrary threshold θ, for example, the mean gray value. Parameters (m_1, σ_1) and (m_2, σ_2) are computed from the gray values to the left and to the right of θ, repsectively. The mean square error between the two normal distributions and the histogram is computed. By systematically increasing and decreasing θ the parameters yielding the smallest mean square error are determined. The corresponding value of θ is selected as a threshold.

A *local threshold* is

$$\theta = \theta(\boldsymbol{f}, \boldsymbol{f}_s) \quad , \tag{3.2.3}$$

where $\boldsymbol{f}_s$ is some suitable subimage of $\boldsymbol{f}$. This approach is often used to separate characters from background or to quantize circuit diagrams. With (3.2.3) it is possible to make θ dependent on local properties of an image and thereby to improve results of thresholding. One example for threshold determination according to (3.2.3) is to choose $\boldsymbol{f}_s$ as a rectangular neighborhood centered at the point under consideration and to adjust θ to the average gray level f_m of $\boldsymbol{f}_s$. A generalization would be to allow a nonlinear dependence between θ and f_m.

A *dynamic threshold* is

$$\theta = \theta(j, k, \boldsymbol{f}, \boldsymbol{f}_s) \quad , \tag{3.2.4}$$

which also depends on the position (j, k) of f_{jk}. Such a dependence has been used in the extraction of boundaries in radiographic images.

An example of the result of a threshold operation is given in Fig. 3.2.

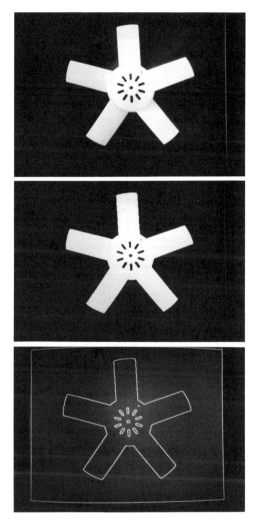

Fig. 3.2. The figure shows, (*from top to bottom*) a gray-level image, the thresholded image, and the contours extracted from the thresholded image

3.2.2 Operations on Binary Images

Once a binary pattern is obtained there are a variety of operations to do further processing. Among them are contour following, noise cleaning, shrinking, skeletonization, and the morphological operations discussed in Sect. 2.4.

Contour following means that the border between object ($f_{jk} = 1$) and background ($f_{jk} = 0$) is traced. This may be done, for instance, by first searching the image line by line until a point with $f_{jk} = 1$ is found. Then the contour is traced by the rule that a left turn is made at a point with $f_{jk} = 1$, and a right turn is made at a point $f_{jk} = 0$.

Noise cleaning is used to eliminate holes in an object or a few isolated points on the background. Logical operations or mask operations may be used to do this or also averaging over small neighborhoods is applicable.

Shrinking and skeletonization is used to reduce an object to a line-like pattern. The line may be the boundary or the medial axis. A variety of further operations is discussed in the references.

3.3 Gray-Level Changes

3.3.1 Differences in Local Neighborhoods

Extraction of *contours* as segmentation objects is motivated by the observation that in many cases contours provide important information about an object or a scene. This is illustrated by Fig. 3.3.

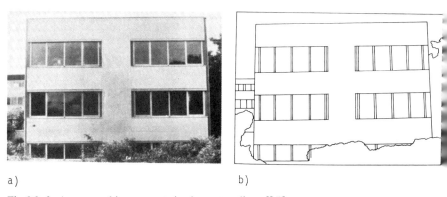

Fig. 3.3a,b. A scene and its representation by contour lines, [3.1]

Much effort has been taken to extract contours. Furthermore, reduction of gray-level images to contours results in significant information reduction and should facilitate further processing. The fundamental aspect of a contour is a change of gray-level. Therefore, a method to detect such changes is required in any case. It has already been mentioned that noise is superimposed on any image. This means that we never encounter an ideal contour, for instance, as shown in Fig. 3.4a, but only "real" contours, for instance, as in Fig. 3.4b. One consequence

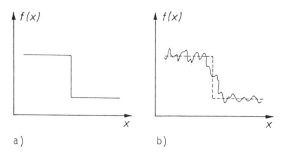

Fig. 3.4a,b. An example of an ideal contour in (**a**) and an example of what is recorded as a real contour due to noise in (**b**)

is that many algorithms for contour extraction tailored for particular task domains have been developed. The other consequence is that contour extraction is done in at least two steps. In the first step points or line segments are determined where gray-level changes occur. In the second step one tries to link a subset of these points (or segments) by a straight or curved line. Usually the second step is further subdivided.

Gray-level changes may be obtained from the derivatives of $f(x, y)$ or differences of the picture matrix $\boldsymbol{f} = [f_{jk}]$. One of the first operators to obtain a differenced image $\boldsymbol{h} = [h_{jk}]$ is the *Robert's cross*

$$f_x = f_{jk} - f_{j+1,k+1},$$
$$f_y = f_{j,k+1} - f_{j+1,k}, \qquad (3.3.1)$$
$$h_{jk} = \sqrt{f_x^2 + f_y^2} \text{ or } h_{jk} = |f_x| + |f_y|.$$

In the above equation and those to follow, the indices j, k are assumed in a standard x, y-coordinate system, that is, starting at the lower left as indicated in (2.1.1). Directional information is available from f_x, f_y. A point $h_{jk} \in \boldsymbol{h}$ is considered to possibly lie on a contour of $\boldsymbol{f}$ if h_{jk} exceeds a threshold, that is, if there is a significant change of gray-level in $\boldsymbol{f}$. Because of Fig. 3.4b it is evident that there will be many points not lying on a contour but indicating a change of gray-level. It is clear that achieving improved insensitivity to noise is of great importance. An example is the *Sobel operator*

$$f_x = (f_{j+1,k+1} + 2f_{j+1,k} + f_{j+1,k-1}) - (f_{j-1,k-1} + 2f_{j-1,k} + f_{j-1,k+1}),$$
$$f_y = (f_{j-1,k+1} + 2f_{j,k+1} + f_{j+1,k+1}) - (f_{j+1,k-1} + 2f_{j,k-1} + f_{j-1,k-1}). \qquad (3.3.2)$$

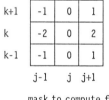

Fig. 3.5. The Sobel operator as a mask operation

A result of this operator is given in Fig. 3.9 below. The operator may be visualized as the masks indicated in Fig. 3.5. The center of the mask is positioned on an image point (j, k); pixels around (j, k) are multiplied by the weight given in the mask and added; the sum is the new pixel value at (j, k) according to (3.3.2).

The idea of (3.3.2) of using more points for better noise immunity may be further developed. Noise is reduced if the image is averaged over a neighborhood

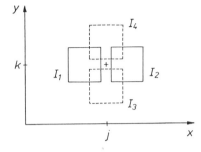

Fig. 3.6. By averaging over intervals $I_\nu, \nu = 1, 2, 3, 4$ the effect of noise on detection of gray-level changes is reduced

of appropriate size. Using the notation of Fig. 3.6 an example of such an operation is

$$h'_{jk} = \left| \sum_{\mu,\nu \in I_1} f_{\mu\nu} - \sum_{\mu,\nu \in I_2} f_{\mu\nu} \right| ,$$

$$h''_{jk} = \left| \sum_{\mu,\nu \in I_3} f_{\mu\nu} - \sum_{\mu,\nu \in I_4} f_{\mu\nu} \right| , \qquad (3.3.3)$$

$$h_{jk} = \max \{h'_{jk}, h''_{jk}\} .$$

Modifications of this operation are determination of h_{jk} according to (3.3.1), inclusion of intervals belonging to the $\pm 45°$ directions around f_{jk}, or variation of interval size until a maximum of h_{jk} is obtained. Averaging may be considered as a low-pass filtering operation as discussed in Sect. 2.3.6. By a combination of filtering and computaton of differences it is obviously possible to generate a variety of other operators. In a particular task domain it will be necessary to experimentally select the best-suited operator. Another example is the *Laplacian operator*, which yields

$$h(x,y) = \frac{\partial^2 f(x,y)}{\partial x^2} + \frac{\partial^2 f(x,y)}{\partial y^2} = \nabla^2 f(x,y) \qquad (3.3.4)$$

and was mentioned already in (2.3.47) for image enhancement; it makes use of second-order differences. A digital version is

$$h_{jk} = \left(f_{j,k-1} + f_{j,k+1} + f_{j+1,k} + f_{j-1,k}\right) - 4f_{jk} . \qquad (3.3.5)$$

For instance, a combination of (3.3.2) and (3.3.5) can be used to increase the reliability of contour point detection. From (3.3.2) a matrix

$$h'_{jk} = \sqrt{f_x^2 + f_y^2}$$

and from (3.3.5) a matrix

$$h''_{jk} = |h_{jk}|$$

are computed. From these a new binary matrix h is obtained by

$$h_{jk} = \begin{cases} 1 & \text{if } h'_{jk} > \theta' \text{ and } h''_{jk} > \theta'' \\ 0 & \text{otherwise} \end{cases}, \quad (3.3.6)$$

A value $h_{jk} = 1$ in this matrix indicates that a contour may exist in this point. In addition, directional information is also available from f_x and f_y of (3.3.2).

It was mentioned in Sect. 2.3.6, that low-pass filtering tends to reduce noise but also to blur details. The above-mentioned averaging operations, particularly (3.3.3), have the same effect. Averaging not only reduces noise but also turns sharp edges into smooth edges. To achieve sharp localization of a contour a nonlinear operator may be used. An integer number q and a number $a = 2^q$ are selected and a difference

$$h^{(a)}_{jk} = \frac{1}{a(2a+1)} \left| \sum_{\mu=-a+1}^{0} \sum_{\nu=-a}^{a} f_{j+\mu,k+\nu} - \sum_{\mu=1}^{a} \sum_{\nu=-a}^{a} f_{j+\mu,k+\nu} \right| \quad (3.3.7)$$

is computed which is similar to (3.3.3). The quantity

$$h_{jk} = \prod_{q=1}^{q_0} h^{(a)}_{jk}, \quad a = 2^q \quad (3.3.8)$$

tends to take large values only at the location of an edge because the product in (3.3.8) is large only if all factors are large. Since (3.3.7) belongs to the x direction, an equivalent computation has to be done for the y direction. It is mentioned that efficient computation of (3.3.8) is possible by using intermediate results of a $2^{q_0} \times 2^{q_0}$ neighborhood.

3.3.2 Contour Filters

Use of concepts from linear filters, as introduced in Sect. 2.3.2, may be further developed by consideration of additive noise. Similar to (2.3.36), let the recorded pattern $f(x,y)$ consist of an ideal signal $s(x,y)$ corrupted by additive noise $n(x,y)$ to yield

$$f(x,y) = s(x,y) + n(x,y) \quad . \quad (3.3.9)$$

One approach is to determine a filter $G_m(\xi, \eta)$ which at its output maximizes the *signal-to-noise ratio*

$$\frac{s_o^2}{E\{n_o^2\}} = \max \quad (3.3.10)$$

if its input is f.

Theorem 3.3.1. The linear filter (known as *matched filter*) maximizing (3.3.10) is given by

$$G_m(\xi, \eta) = \frac{S^*(\xi, \eta) \exp[-i(\xi x_0 + \eta y_0)]}{S_{nn}(\xi, \eta)}, \quad (3.3.11)$$

where $S^*(\xi, \eta)$ is the complex conjugate of the Fourier transform of $s(x, y)$, $S_{nn}(\xi, \eta)$ is the spectral density of homogeneous noise $n(x, y)$, and (x_0, y_0) is the point where the maximum occurs.

Proof. See, for instance, [3.2], pp. 553–558. □

The above result may be used for contour detection by identifying s with contour segments of fixed shape, orientation, and location (contour masks). If in addition n is white noise, (3.3.11) reduces to

$$G = S^* \exp[-i(\xi x_0 + \eta y_0)] , \quad g(x, y) = s(x_0 - x, y_0 - y) . \qquad (3.3.12)$$

The impulse response is the signal rotated by π and shifted by (x_0, y_0). Convolution of g and f is the same as cross correlation of s and f. If s is some contour segment, cross correlation of s and f will yield high output at locations (x_0, y_0) where an edge similar to s is present. A generalization to pth order gradient of this approach is possible.

Another approach is to use a minimum mean-square-error criterion to derive a recursive filter for contour detection. It is intended to obtain the Laplacian operator of the ideal image $s(x, y)$ in (3.3.9). The Laplacian in (3.3.4) is attenuated at high frequencies to give

$$G(\xi, \eta) = (\xi^2 + \eta^2) \exp(-\xi^2 - \eta^2) \qquad (3.3.13)$$

as the filter characteristic. Let $h(x, y)$ be the result of passing $s(x, y)$ through a filter with impulse response $g(x, y)$ [the FT of g being G in (3.3.13)], as shown in Fig. 3.7. Then a filter with impulse response g_0 is desired such that passing $f(x, y)$ through g_0 yields an estimate $\hat{h}(x, y)$ of $h(x, y)$ which minimizes mean-square error

$$\varepsilon = E\left\{[h(x, y) - \hat{h}(x, y)]^2\right\} . \qquad (3.3.14)$$

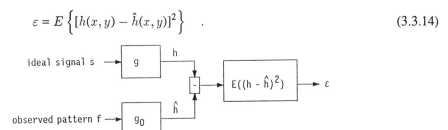

Fig. 3.7. Mean square error criterion for edge detection

Theorem 3.3.2. *The filter characteristic G_0 minimizing (3.3.14) is*

$$G_0(\xi, \eta) = \frac{G(\xi, \eta) S_{ss}(\xi, \eta)}{S_{ss}(\xi, \eta) + S_{nn}(\xi, \eta)} . \qquad (3.3.15)$$

Spectral densities of $s(x, y)$ and (homogeneous) $n(x, y)$ are denoted by S_{ss} and S_{nn} in (3.3.15), and s and n are assumed to be uncorrelated.

Proof. Follows as a generalization of Wiener filtering for an estimate $\hat{s}$ of s; see, for instance, [3.3] . □

To obtain S_{ss} a model of the process "images with contours" has to be developed. With this a *recursive digital filter* is obtained to compute the estimate $\hat{h}$ from

$$\hat{h}_{jk} = \sum_{\mu=0}^{m_1} \sum_{\nu=0}^{n_1} b_{\mu\nu} f_{j-\mu,k-\nu} - \sum_{\substack{\mu=0 \\ \mu=\nu\neq 0}}^{m_2} \sum_{\nu=0}^{n_2} a_{\mu\nu} \hat{h}_{j-\mu,k-\nu}, \quad j,k > 0 \quad . \quad (3.3.16)$$

Symmetric results are achieved by repeated application of the filter from each of four corners of the image. Tables of parameters $a_{\mu\nu}, b_{\mu\nu}$ are given in [3.4].

A recent result along this line is the simultaneous optimization of three criteria in the *Canny operator*. Its derivation starts from a one-dimensional observation $f(x)$ of the type (3.3.9) which consists of an ideal edge $s(x)$ superimposed by white noise $n(x)$ to yield

$$f(x) = s(x) + n(x) \quad . \quad (3.3.17)$$

It is assumed that the center of the edge is located at $x = 0$. The observation is passed through a linear filter $g(x)$ taking nonzero values in the finite interval $[-W, +W]$. An edge is to be marked by a local maximum in the filter output.

Three performance criteria are stated for the filter. The first is conventional signal-to-noise ratio as above and is called the *detection criterion*; it is defined by

$$\Sigma = \frac{\left| \int_{-W}^{W} s(-x) g(x) dx \right|}{n_0 \sqrt{\int_{-W}^{W} g^2(x) dx}} \quad , \quad (3.3.18)$$

where n_0 is the mean squared noise amplitude per unit length. The second criterion is a *localization criterion*. It measures the reciprocal of the root-mean-squared distance of the detected edge from the center of the true edge and is approximated by

$$\Lambda = \frac{\left| \int_{-W}^{W} s'(-x) g'(x) dx \right|}{n_0 \sqrt{\int_{-W}^{W} g'^2(x) dx}} \quad , \quad (3.3.19)$$

where s' and g' denote the first derivative of s and g, respectively. A useful filter should maximize both criteria. Therefore, one seeks a filter g yielding

$$(\Sigma \Lambda)_{\max} = \max_{\{g\}} \{\Sigma \Lambda\} \quad . \quad (3.3.20)$$

It follows from the Schwarz inequality that

$$g(x) = s(-x) \quad (3.3.21)$$

maximizes (3.3.20). This result coincides with (3.3.12) if we consider only one dimension and let $x_0 = 0$.

It is important to note that (3.3.20) optimizes the filter output at $x = 0$, but nothing is required about neighboring points. It is possible that additional relative maxima occur in the vicinity of the true edge. This effect would result in the detection of several edges in addition to the true edge. Therefore, it is desirable to introduce a third criterion, the *elimination of multiple responses*. A reasonable measure is the mean distance between maxima in the filter output if noise is the filter input. This distance is

$$\Delta_{\max} = 2\pi \sqrt{\int_{-\infty}^{\infty} g'^2(x)dx \Big/ \int_{-\infty}^{\infty} g''^2(x)dx} \quad . \tag{3.3.22}$$

It is reasonable to require that this distance should be some fraction k of the width W of the filter impulse response

$$\Delta_{\max} = kW \quad , \tag{3.3.23}$$

since the filter output will be concentrated in a region of width $2W$.

Now the problem is to find a filter $g(x)$ maximizing (3.3.20) subject to the multiple response constraint (3.3.22) and perhaps additional constraints; one such additional constraint is that the filter should have zero output for constant input. So far only numerical optimization seems possible, but no closed form solution. Examples of optimal filters for three edge types are given in Fig. 3.8. A comparison of Fig. 3.8f,g shows that the optimal filter for a step edge is closely approximated by the first derivative $\varphi'(x)$ of the Gaussian function

$$\varphi(x) = \exp\left[-x^2/(2\sigma^2)\right] \quad . \tag{3.3.24}$$

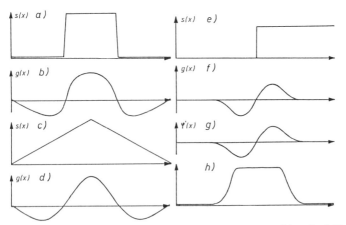

Fig. 3.8a–h. Examples of optimal edge filters: (a) ridge edge, (b) optimal filter for ridge edge, (c) roof edge, (d) optimal filter for roof edge, (e) step edge, (f) optimal filter for step edge, (g) first derivative of Gaussian function, (h) example of a projection function. Results according to [3.5]

3.3 Gray-Level Changes

This is also confirmed by a numerical analysis.

In two dimensions an edge has a certain direction which is defined here as the direction of the tangent to the edge contour. From the above discussion a two-dimensional edge operator can be derived by convolving a (one-dimensional) detection filter aligned normal to the edge direction with a projection function parallel to this direction. In the case of a step edge an efficient solution can be obtained if the projection function is also a Gaussian function having the same σ as the detection function. This gives a two-dimensional Gaussian function

$$\varphi(x,y) = \exp\left[-(x^2+y^2)/(2\sigma^2)\right] \quad . \tag{3.3.25}$$

The image $f(x,y)$ is convolved with $\varphi(x,y)$ to obtain

$$h(x,y) = f(x,y) \otimes \varphi(x,y) \quad . \tag{3.3.26}$$

From Sect. 2.3.4 it can be seen that this can be done efficiently since φ is a separable function. This convolution also reduces noise because it is a low-pass operation. In the filtered image an estimate $\hat{n}$ of the normal n to the edge direction is computed by

$$\hat{n}(x,y) = \frac{\nabla h(x,y)}{|\nabla h(x,y)|} \quad . \tag{3.3.27}$$

The local maxima of $\partial h(x,y)/\partial \hat{n}$ indicate edge locations. They may also be computed from the locations where

$$\frac{\partial^2 h(x,y)}{\partial \hat{n}^2} = 0 \quad . \tag{3.3.28}$$

This sequence of operations is called a *simple edge operator*. It is equivalent to convolving $f(x,y)$ with $\partial \varphi(x,y)/\partial \hat{n}$ by (2.3.26).

Since the signal-to-noise ratio of edges will vary from point to point in the image, it is useful to have simple edge operators of different width or different σ in (3.3.25). Since both Σ and Λ in (3.3.20) increase with the length of the projection function, it is useful to have operators with a certain direction. This can be achieved by adding the outputs of several simple edge operators having the same direction. Addition is performed in equal intervals on a straight line parallel to the edge direction. If the interval length is less than 2σ, the projection function is almost constant and gradually tends to zero at the ends. In an actual implementation several design choices are necessary. They concern the operator width, the length of the projection function, the directions of the operators and the thresholding for edge detection. An example of detected edges in a particular implementation is given in Fig. 3.9.

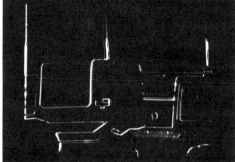

Fig. 3.9. An example of edge detection showing (*from top to bottom*) a gray-level image, the result of the Sobel operator, and the result of the Canny operator, [3.6]

3.3.3 Zero Crossings

Figure 3.10 shows that the second derivative of a function has a zero crossing at the position of maximal slope. Even if the increase of $f(x)$ is gradual and the first derivative has a broad maximum, the zero crossing is precisely located. Therefore, instead of using the maximum of the first derivative as in Sect. 3.3.1 it may be advantageous to use the zero crossing of the second derivative. It is necessary to take into account the effect of noise as illustrated in Fig. 3.4. An operator for detecting gray-level changes should meet three criteria. In order to reduce noise a smoothing operation should be applied which is band-limited in order to limit edge detection to a certain frequency interval. In addition, the smoothing operators should also be limited in the space domain in order not to

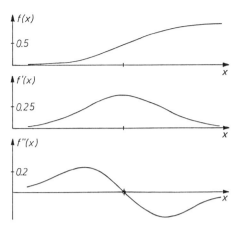

Fig. 3.10. Edge detection at zero crossings of the second derivative

blur the image too much. As a third criterion the operator should not be direction dependent in order to reduce computational effort.

There is no function which simultaneously is arbitrarily concentrated in the space and frequency domains. It may be shown that the Gaussian function gives the best compromise between limitation in both domains. Therefore, the gray-value function $f(x,y)$ is smoothed by a low-pass filter having a Gaussian impulse response

$$g(x,y) = (2\pi\sigma^2)^{-1} \exp\left[-(x^2+y^2)/(2\sigma^2)\right] \quad . \tag{3.3.29}$$

Then the Laplacian operator (3.3.4) is used to compute second derivatives. The result is

$$h(x,y) = \nabla^2[f(x,y) \otimes g(x,y)] = \left[\nabla^2 g(x,y)\right] \otimes f(x,y) \quad . \tag{3.3.30}$$

The last equation results from (2.3.26). It shows that smoothing and differentiation is achieved by convolution of f with the function

$$\nabla^2 g(x,y) = (2\pi\sigma^6)^{-1}(x^2+y^2-2\sigma^2)\exp\left[-(x^2+y^2)/(2\sigma^2)\right] \quad . \tag{3.3.31}$$

A plot of this circularly symmetric function is shown in Fig. 3.11a. Its Fourier transform is

$$\begin{aligned} \text{FT}\{\nabla^2 g(x,y)\} &= -(\xi^2+\eta^2)\,\text{FT}\,\{g(x,y)\} \\ &= -(\xi^2+\eta^2)\exp\left[-\sigma^2(\xi^2+\eta^2)/2\right] \\ &= -\omega^2 \exp\left(-\omega^2\sigma^2/2\right) \,. \end{aligned} \tag{3.3.32}$$

A plot of its absolute value is shown in Fig. 3.11b. It is a band-pass filter whose center frequency depends on σ.

The gray-level changes of $f(x,y)$ may thus be detected by passing f through a band-pass filter (3.3.32) and determining zero crossings of the result. Doing this for several values of σ gives a hierarchy of resolutions, details, and noise reduction. A small value of σ gives a low-pass filter in (3.3.29) with a high

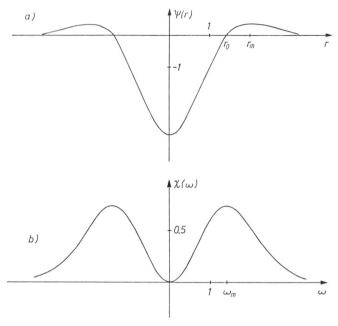

Fig. 3.11. (a) Plot of $\psi(r) = \nabla^2 g(r)$, $r^2 = x^2 + y^2$, in (3.3.31) with $\sigma = 1$. $\psi(0) = -1/(\pi\sigma^6)$, $r_0^2 = 2\sigma^2$, $r_m = 2\sigma$, $\psi(r_m) = [\pi\sigma^4 \exp(2)]^{-1}$. (b) Plot of $\chi(\omega) = |\text{FT}\{\nabla^2 g(r)\}|$ in (3.3.32) with $\sigma = 1$. $\chi(0) = 0$, $\chi(\omega_m) = (2/\sigma^2)\exp(-1)$, $\omega_m^2 = 2/\sigma^2$

cut-off frequency or a band-pass filter in (3.3.32) with a high center frequency; this gives fine details in gray-level changes and large sensitivity to noise. Noise sensitivity may be reduced by choosing a larger value of σ; it may also be reduced by discarding zero crossings where the difference in values of f across the zero crossing is too small. It is mentioned that (3.3.31) may be closely approximated by the difference of two suitably chosen Gaussian functions. Since these functions are separable, computational complexity is reduced according to (2.3.25).

3.3.4 Line and Plane Segments

Fitting a small segment of a line or a plane to a part of an image is a technique which also allows noise reduction. The approach of the *Hueckel operator* is to superimpose a circular window $f_0(x, y)$ onto the image $f(x, y)$. Inside this circular window $f(x, y)$ is approximated by an idealized edge, for example, an edge of the form

$$g(x, y; a_0, a_1, a_2, b, d) = \begin{cases} b & \text{if } a_1 x + a_2 y \leq a_0 \\ b + d & \text{otherwise.} \end{cases} \quad (3.3.33)$$

As shown in Fig. 3.12a this is an edge having intensity b on one side, intensity $b + d$ on the other, and orientation given by the line $a_1 x + a_2 y = a_0$. The edge g in (3.3.33) contains parameters a_0, a_1, a_2, b, d which are adjusted such that g fits f optimally inside f_0 in the sense

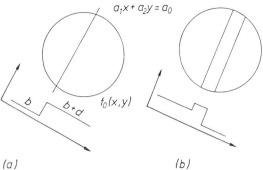

Fig. 3.12a,b. A contour is modeled as shown in (a), a modification is given in (b)

$$\varepsilon = \min_{\{a_0,a_1,a_2,b,d\}} \iint_{(x,y)\in f_0} (f-g)^2 dx\, dy \quad . \tag{3.3.34}$$

An efficient solution of (3.3.34) is given in the references. It is based on an expansion of f and g into orthogonal basis functions $\varphi_\nu(x,y)$, truncating the series to 8 terms, and considering ε in (3.3.34) only for the 8 terms of the expansion. It may be shown that minimization of this ε is equivalent to minimization of a function of only a_1, a_2, which in addition can be normalized to $a_1^2 + a_2^2 = 1$. With these the remaining coefficients can be computed. The quality of the edge g determined by minimization of ε is measured by the angle between the coefficient vectors of f (inside f_0) and g. If this angle is too large the edge is rejected. A modification of this approach to an edge of the form shown in Fig. 3.12b is possible. Since operators like Hueckel or Canny are computationally fairly expensive, they may be applied only at points where a simple operator such as (3.3.1) indicates sufficiently high gray-level change.

A plane

$$z = ax + by + c \tag{3.3.35}$$

may be used to approximate the gray values of an image within a rectangular region centered at f_{jk}. A vector normal to this plane gives an indication of gray-level changes. Parameters a, b, c are determined to minimize

$$\varepsilon = \sum_{\mu=-m}^{m} \sum_{\nu=-m}^{m} \left[f_{j+\mu,k+\nu} - (ax_{j+\mu} + by_{k+\nu} + c) \right]^2 \quad . \tag{3.3.36}$$

Taking partial derivatives with respect to a, b, c and equating them to zero gives three equations for the parameters. These may be simplified by shifting f_{jk} to the origin to get f_{00} and using $x_j = j\Delta x = j$. Then one obtains

$$a = \alpha \sum_{\mu,\nu} f_{\mu\nu} x_\mu ,$$

$$b = \alpha \sum_{\mu,\nu} f_{\mu\nu} y_\nu ,$$

$$c = (2m+1)^{-2} \sum_{\mu,\nu} f_{\mu\nu} ,$$

$$\alpha = 3/[m(m+1)(2m+1)^2] ,$$

(3.3.37)

with $\sum_{\mu,\nu}$ indicating the double sum as in (3.3.36). Values of m may be, for instance, $m = 1$ or $m = 2$. The angle between a normal on the (x, y) plane and a normal on the plane in (3.3.35) is

$$\gamma = \text{arc cos } [(1 + a^2 + b^2)^{-1/2}] , \qquad (3.3.38)$$

with $\gamma = 0$ corresponding to constant $f(x,y)$, that is, no change of gray level at f_{jk}.

3.3.5 Statistical and Iterative Methods

Since in contour detection a choice has to be made between the alternatives "an edge is present in some neighborhood" and "no edge is present", it seems natural to use an approach based on hypothesis testing. For example, it has to be decided whether a boundary is present between two points (j, k) and $(j + 1, k)$. To achieve noise immunity, intervals I_1, I_2 around the two points are considered; the intervals may look like those in Fig. 3.6. The model of a contour is a step between two intervals of constant gray value, similar to Fig. 3.12. Gray values are disturbed by additive white noise. There are two hypotheses:

$$\begin{aligned} H_0 &: I_1 \text{ and } I_2 \text{ belong to one object} , \\ H_1 &: I_1 \text{ and } I_2 \text{ belong to different objects} . \end{aligned} \qquad (3.3.39)$$

In both cases the probability density of gray values is assumed to be normal. Let there be N_1 samples $f_{i,j} \in I_1$ and N_2 samples $f_{i,j} \in I_2$ which for simplicity are indexed by $f_i, i = 1, \ldots, N_1$ or N_2. Then

$$\begin{aligned} p(f_i|H_0) &= N(\mu_0, \sigma_0) , \\ p(f_i|H_1, I_\nu) &= N(\mu_\nu, \sigma_\nu) \quad \nu = 1, 2 , \end{aligned} \qquad (3.3.40)$$

with

$$\begin{aligned} \mu_\nu &= N_\nu^{-1} \sum_{i \in I_\nu} f_i , & \nu &= 1, 2 , \\ \sigma_\nu^2 &= N_\nu^{-1} \sum_{i \in I_\nu} (f_i - \mu_\nu)^2 , & \nu &= 1, 2 , \end{aligned} \qquad (3.3.41)$$

$$\mu_0 = (N_1 + N_2)^{-1} \sum_{i \in I_1 \cup I_2} f_i ,$$
$$\sigma_0^2 = (N_1 + N_1)^{-1} \sum_{i \in I_1 \cup I_2} (f_i - \mu_0)^2 .$$
(3.3.42)

Assuming independent samples f_i the density of the sample is, for instance,

$$p(f_i \in I_1 | H_1) = \prod_{i \in I_1} \left(2\pi\sigma_1^2\right)^{-1/2} \exp\left[-(f_i - \mu_1)^2/(2\sigma_1^2)\right] , \quad (3.3.43)$$

and similar expressions hold for $p(f_i \in I_2 | H_1)$ and $p(f_i \in I_1 \cup I_2 | H_0)$. With (3.3.41) the simple expression

$$p(f_i \in I_1 | H_1) = \left[\left(2\pi\sigma_1^2\right)^{N_1/2} \exp\left(N_1/2\right)\right]^{-1} \quad (3.3.44)$$

for (3.3.43) results. Deciding between H_0 and H_1 by use of the maximum likelihood ratio yields

$$p \frac{(f_i \in I_1 | H_1) p(f_i \in I_2 | H_1)}{p(f_i \in I_1 \cup I_2 | H_0)} > \Theta ,$$
$$\sigma_1^{N_1 + N_2} / \left(\sigma_1^{N_1} \sigma_2^{N_2}\right) \begin{cases} > \Theta \to H_1 \\ < \Theta \to H_0 \end{cases}$$
(3.3.45)

The variances of gray values in intervals I_1, I_2, and $I_1 \cup I_2$ are computed and compared according to (3.3.45) in order to decide for H_0 or H_1.

Iterative methods are used to enhance or attenuate gray-level changes depending on evidence from neighboring points. In general, if there are several points with gray-level changes of similar value and direction in a line, this is taken as evidence to enhance these changes; if there are only very few points, changes are attenuated. A variety of iterative schemes have been developed out of which only one is taken as an example. The original image f is transformed to an image h containing magnitude p and orientation q of edge elements. They are quantized to integer values. A point h_{jk} of h with magnitude p and orientation q is superimposed by a mask of points such that the n points have a certain orientation; some examples of masks are given in Fig. 3.13. The point h_{jk} is inspected by the mask corresponding to its orientation q. With the magnitudes and orientations of points of h inside the mask denoted by $p_i, q_i, i = 1, \ldots, n$ a parameter

$$r = \sum_{i=1}^{n} w_{|q_i - q|} p_i \quad (3.3.46)$$

is computed. This means that the magnitudes p_i of points of h inside the mask are weighted and summed. If the weights w_m, $m = |q_i - q|$, are positive for small m and negative for large m, this amounts to a large value of r if points with orientations similar to q are inside the mask. Therefore, the magnitude p of h_{jk} is increased if $r > 0$, and decreased if $r < 0$. This process is iterated. Orientation

94   3. Segmentation

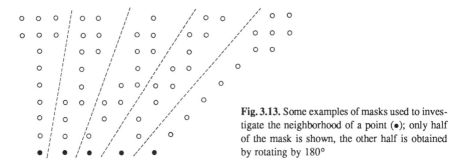

Fig. 3.13. Some examples of masks used to investigate the neighborhood of a point (•); only half of the mask is shown, the other half is obtained by rotating by 180°

q of h_{jk} is adjusted in an independent algorithm. It is based on the principle that orientations q_i of points within the mask corresponding to orientation q of h_{jk} should not differ too much from q. If too many points in the mask have significantly different orientations, the existence of a line at h_{jk} is questionable. If many points in the mask have only slightly different orientations, the value q is slightly changed.

3.3.6 Generalization to Several Spectral Channels

In the foregoing we only considered gray-value functions. It is possible to generalize the methods based on computation of differences also to color or multispectral images. Instead of considering a scalar value f_{jk} a vector $\boldsymbol{f}_{jk}$ of values in different spectral channels is considered. This can be done in (3.3.1 or 2). Suitable measures of color difference were given in Sect. 2.2.5.

3.4 Contour Lines

3.4.1 Computing Curved and Straight Lines

So far we have discussed only operators which return a set S of points where significant changes of gray values occur, possibly together with measures of magnitude and direction of these changes. The next important step now is to find subsets of points lying on the same contour line, and to fit the line through these points. Usually a parametric family of curves in the x, y plane given by

$$y = g(x, \boldsymbol{a}) \quad \text{or} \quad 0 = g(x, y, \boldsymbol{a}) \qquad (3.4.1)$$

is assumed, where $\boldsymbol{a}$ is a parameter vector. It is obvious (for instance, from Fig. 3.3) that often it will be neither possible nor meaningful to fit one curve to the whole set S of points. Rather, a piecewise approximation by several functions will be attempted. So the following problems have to be solved:

1) A family of functions in (3.4.1) has to be selected.
2) A criterion to evaluate the quality of approximation has to be chosen.

3) The set of points to be approximated must be partitioned into subsets belonging to the same contour line.

Some approaches to these problems will be discussed. It is mentioned that (3.4.1) may be used, at least in principle, for one- and two-dimensional patterns. In the one-dimensional case we have a function (pattern) $y = f(x)$ and try to approximate it by (3.4.1). In the two-dimensional case we have $z = f(x,y)$ with z being the gray value at location (x, y). But the purpose of contour extraction is to get rid of gray values except zero and one. So it is not intended to approximate the surface z (for instance, this is of interest in CAD), but only a line in the x, y plane. There may be some differences in the appearance of a function $y = f(x)$ and a contour line extracted from $z = f(x, y)$ as shown in Fig. 3.14.

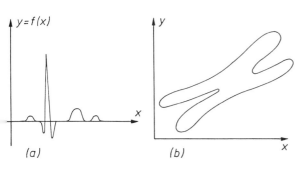

Fig. 3.14a,b. An example of a one-dimensional function in (a) and contour lines of a two-dimension function in (b)

Examples of families of functions in (3.4.1) are

$$y = a_1 x + a_2 , \tag{3.4.2}$$

$$y = \sum_{\nu=1}^{n} a_\nu \varphi_\nu(x) , \tag{3.4.3}$$

$$0 = a_1 x + a_2 y + a_3 x^2 + a_4 x y + a_5 y^2 + a_6 , \tag{3.4.4}$$

$$y = a_1 x^2 + a_2 x . \tag{3.4.5}$$

In many cases a piecewise linear approximation (3.4.2) is used, quadratic functions (3.4.4, 5) are also in use occasionally.

Two criteria for measuring the quality of approximation are mean-square error and maximal absolute difference. Let

$$S = \{(x_\nu, y_\nu) \,|\, \nu = 1, \ldots, N\} \tag{3.4.6}$$

be the set of N points which have been selected by one of the above methods as candidates for contour points. Let

$$S_i = \{(x_\nu, y_\nu) \,|\, \nu = 1, \ldots, N_i\} , \quad S \supset S_i \tag{3.4.7}$$

be a set of points to be approximated by a segment of a function (3.4.1), for instance, a straight line segment (3.4.2). To avoid useless notational complications points $(x_\nu, y_\nu) \in S_i$ taken from S are assumed to be reindexed from $\nu = 1, \ldots, N_i$. Then the *mean-square error* is given by

$$\varepsilon_m = N_i^{-1} \sum_{\nu=1}^{N_i} [y_\nu - g(x_\nu, \boldsymbol{a})]^2 \,, \tag{3.4.8}$$

and the *maximum absolute error* by

$$\varepsilon_a = \max_{\{\nu\}} |y_\nu - g(x_\nu, \boldsymbol{a})| \,. \tag{3.4.9}$$

A disadvantage of both definitions is that they measure error in the y direction although it would be more desirable to measure the error orthogonal to the line. In general this is more complicated and will be omitted here. In any case a parameter vector $\boldsymbol{a}$ is looked for which minimizes ε_m or ε_a. This is fairly easy for (3.4.8) and more difficult for (3.4.9). The main difference is that a large error in a single point only slightly affects ε_m, but greatly affects ε_a. Assuming $g(x, \boldsymbol{a})$ to be linear in the parameter $\boldsymbol{a}$, that is to have the general form (3.4.3), the solution for $\boldsymbol{a}$ is given by Theorem 2.2.1 of Sect. 2.2.7. This is seen by introducing vectors

$$\begin{aligned} \boldsymbol{y} &= (y_1, \ldots, y_{N_i})_t \,, \\ \boldsymbol{a} &= (a_1, \ldots, a_n)_t \end{aligned} \tag{3.4.10}$$

and the $N_i \times n$ matrix $\boldsymbol{Z}$ with elements

$$z_{jk} = \varphi_k(x_j) \,, \quad k = 1, \ldots, n \,, \quad j = 1, \ldots, N_i \,, \tag{3.4.11}$$

and j indicating the jth row of $\boldsymbol{Z}$. Then, as in (2.2.41),

$$\boldsymbol{y} = \boldsymbol{Z}\boldsymbol{a} \,, \tag{3.4.12}$$

and the parameter $\boldsymbol{a}$ minimizing (3.4.8) is from (2.2.43)

$$\boldsymbol{a} = (\boldsymbol{Z}_t \boldsymbol{Z})^{-1} \boldsymbol{Z}_t \boldsymbol{y} \,. \tag{3.4.13}$$

For the linear function (3.4.2) closed-form solutions for a_1, a_2 are available and also solutions for minimization of normal distances are possible. Minimization of ε_a in (3.4.9) amounts to solving a linear programming problem. An algorithm for this case is given in the references.

The crucial point that remains is selection of subsets S_i in (3.4.7) which are to be approximated in the sense of (3.4.8) by a function of the type (3.4.3). A useful technique is combination of approximation and subset selection by requiring that a certain error should not be exceeded when approximating points in S_i. So the subsets are determined from selection of an error criterion, a maximum tolerable error, and a family of functions. An outline of one approach to solving this problem is the following *split-and-merge* procedure:

1. Assume that S is an ordered set of boundary points, $S_i^{(0)}, i = 1, \ldots, m$, an initial partition of S, $S_i^{(\nu)}$ the partition in the νth iteration, and $\varepsilon_i^{(\nu)}, \varepsilon^{(\nu)}$ the error of approximating S_i, S, respectively, in the νth iteration (the error ε may be (3.4.8 or 9) or some other criterion). Select a maximum tolerable error $\varepsilon_{\max}$.
2. Compute $\varepsilon^{(0)}$ for the initial partition.
3. If $\varepsilon^{(\nu)} > \varepsilon_{\max}$ do step 4, otherwise do step 5.
4. Split the subsets $S_i^{(\nu)}$ with largest $\varepsilon_i^{(\nu)}$ into two new subsets; increment ν; update parameters; do step 3.
5. Find a pair $S_i^{(\nu)}, S_{i+1}^{(\nu)}$ whose mergence to one new subset causes the smallest increase in $\varepsilon^{(\nu)}$. Do the mergence if afterwards $\varepsilon^{(\nu)} < \varepsilon_{\max}$ still holds. Repeat step 5 until no more mergence occurs; update parameters.
6. Do an iteration of 8.
7. If no changes occurred above, END the procedure, otherwise do step 3.
8. Reduce ε for fixed m.
    8.1 Choose $S_i^{(\nu)}, S_{i+1}^{(\nu)}$, for instance, by trying $i = 1, \ldots, m - 1$, and obtain the sum s of their errors.
    8.2 Make the rightmost point of $S_i^{(\nu)}$ tentatively be the leftmost point of $S_{i+1}^{(\nu)}$ and obtain the sum s' of errors in modified subsets.
    8.3 Make the leftmost point of $S_{i+1}^{(\nu)}$ tentatively be the rightmost point of $S_i^{(\nu)}$ and obtain the sum s'' of errors in modified subsets.
    8.4 Select subsets belonging to min $\{s, s', s''\}$.
    8.5 Repeat until no more changes occur.

3.4.2 Hough Transform

Another method of subset selection is the *Hough transform*, which does not require an ordered set of boundary points. It is useful for functions with few, say two or three, parameters, for instance, as given in (3.4.2, 5). Instead of searching the set S of boundary points for a subset lying on a line the parameter space is searched. The principle may be seen from the parametric representation of a straight line; parameters α, $0 \leq \alpha < 2\pi$, and r, $r \geq 0$, are illustrated in Fig. 3.15 (alternatively, β may be used). To a straight line in the x, y plane corresponds one point in the α, r plane. Alternatively, one point (x_i, y_i) of the x, y plane may be viewed as defining a curve

$$r = x_i \sin \alpha + y_i \cos \alpha \tag{3.4.14}$$

in the α, r plane. It is obvious that this dual viewpoint applies to any parametric family of functions, not only to (3.4.14). For example, if (3.4.2) is used, the point (x_i, y_i) defines the line $a_2 = y_i - a_1 x_i$ in an a_1, a_2 plane. The advantage of using (3.4.14) instead of (3.4.2) is that the parameter space of the former is finite since r will not exceed the image diagonal. The disadvantage is that the evaluation of (3.4.14) is more time consuming than (3.4.2). If the two cases "line with slope less than 45°" and "line with slope more than 45°" are distinguished, (3.4.2) can be used and a finite parameter space is still guaranteed. Several points in the

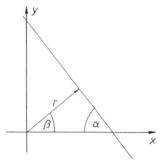

Fig. 3.15. Parameters r and α of the representation (3.4.14) of a straight line

x, y plane are on the same straight line if the corresponding curves in the α, r plane (or a_1, a_2 plane) intersect in one point, this point giving the parameters of the straight line. To obtain lines from a set of points the α, r plane is suitably quantized, say with $p \times q$ elements. Here it is advantageous to have a finite parameter space. An accumulator array $H(\alpha, r)$ is initialized with zero; it can be viewed as attributing a counter to any point α_j, r_k of the quantized α, r plane. For any point $x_i, y_i \in S$ compute

$$r_k = x_i \sin \alpha_j + y_i \cos \alpha_j, \quad j = 1, \ldots, p \tag{3.4.15}$$

and increase the counter of each α_j, r_k by one. When this has been done for all $x_i, y_i \in S$ the value of the counter of α_j, r_k gives the number of points on a line with these parameters. Ideally, all points on a line will increase only the same counter, but due to noise they may increase several neighboring counters. Thus points in S on the same line give rise to a cluster in the α, r plane.

A modification of this approach avoiding the evaluation of (3.4.15) is possible for two-parameter curves if in addition to x, y the derivative $y' = \partial y / \partial x$ is known. An estimate of this may be obtained from the above operators for detection of gray-level changes. Let

$$y = g(x, a_1, a_2), \quad y' = \partial g / \partial x \tag{3.4.16}$$

be the two-parameter family of curves and the triple $(x_i, y_i, y'_i), i = 1, \ldots, N$ be given. If it is possible to solve (3.4.16) for a_1, a_2, unique values of the parameters

$$a_1 = g_1(x, y, y'), \quad a_2 = g_2(x, y, y') \tag{3.4.17}$$

can be computed for any of the (x_i, y_i, y'_i). A simple example is (3.4.5) with

$$y = a_1 x^2 + a_2 x \quad \text{and} \quad y' = 2a_1 x + a_2 \tag{3.4.18}$$

giving the closed-form solution

$$a_1 = \frac{y'}{x} - \frac{y}{x^2}, \quad a_2 = \frac{2y}{x} - y'. \tag{3.4.19}$$

Again, noise points lying on the same curve will give a cluster in parameter space. In general, if there are m parameters, the use of directional information reduces them to $m - 1$ parameters.

Since the computational effort for the Hough transform is fairly large, it is interesting to explore methods for reducing it. One approach is to use a *resolution hierarchy*. The parameter space is first searched in a coarse resolution giving coarse initial parameter estimates. These are used to constrain the interval of search in the next finer resolution where improved parameter estimates are computed. This process is iterated until the finest resolution is obtained.

3.4.3 Linking Contour Points

Of course, other methods of subset selection are possible and have been used. Some are briefly mentioned. A set of points on a contour may be determined first. The next step is forming straight line segments by starting with two neighboring points and computing the parameters of the corresponding line similar to (3.4.13). A new point is merged to the line segment if its distance is less than a threshold. In this case parameters are updated, otherwise a new segment is introduced. Selecting a low threshold gives a large number of short line segments. After this, two adjacent segments are tentatively joined to give one longer straight line. Joining is considered successful if the mean-square error of the approximation is less than another threshold, and then the next adjacent segment is tried. This process is continued until lines cannot be extended any more.

Another possibility is usage of masks or search regions. A point on a contour with a certain orientation is inspected by slitlike masks of similar orientation to see whether more points are within the mask. If a segment of a straight or curved line is available, a search region for the next point can be determined by extrapolation of the line. Within the mask or search region a "good" contour point is determined; this is a point having sufficient contrast and allowing an extension of the line within certain continuity constraints.

Finally, several search methods for finding "optimal" contours have been developed. Here the approach is to find a contour which maximizes a goodness function or to find an optimal path in a weighted graph. One approach is contour following by dynamic programming (DP). A line consisting of N points is given a score, for example, the weighted sum of contour contrast and straightness, where straightness may be enforced by subtracting curvature or by constraining the set of allowed successor points. From among the set of contour points whose contrast exceeds a threshold, a subset of adjacent points maximizing the score is computed by DP. A special case arises if a nearly elliptical closed contour is to be determined. If a point P near to the center of the region enclosed by the contour can be found, it is useful to transform the image to polar coordinates (r, ψ) centered at P. A contour of nearly elliptical shape transforms to a line of low curvature; the start point $(r_s, 0)$ and the end point $(r_e, 2\pi)$ of the closed contour in polar coordinates have to meet the constraint $r_e \in \{r_s, r_s + 1, r_s - 1\}$.

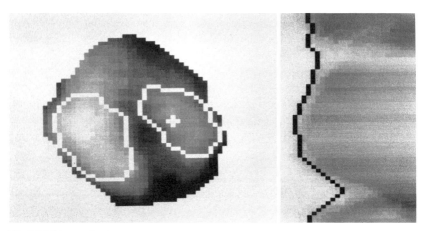

Fig. 3.16. The result of contour following using dynamic programming. (*Left*) Contours of left and right ventricles appear on the right and left part of this image, respectively. (*Right*) contour of the left ventricle in polar coordinates [3.7]

Again, the best sequence of points maximizing the score is determined by DP. Figure 3.16 shows an example of contour following in a scintigraphic image.

3.4.4 Characterization of Contours

Characterization (or representation, description) of contours (or shape) is understood here as giving a means of classification and analysis. Therefore, the main purpose is not to include as much information as possible (for instance, to establish a one-to-one relation between the original contour and its representation), but to select as little as necessary to allow successful analysis. To distinguish the two shapes of Fig. 3.17a just a few parameters are sufficient, but to describe Fig. 3.17b and to isolate and name the objects, fairly complete lists of lines and junctions are required. This indicates that according to the task domain quite different methods are useful, and even for the same task domain several different methods may be needed. Two types of methods are the local *space domain techniques* and the global *scalar transform techniques*.

Scalar transform techniques are, for instance, shape factors, moments, Fourier transform or other series expansions, and masks. A further distinction can be made between internal and external transforms. The former make use of the interior of the contour, for instance, by setting the points inside the contour equal to one and those outside equal to zero, or by using the gray value $f(x, y)$ at point (x, y) inside the contour; the latter only use the contour line. It will become evident from the following which type applies.

Moments are used to normalize parameters and to get features for classification. For a function $f(x, y)$ of two variables the *moment* of order $(m + n)$ is

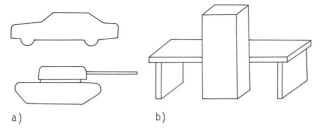

Fig. 3.17a,b. In (a) classification of shapes is adequate, in (b) a description localizing and naming objects is required

$$m_{mn} = \int_{-\infty}^{\infty} \int_{-\infty}^{\infty} x^m y^n f(x,y) dx\, dy \ . \tag{3.4.20}$$

There is a one-to-one relation between the set $\{m_{mn}|m,n = 1,2,\ldots\}$ and $f(x,y)$. With

$$x_g = m_{10}/m_{00}\ , \quad y_g = m_{01}/m_{00}\ , \tag{3.4.21}$$

and the coordinate transformation

$$x = x' + x_g\ , \quad y = y' + y_g\ , \tag{3.4.22}$$

the translation-invariant *central moments*

$$\mu_{mn} = \int_{-\infty}^{\infty} \int_{-\infty}^{\infty} x'^m y'^n f(x,y) dx'\, dy' \tag{3.4.23}$$

may be defined. Various other invariant parameters can be derived from moments.

Coefficients of the discrete Fourier transformation (2.3.13) give another example of a one-to-one relation between a pattern and a set of parameters. In this case the function $f(x,y)$ (or rather its sampled version) is used. But it is also possible to obtain Fourier descriptors of the contour line itself. Assume a set of points $(x_i, y_i), i = 0, 1, \ldots, N$, $(x_0, y_0) = (x_N, y_N)$ to be given which describes a simple closed curve of circumference L in clockwise direction. Two representations are in use. The first is based on arc length l_i between (x_0, y_0) and (x_i, y_i) and angle $\alpha'(l_i)$ between a tangent at (x_0, y_0) and a tangent at (x_i, y_i). This angle is normalized to

$$\alpha(t) = \alpha'\left(\frac{Lt}{2\pi}\right) + t\ , \quad 0 \leq t \leq 2\pi\ , \tag{3.4.24}$$

which is invariant under translation, rotation, and scale change, and maps simple closed curves to periodic functions on $(0, 2\pi)$. The complex Fourier descriptors are

$$c_\nu = \frac{1}{2\pi} \int_0^{2\pi} \alpha(t) \exp(-i\nu t) dt\ . \tag{3.4.25}$$

The other approach is based on a parametric representation conceived as a complex function

$$\varphi(l) = x(l) + iy(l) \tag{3.4.26}$$

of the boundary $x(l), y(l)$. In this case Fourier descriptors

$$c'_\nu = L^{-1} \int_0^L \varphi(l) \exp(-i2\pi nl/L) dl \ . \tag{3.4.27}$$

Space domain techniques are similarly divided into internal and external transforms. Among the former are projections, medial axis transformation, and decomposition of the shape; among the latter, chain code (Sect. 2.1.5), approximation of contours (Sect. 3.4.1), and symbolic description of contours.

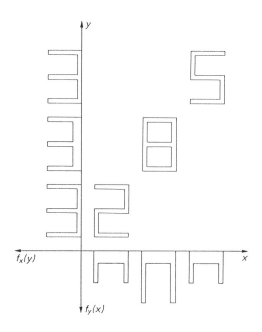

Fig. 3.18. Two-dimensional shapes and their projections onto the x and y axis

Projections may be defined in various ways. An example is the projection of a function $f(x, y)$ onto the y axis given by

$$f_x(y) = \int_{-\infty}^{\infty} f(x, y) dx \ , \tag{3.4.28}$$

and analogously for the x axis; the appearance of such projections is illustrated in Fig. 3.18. It is clear that $f_x(y)$ and $f_y(x)$ may be identical for different shapes. A generalization of (3.4.28) is obtained by using projections along an arbitrary direction (*Radon transform*). The medial-axis transformation converts a shape to a set of lines as illustrated in Fig. 3.19. A point on the medial axis is one having more than one nearest neighbor on the boundary.

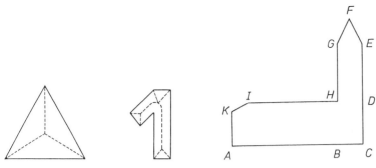

Fig. 3.19. Medial axis of some simple patterns

Fig. 3.20. The above shape may be decomposed, for instance, into the convex subsets (ABHIK) ∪ (BCDEFGH) or (ABCDHIK) ∪ (HDEG) ∪ (EFG)

Decomposing a shape is an attempt to find conceptually simple components which together make up the original shape. If a boundary is approximated by piecewise linear segments, it is reasonable to only consider polygons. Convex subsets of a polygon may then be used as simple components to decompose the original object. An example is given in Fig. 3.20.

Approximation of a contour as described in Sect. 3.4.1 is itself a characterization which may be used as a starting point for further processing. For instance, a symbolic labeling of boundary segments can be obtained for syntactic pattern recognition. A chain-coded version can be used and "critical" points such as intersections, discontinuities of curvature, and endpoints extracted.

3.5 Regions

3.5.1 Homogeneity

Extraction of regions is motivated by the observation that contour detection is very difficult in the presence of noise and that there are cases where contours alone do not give sufficient information. For instance, in Fig. 3.3 or 3.17 contours seem to be quite sufficient. On the other hand, the "shape" of the rectangular region in Fig. 3.21 says little about the "object"; it may be forest, grassland, tilled land, and so on if no additional information is given. In this case it is essential to know something about the interior of the contour, for instance, color of the surrounded region. The fundamental aspect of a region is homogeneity with respect to suitably chosen parameters or properties. In this sense it is dual to a contour. Of course, a closed contour usually surrounds a region, and between two neighboring regions there is a border (or contour). Thus, one stipulates the other.

Segmentation of an image into regions means that the image array f is partitioned into connected arrays f_ν

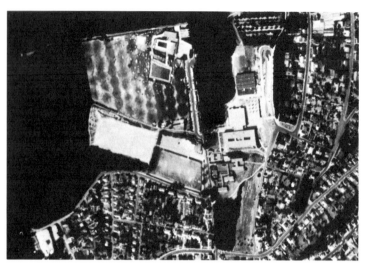

Fig. 3.21. An aerial photo with a distinct rectangular region (record from DFVLR-Forschungszentrum Oberpfaffenhofen)

$$f \to \{f_\nu | \nu = 1, \ldots, N\} , \qquad (3.5.1)$$

such that

$$\bigcup_{\nu=1}^{N} f_\nu = f \quad \text{and} \quad f_\mu \cap f_\nu = \phi \quad \text{for} \quad \mu \neq \nu . \qquad (3.5.2)$$

The criterion of homogeneity is denoted by H and $H(f_\nu)$ is a predicate which is one if f_ν satisfies H, and zero otherwise. An example would be

$$H(f_\nu) = 1 \quad \text{if} \quad |f_{ij} - f_{kl}| < \Theta \quad \text{for all} \quad f_{ij}, f_{kl} \in f_\nu . \qquad (3.5.3)$$

Then it is required that

$$H(f_\nu) = 1 \quad \text{for} \quad \nu = 1, \ldots, N , \qquad (3.5.4)$$

$$H(f_\mu \cup f_\nu) = 0 \quad \text{for} \quad \mu \neq \nu \quad \text{and adjacent} \quad f_\mu, f_\nu . \qquad (3.5.5)$$

Conditions (3.5.2, 4, 5) mean that every point $f_{ij} \in f$ must belong to one and only one region, that all regions are homogeneous in the sense of H, and that merging two regions with a common boundary causes a violation of H. Thresholding (Sect. 3.2) is a special case of region extraction with

$$H(f_\nu) = 1 \quad \text{if} \quad \Theta_{\nu-1} \leq f_{ij} \in f_\nu \leq \Theta_\nu , \quad \nu = 1, \ldots, N \qquad (3.5.6)$$

and, usually, N equal to two or three. Although (3.5.3, 6) are fairly simple, it is stressed that this need not be the case in general. In fact, the predicate H may be completely arbitrary (provided it is useful for a task domain). It may be defined on a set of points, it may use gray levels, color, depth, motion, or some symbolic representation of a set of gray levels, it may be a combination of several predicates, and so on.

3.5.2 Merging

One possibility is to start, in the extreme case, with $N = M^2$ regions

$$\boldsymbol{f}_\nu = \{f_{jk}\}, \quad \nu = 1, \ldots, N; \quad j, k = 1, \ldots, M \tag{3.5.7}$$

if the image is sampled with M^2 points. This fulfills (3.5.2,4), but usually will violate (3.5.5). By successively merging regions one tries to obtain a partition satisfying (3.5.2, 4, 5).

In one approach the whole image is initially segmented into cells of size $2 \times 2, 4 \times 4$ or 8×8 sample values. A statistic of the gray levels of cells is computed next; an example is computation of the gray-level histograms. The statistic of the first cell is compared to that of a neighboring cell. If the two statistics are not similar, the cell is labeled as dissimilar. If the statistics are similar, the two cells are merged, forming one new cell with a newly computed statistic. Then attempts are made to extend the new cell further by examining all of its neighboring cells and proceeding as indicated. The new cell is labeled finished if no more merging is possible. Then the next cell which is not yet finished is treated in the same way. The process stops when all cells are labeled finished. Similarity of statistics is measured by a statistical test.

An algorithm incorporating more global information is described next. Initially the image is segmented into atomic regions which contain only points of equal gray value. These regions will be fairly small because due to noise even a "homogeneous" surface will contain different gray levels. A boundary segment is introduced between two neighboring picture points belonging to different regions. The strength of a boundary segment is the difference between the gray values of the two neighboring points. The boundary segment is weak if its strength is less than a threshold Θ. The length l of the weak part of the boundary is the number of its weak boundary segments. Two adjacent regions $\boldsymbol{f}_\mu, \boldsymbol{f}_\nu$ with perimeters l_μ, l_ν are merged if

$$l/l_{\min} > \Theta_1, \quad l_{\min} = \min\{l_\mu, l_\nu\}. \tag{3.5.8}$$

This requirement allows one to control the length l_r of the perimeter resulting from merging regions $\boldsymbol{f}_\mu, \boldsymbol{f}_\nu$, by adjusting Θ_1. The resulting perimeter must be smaller than $\max\{l_\mu, l_\nu\}$ if $\Theta_1 > 0.5$; it may be larger if $\Theta_1 < 0.5$. In addition to (3.5.8) a second heuristic is used to merge two adjacent regions with common boundary of length l_c if

$$l/l_c > \Theta_2. \tag{3.5.9}$$

The first heuristic is more global than the second. Experiments show that it is not possible to use only the second heuristic.

The technique of piecewise approximation of a function $z = f(x, y)$ can be used to obtain a segmentation and to merge similar segments. The two-dimensional image is scanned at lines $z = f(x, y = y_k)$ to obtain one-dimensional functions which are approximated by m segments. The overlapping segments of different lines are merged if they are similar. An extension of this to a two-dimensional split-and-merge procedure is given in Sect. 3.5.4. Merging of similar regions can also be based on the criterion

$$|m_\mu - m_\nu|\,(A_\mu + A_\nu) = \min \,. \tag{3.5.10}$$

In the above equation m_μ, m_ν are mean gray levels of adjacent regions $\boldsymbol{f}_\mu, \boldsymbol{f}_\nu$ and A_μ, A_ν are their areas. This gives precedence to merging smaller regions.

The basic merge procedure, therefore, is:

1. Obtain an initial segmentation containing a large number of very small regions.
2. Define criteria for merging two adjacent regions.
3. Successively merge regions until no more mergences are possible without violation of (3.5.4).

3.5.3 Splitting

Another possibility is to start with one region

$$\boldsymbol{f}_1 = \boldsymbol{f} \tag{3.5.11}$$

which is the whole image. Usually this will violate (3.5.4), and by introduction of new regions one tries to obtain a partition satisfying (3.5.2, 4, 5).

An example is an approach which uses the whole image as the first region. A region is defined to be homogeneous if the mean gray value of any of its subregions is equal to the mean gray value of the region. The equality of the means is evaluated by a statistical test. For m-channel multispectral images a vector of m mean values is used. As long as there are inhomogeneous regions these are subdivided to obtain more homogeneous regions. It can be shown that in order to test a region for homogeneity it is sufficient to test two arbitrary subregions for equal mean gray level. Several partitions of an inhomogeneous region are tried to find good subregions. Partition error for the image is defined by

$$\varepsilon = \frac{\sum_{\nu=1}^{N} A_\nu \sigma_\nu^2}{\sum_{\nu=1}^{N} A_\nu}\,, \tag{3.5.12}$$

where A_ν is as in (3.5.10) and σ_ν^2 is the variance of gray levels in A_ν.

Successive splittings can also be used to obtain regions in gray-level and color images. The criteria for splitting are based on thresholds which are determined from one-dimensional histograms. Once the whole image has been designated as the first region to be segmented, the following steps are performed:

1. Take the next region to be segmented, compute histograms of feature values and smooth them. If there is no unfinished region then END. Histograms are taken from the red, green, and blue spectral channel and some linear combinations (e.g. Y, I, Q, see Sect. 1.7). Modifications are possible by taking

the number of edge elements per unit area, the variance inside windows, or other textural features.
2. The histograms are searched for the best peak. This should be well separated and have definite minima. If no prominent peak is found, the region is considered to be segmented (finished).
3. The peak is used to determine upper and lower thresholds which are applied to the region under consideration (not to the whole image). This way a binary image is generated; see (3.5.6). The binary image is smoothed by averaging and rounding operations. This removes small holes in the regions.
4. Connected regions are extracted from the binary image, and the area covered by these is removed from the area currently considered for segmentation. The regions are saved to be checked for further segmentation. Segmentation is continued on the remainder of the region.

The speed of the process is increased considerably through a resolution hierarchy. An initial segmentation is made on an image of reduced resolution and is then used to guide the final segmentation of the full size image.

The above-mentioned idea of splitting an image according to the maxima of a suitable function may be modified in various ways. For example, a function may be used which takes large values in regions where contours have sudden turns. A direct generalization is the use of two-dimensional histograms or, in general, selection of clusters in the m-dimensional vectors of spectral measurements in a multispectral image. The selection of such clusters may be made with unsupervised learning or with the mapping of high-dimensional data onto a plane. This amounts to a classification of multispectral image points by numerical classification methods (Chap. 4).

3.5.4 Split and Merge

It seems natural to combine split and merge operations in one algorithm which, hopefully, will exhibit the advantages of both methods. As an example we consider the piecewise constant approximation of the two-dimensional function $f(x, y)$. This is represented by the predicate

$$H(\boldsymbol{f}_\nu) = \begin{cases} 1 & \text{if } |f_{jk} - a| \geq \varepsilon \text{ for } f_{jk} \in \boldsymbol{f}_\nu \\ 0 & \text{otherwise.} \end{cases} \quad (3.5.13)$$

Modifications by use of other predicates are possible. Merge and split operations are carried out within a particular data structure, the *quad tree* (or segmentation tree, quartic picture tree). The root of the tree is the whole image f, the leaves are single sample values f_{jk}, and each node corresponds to a square picture region. The successors of each node correspond to a partition of the region into four equal subregions as shown in Fig. 3.22. In the processing of volumetric data the quad tree is modified to an *oct tree*, meaning that a certain volume is split into eight subvolumes of equal size. A segmentation of the image f in the quad tree corresponds to a node cutset of the tree. Merging and splitting then means moving up and down in the tree. During execution of the algorithm

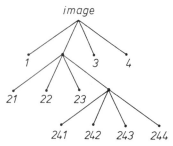

Fig. 3.22. Segmentation of a quadratic image into four regions of equal size. Region 2 is further subdivided into four regions of equal size and so on. A particular segmentation is represented by a cutset of the quad tree (or segmentation tree)

only the respective cutset is stored, not the tree. If in some state the cutset is as in Fig. 3.22, a merge operation on nodes {241, 242, 243, 244} would remove these four nodes from the cutset and replace them by one node {24}. Similarly splitting node 4 would remove this node from the cutset and replace it by four nodes {41, 42, 43, 44}, which corresponds to a division of region 4 into regions of equal size. The algorithm performs the following sequence of steps without any possibility of cycling between steps:

1. Obtain an initial segmentation of the image. This may be chosen at random or based on initial information, for instance, the probable location of an object.
2. Perform all possible merges. Since (3.5.13) is equivalent to the requirement that maximal and minimal gray values in f_ν differ by at most 2ε, it is sufficient to have only these two values available. If a merge of 4 nodes of one predecessor is possible without violation of H, the nodes are merged to one node.
3. Perform all necessary splits. A node not satisfying H is split into four successor nodes.
4. Group adjacent regions to one new region if H remains true after grouping. These operations abandon the tree structure and allow the removal of unnatural boundaries imposed by the 2×2 subdivision. Note that steps 2 and 3 above never would allow the merging of, for instance, regions 21 and 22 in Fig. 3.22 or the merging of regions 23 and 3. Step 2 only might yield a merging of 21, 22, 23 and 24.
5. Eliminate very small regions by combining them with a nearest neighbor, that is, with an adjacent region having the most similar average gray value.

Figure 3.23 shows an example of the application of this split-and-merge algorithm.

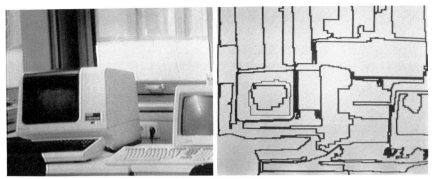

Fig. 3.23. Result of region growing

3.5.5 Remarks

It is apparent that there is a strong analogy between the determination of regions in the analysis of complex patterns and unsupervised learning in the classification of simple patterns. In the former case, a set of homogeneous points of image f are sought which satisfy (3.5.2, 4, 5); in the latter case, a set of similar patterns $f^r(x) \in \omega$ are sought which belong to a given sample ω of patterns. The analogy is seen to be almost an identity in the case of multispectral images. Here the (j, k)th sample point itself consists of m spectral measurements and is a vector $f_{jk} \in R^m$. This may be seen as equivalent to the feature vector $c^r \in R^n$ extracted from a simple pattern $f^r(x)$. A significant difference is the fact that the sample ω is an unordered set of simple patterns, whereas the image array f is an ordered set of sample points. Otherwise (3.5.5) would be meaningless. Another significant difference is the fact that in many cases not only the properties of an image point are considered to decide on its membership of a region, but also the properties of other neighboring points. Although there are examples of this in the above methods, this is the case to a larger extent in Sect. 3.6 where textural properties are taken into account.

Again, only methods which do not require information about structural properties or the task domain have been discussed. It was already mentioned in Sect. 3.1 that in many cases such information is indispensible to obtain a satisfactory segmentation.

It seems reasonable to introduce a boundary (or a contour) between two different adjacent regions which were determined by one of the above methods. It also may be expected that a closed contour, which was obtained by one of the methods in Sect. 3.3, 3.4, will enclose a region of points which are homogeneous according to some criterion. In this sense determination of regions is also a method of contour extraction and vice versa. Mutual support of both approaches seems natural. A combination of results from line extraction and region growing may be achieved by a rule-based system (Sect. 7.3). An example of such a rule might be

```
IF   (the size of the region is not small
 AND there is a line inside the region
 AND the length of the line is not small
 AND the average gradient of the line is large),
THEN (split the region along the line).                    (3.5.14)
```

3.6 Texture

3.6.1 The Essence of Texture

It was mentioned already in Sect. 3.5.1 that there are cases where it is necessary to know something about the interior of a contour. In Sect. 3.5 only fairly simple properties such as color or gray level were mentioned. The properties of an area or surface may be more complicated, as becomes evident from Fig. 3.24. Although the areas shown in this figure are inhomogeneous on the basis of pointwise gray values, it is apparent that on a larger basis they exhibit homogeneous properties. The predicate best suited for measuring this kind of homogeneity is not as easy to find as, for example, that of a monochromatic area. But nevertheless, the homogeneity is obvious to a human observer. These more complicated properties of a surface are referred to as *texture* or textural properties. Some aspects of human texture discrimination have been investigated experimentally. According to these experiments, the textures in gray-level images are discriminated spontaneously only if they differ in second-order moments. Textures with equal second-order moments, but with different third-order moments require a deliberate cognitive effort. This is an indication that, for automatic processing, statistics up to second order may be most important (if statistical methods are used at all). In addition color, blobs, and line endings are cues for spontaneous texture discrimination.

Intuitively, texture is an obvious phenomenon. The question remains how to characterize it in a quantitative manner. An intuitive model starts with a textural primitive which is repeated to obtain an ideal (but unobservable) texture. After some transformations an observable surface texture results. This idea is illustrated in Fig. 3.25. A modification might be to start with one or more texture primitives which are subjected to deterministic or random perturbations. An ideal texture is formed by deterministic or random repetition of the primitives; from the ideal texture the observable texture is obtained after deterministic or random transformations. Although one may imagine that this is adequate in most cases, the idea is hard to formalize.

3.6 Texture   111

Fig. 3.24. Examples of textures; (*upper left*) grass lawn, (*upper right*) beach pebbles, (*lower left*) lizard skin, (*lower right*) bark of tree; from [Ref. 3.8; Figs. D9, 13, 23, 35]

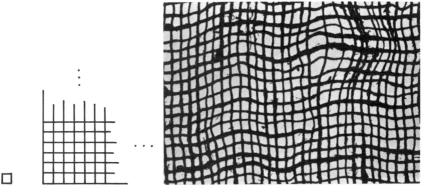

Fig. 3.25. The square is used as a texture primitive to build an ideal texture. Transformations yield an observable texture, in this example loose burlap; from [Ref. 3.8; Fig. D103]

3.6.2 Numerical Characterization

Models developed for texture include, for instance, stationary random fields, spatial filters driven by a stochastic process, fractal geometry, and results from human vision. Using an approach based on such results, a set of 27 features for characterization of a texture can be derived by the following steps. The image $f(x, y)$ is normalized to

$$f_n(x, y) = \frac{f(x, y) - m}{\sigma} \tag{3.6.1}$$

where m and σ^2 are the mean and variance of image intensity. The normalized image f_n is filtered by band-pass filters given in polar coordinates ω, α by

$$\begin{aligned} G(\omega, \alpha) &= G(\omega)G(\alpha) \,, \\ G(\omega) &= [1 + (\omega - \omega_0)^2/\omega_b]^{-1/2} \,, \\ G(\alpha) &= \exp\left[-(\alpha - \alpha_0)^2/(2\alpha_b)\right] + \exp\left[-(\alpha - \alpha_0 - \pi)^2/(2\alpha_b)\right] \,. \end{aligned} \tag{3.6.2}$$

The central frequencies are defined by ω_0 and α_0. Three values for ω_0 and nine for α_0 are selected, resulting in 27 filters. The radial and angular bandwidth is determined by ω_b and α_b where reasonable values are $\omega_0/\omega_b = 2$ and $\alpha_0/\alpha_b = 10$. Let the result of filtering f_n by the νth filter, $\nu = 1, \ldots, 27$, be $h_\nu(x, y)$. The νth textural feature c_ν is defined by

$$c_\nu = \iint_{\text{area}} h_\nu^6(x, y) dx \, dy \,, \tag{3.6.3}$$

with "area" indicating that windows may be specified for integration. The 6th power in (3.6.3) is taken as a compromise between square norm and maximum norm to avoid too much and too little averaging of details.

The idea of repetition of textural primitives is the basis for defining $L \times L$ *spatial co-occurrence matrices* $P(d, \alpha)$ where L is the number of gray levels. The matrix

$$P(d, \alpha) = [p_{\mu\nu}(d, \alpha)] \tag{3.6.4}$$

has elements $p_{\mu\nu}(d, \alpha)$ which give the number of sample point pairs $f_{ij} = \mu$ and $f_{kl} = \nu$, where f_{ij} and f_{kl} have distance d and orientation α. The matrix may be computed for different values of d and α and used to obtain textural features. Four of the fourteen features are given as an example. With the normalized elements

$$p_{\mu\nu} = \frac{p_{\mu\nu}(d, \alpha)}{\sum_{\mu=0}^{L-1} \sum_{\nu=0}^{L-1} p_{\mu\nu}(d, \alpha)} \tag{3.6.5}$$

these features are

$$\begin{aligned} c_1 &= \sum_{\mu,\nu} p_{\mu\nu}^2, \\ c_2 &= \sum_{l=0}^{L-1} l^2 \sum_{|\mu-\nu|=l} p_{\mu\nu}, \\ c_3 &= \frac{\sum_{\mu,\nu} \mu\nu p_{\mu\nu} - m_\mu m_\nu}{\sigma_\mu \sigma_\nu}, \\ c_4 &= -\sum_{\mu,\nu} p_{\mu\nu} \log p_{\mu\nu}. \end{aligned} \tag{3.6.6}$$

In the above equation m_μ and σ_μ are the mean and standard deviation of the marginal distribution

$$p_\nu^{(\mu)} = \sum_\mu p_{\mu\nu}, \tag{3.6.7}$$

and m_ν, σ_ν are defined analogously. The features in (3.6.6) are termed angular second moment, contrast, correlation, and entropy. Similarity of matrices $P(d, \alpha)$ in (3.6.4) may be used as a predicate H for homogeneity in a split-and-merge algorithm of the type described in Sect. 3.5.4.

The computation of statistical parameters may be varied. For instance, instead of taking gray levels μ and ν in (3.6.4) only the difference $l = |\mu - \nu|$ may be considered. Also the *gray-level run-length matrix*

$$G(\alpha) = [g_{\mu l}(\alpha)] \tag{3.6.8}$$

may be used. The elements $g_{\mu l}(\alpha)$ are the frequency of a run length l at gray level μ in direction α. Run length l is the number of collinear picture points with constant gray level μ which may be determined in different directions α. Omitting α for brevity and with

$$N_t = \sum_\mu \sum_l g_{\mu l} \tag{3.6.9}$$

five textural features are defined. They are termed short-run emphasis, long-run emphasis, gray-level nonuniformity, and run percentage and are given by

$$c_1 = N_t^{-1} \sum_\mu \sum_l g_{\mu l}/l^2,$$

$$c_2 = N_t^{-1} \sum_\mu \sum_l l^2 g_{\mu l},$$

$$c_3 = N_t^{-1} \sum_\mu \left(\sum_l g_{\mu l}\right)^2, \qquad (3.6.10)$$

$$c_4 = N_t^{-1} \sum_l \left(\sum_\mu g_{\mu l}\right)^2,$$

$$c_5 = N_p^{-1} \sum_\mu \sum_l g_{\mu l}.$$

N_p denotes the number of possible runs if all had length one.

Texture discrimination can also be based on histograms of local properties. An example of a property is a spot detector giving

$$h_{jk} = \left| \frac{1}{(2a+1)^2} \sum_{\mu=j-a}^{j+a} \sum_{\nu=k-a}^{k+a} f_{\mu\nu} - \frac{1}{(2b+1)^2} \sum_{\mu=j-b}^{j+b} \sum_{\nu=k-b}^{k+b} f_{\mu\nu} \right|. \qquad (3.6.11)$$

It is the difference in the mean gray value of two square regions of size $(2a+1)^2$ and $(2b+1)^2$, $b > a$, which are centered at (j, k). A value h_{jk} is suppressed if there is another, h_{lm} with $h_{lm} > h_{jk}$, inside the square of size $(2a+1)^2$. This *nonmaximum suppression* is used to obtain bimodal histograms. The two modes then correspond to two different textures which thus may be discriminated. The parameters a, b are chosen so as to yield the strongest bimodality.

In addition, other texture measures have been developed. Spectral features obtained from $|F_w|^2$, with F_w being the Fourier transform computed on a window w (or subimage) of the image $f(x, y)$ with (2.3.13), can be used. These are low-order Fourier coefficients or averages over bar-, ring-, or wedge-shaped areas of the spatial frequency plane as indicated in Fig. 3.26. Another example is the determination of differences in the autocorrelation functions of different regions where the (normalized) *autocorrelation* is computed from

$$R_{jk} = \frac{\sum_{\mu=l-a}^{l+a} \sum_{\nu=m-a}^{m+a} f_{\mu\nu} f_{\mu-j,\nu-k}}{\sum_{\mu,\nu} f_{\mu\nu}^2}. \qquad (3.6.12)$$

The textural features described above may be used, for example, to detect edges between regions of different texture, to classify textured regions such as grass, water, and so on, to segment an image into regions which are homogeneous with respect to texture, and to obtain depth cues in two-dimensional images of three-dimensional scenes. Experimental evidence indicates that spectral features are less suited.

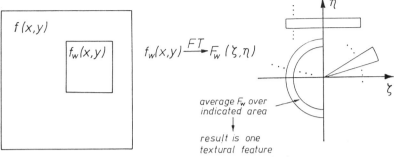

Fig. 3.26. Computation of spectral features for texture characterization

3.6.3 Syntactic Characterization

Besides the numerical characterization of textures a syntactical approach is also possible as shown briefly in the following. As a textural primitive in the sense of Sect. 3.6.1 a window of fixed size $M_t \times M_t$ is chosen. The gray values inside the window are represented by a tree as indicated in Fig. 3.27. It is apparent that the tree structure may be chosen differently, but it is then fixed for all the windows of an image. Different textural primitives have different labeling of nodes in the tree and are generated by a tree grammar. Textural primitives (windows) are grouped by placement rules expressed in another grammar. With stochastic grammars and error-correcting parsing, irregular textures can also be discriminated.

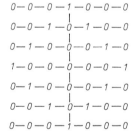

Fig. 3.27. Example of a textural primitive and its representation by a tree

3.7 Motion

3.7.1 Approaches to Motion Estimation

We assume a camera is centered on a coordinate system as shown in Fig. 2.12. If a pattern $f(x,y)$ is moving in the image plane of this system with respect to the background or if the camera is moving with respect to the pattern, the gray-value function becomes a time-dependent function $f(x,y,t)$. In general, a point P having coordinates (x,y,z) in R^3 at time t will have coordinates

$(x+dx, y+dy, z+dz)$ at time $t+dt$. This results in a velocity vector $\boldsymbol{v}(x,y,z,t)$ with components

$$u(x,y,z,t) = \frac{dx}{dt}, \quad v(x,y,z,t) = \frac{dy}{dt}, \quad w(x,y,z,t) = \frac{dz}{dt}. \qquad (3.7.1)$$

Since the z-coordinate is lost when projecting P on the image plane, it is not possible to recover w if only one point is considered. If two or more points of a rigid body are observed, a time change in the z-coordinate will cause a time change in the relative (x,y) position of the points and this can be used to recover the component w of the velocity. In (3.7.1) only the motion of a point is considered. The motion of a rigid body usually is described in terms of a translation along the three axes x, y, and z and a rotation consisting of three rotations around the three axes. In the following we will restrict the discussion to an estimation of the components u and v of the velocity vector in the image plane. From (3.7.1) it is apparent that motion estimation requires at least two images recorded at two different times t and $t + dt$, where the time separation dt is large enough to ensure a reasonably large dx and dy, but small enough to ensure almost constant velocities u and v.

There are two basic approaches to motion estimaton : *the computation of optical flow* and *the identification of corresponding image points* in the two images. A brief account of these two approaches is given below.

3.7.2 Optical Flow

Figure 3.28 shows a cross section of a gray-level edge which is constant along the y direction and changes linearly along the x direction. If the illumination is constant during the time between t and $t + dt$ and if the edge moves in the x direction, the observed change of gray value is due to the motion of the edge. In general, in addition to the requirement of constant illumination the orientation of the surface with respect to the observer should not change or the gray value should be independent of it. From the figure it is evident that in this case

$$f(x,t) = f(x+dx, t+dt). \qquad (3.7.2)$$

Furthermore, we observe that

$$\begin{aligned} &f(x,t+dt) - f(x,t) = df = -m\,dx = -f_x\,dx, \\ &m = \partial f/\partial x = f_x, \\ &[f(x,t+dt) - f(x,t)]/dt = f_t = -f_x dx/dt = -f_x u, \\ &f_t = \partial f/\partial t. \end{aligned} \qquad (3.7.3)$$

This gives in the one-dimensional case the constraint equation

$$f_t + f_x u = 0. \qquad (3.7.4)$$

If only a motion in the x direction is possible, this equation is sufficient to compute the velocity.

In the two-dimensional case a gray value function is observed at times t and $t + dt$. If the illumination is constant, in analogy to (3.7.2) the relation

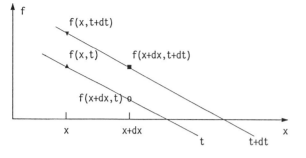

Fig. 3.28. Gray level change caused by the motion in x direction of a gray level edge

$$f(x, y, t) = f(x + dx, y + dy, t + dt) \qquad (3.7.5)$$

holds; this can also be seen from Fig. 3.29. The tuple (dx, dy) is called the *displacement vector*. It is a constant on an object if only one rigid body under translation is observed. If rotation is also possible or if more than one object is observed, the displacement vector is a function of x and y. The vectors $[dx(x, y), dy(x, y) \mid 0 \leq x \leq x_{\max}, 0 \leq y \leq y_{\max}]$ make up the *displacement vector field*. The corresponding velocity vectors ($u = dx/dt, v = dy/dt$) constitute the *optical flow*.

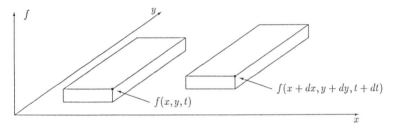

Fig. 3.29. Motion of a rigid body in a plane

Expanding the right side of (3.7.5) into a Taylor series yields

$$\begin{aligned} f(x, y, t) &= f(x + dx, y + dy, t + dt) \\ &= f(x, y, t) + \frac{\partial f}{\partial x} dx + \frac{\partial f}{\partial y} dy + \frac{\partial f}{\partial t} dt + \{\text{rest}\} . \end{aligned} \qquad (3.7.6)$$

Neglecting the {rest} consisting of higher order terms this reduces to a constraint equation which is in complete analogy to (3.7.4)

$$\begin{aligned} 0 &= f_x dx + f_y dy + f_t dt \\ 0 &= f_x u + f_y v + f_t \end{aligned} \qquad (3.7.7)$$

This is called the motion constraint equation. Since there are two velocity components u and v but only one equation, (3.7.7) is not sufficient to compute the velocity. An additional constraint is necessary and this results from the *smoothness constraint* introduced below.

In most practical situations one will not observe a set of randomly moving points but a set of rigid objects. Even if a set of non-rigid objects is observed (e.g.

118   3. Segmentation

a human body), there will be some coherence among the points of this object. Therefore, it is reasonable to assume that the optical flow field is "smooth" in some sense, that is there are regions of constant velocities or only gradual changes of the velocities. A severe deviation from this assumption are borders between a moving object and the background where discontinuities of velocity occur. This phenomenon is neglected for the moment. One possibility to measure the smoothness of a velocity field is the criterion

$$E_s^2 = \left(\frac{\partial u}{\partial x}\right)^2 + \left(\frac{\partial u}{\partial y}\right)^2 + \left(\frac{\partial v}{\partial x}\right)^2 + \left(\frac{\partial v}{\partial y}\right)^2 . \tag{3.7.8}$$

In addition there is the constraint (3.7.7) on the gray values f

$$E_f = f_x u + f_y v + f_t . \tag{3.7.9}$$

The problem now is to compute an optical flow field minimizing E_s^2 and yielding $E_f \approx 0$. This is summarized by the criterion

$$E = \iint \left[\alpha^2 E_s^2 + E_f^2\right] dx\, dy , \tag{3.7.10}$$

where the double integral is over the area of the image. Setting $u_x = \partial u/\partial x$ and so on, (3.7.10) can be written as

$$E = \iint \left[\alpha^2 \left(u_x^2 + u_y^2 + v_x^2 + v_y^2\right) + \left(f_x u + f_y v + f_t\right)^2\right] dx\, dy$$

$$= \iint F\left(u, u_x, u_y; v, v_x, v_y\right) dx\, dy . \tag{3.7.11}$$

The solution to (3.7.11) is known from variational calculus.

Theorem 3.7.1. A necessary condition for an extremum of F are the equations

$$F_u - \frac{\partial F_{u_x}}{\partial x} - \frac{\partial F_{u_y}}{\partial y} = 0 , \quad F_v - \frac{\partial F_{v_x}}{\partial x} - \frac{\partial F_{v_y}}{\partial y} = 0 , \tag{3.7.12}$$

where F_u and so on denote partial derivates with respect to u and so on.

Proof. See, for example, [Ref. 3.9, pp. 173, 174]. □

Using (3.7.11, 12) then gives the equations

$$f_x^2 u + f_x f_y v = \alpha^2 \nabla^2 u - f_x f_t , \quad f_x f_y u + f_y^2 v = \alpha^2 \nabla^2 v - f_y f_t . \tag{3.7.13}$$

In (3.7.13) there are two equations for the two components u and v of the velocity field. The operation $\nabla^2(.)$ is the Laplacian operator defined in (3.3.4).

3.7.3 Computation of Optical Flow

In order to compute an optical flow field from (3.7.13) discrete versions of the different partial derivatives have to be provided first; then an algorithm for solving the system of linear equations has to be selected. Considering an x, y, t coordinate system having discrete coordinate points i, j, k as shown in Fig. 3.30, a discrete approximation of derivatives f_x, f_y, f_t can be computed using a $2 \times 2 \times 2$ neighborhood of image points from

$$
\begin{aligned}
f_x &= \left[\left(f_{i+1,j,k} + f_{i+1,j+1,k} + f_{i+1,j,k+1} + f_{i+1,j+1,k+1}\right) \right. \\
&\quad \left. - \left(f_{i,j,k} + f_{i,j+1,k} + f_{i,j,k+1} + f_{i,j+1,k+1}\right)\right] / 4 \,, \\
f_y &= \left[\left(f_{i,j+1,k} + f_{i+1,j+1,k} + f_{i,j+1,k+1} + f_{i+1,j+1,k+1}\right) \right. \\
&\quad \left. - \left(f_{i,j,k} + f_{i+1,j,k} + f_{i,j,k+1} + f_{i+1,j,k+1}\right)\right] / 4 \,, \\
f_t &= \left[\left(f_{i,j,k+1} + f_{i+1,j,k+1} + f_{i,j+1,k+1} + f_{i+1,j+1,k+1}\right) \right. \\
&\quad \left. - \left(f_{i,j,k} + f_{i+1,j,k} + f_{i,j+1,k} + f_{i+1,j+1,k}\right)\right] / 4 \,.
\end{aligned}
\tag{3.7.14}
$$

The $2 \times 2 \times 2$ neighborhood gives some noise reduction as discussed in Sect. 3.3.1. For better noise reduction either a larger neighborhood may be used or the same neighborhood is used after appropriate low-pass filtering.

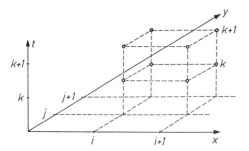

Fig. 3.30. Indices in a $2 \times 2 \times 2$ neighborhood of image points

A discrete version of the Laplacian was given already in (3.3.5). There are other discrete versions which may be summarized by

$$\nabla^2 u_{ijk} \approx \beta \left(\overline{u}_{ijk} - u_{ijk}\right) \,. \tag{3.7.15}$$

The version (3.3.5) has $\beta = 4$ and obvious $\overline{u}_{ijk}$. Another version is $\beta = 3$ and

$$
\begin{aligned}
\overline{u}_{ijk} &= \left(u_{i-1,j,k} + u_{i+1,j,k} + u_{i,j-1,k} + u_{i,j+1,k}\right) / 6 \\
&\quad + \left(u_{i-1,j-1,k} + u_{i-1,j+1,k} + u_{i+1,j+1,k} + u_{i+1,j-1,k}\right) / 12 \,.
\end{aligned}
\tag{3.7.16}
$$

Using (3.7.15), $a = \alpha^2 \beta$, and solving (3.7.13) for u and v gives

$$
\begin{aligned}
u &= \overline{u} - \frac{f_x \left(f_x \overline{u} + f_y \overline{v} + f_t\right)}{f_x^2 + f_y^2 + a} \,, \\
v &= \overline{v} - \frac{f_y \left(f_x \overline{u} + f_y \overline{v} + f_t\right)}{f_x^2 + f_y^2 + a} \,.
\end{aligned}
\tag{3.7.17}
$$

Every term in the above equations, except a, depends on the position i, j, k in the image sequence; this was not added explicitly in order to avoid the many indices. Since $\overline{u}$ depends on a certain neighborhood, (3.7.17) is a system of linear equations having $2M^2$ unknowns if the image has M^2 pixels. This system may be solved by the iterative Gauss-Seidel algorithm. If in step number N of the iteration values u_N, v_N of the velocity components have been computed for every image point, new values in step $N + 1$ are computed by

$$u_{N+1} = \overline{u}_N - \frac{f_x \left(f_x \overline{u}_N + f_y \overline{v}_N + f_t \right)}{f_x^2 + f_y^2 + a}$$

$$v_{N+1} = \overline{v}_N - \frac{f_y \left(f_x \overline{u}_N + f_y \overline{v}_N + f_t \right)}{f_x^2 + f_y^2 + a} \ . \tag{3.7.18}$$

If no a priori information is available about the optical flow field, the iteration is started using $u = v = 0$ in every image point. The iteration may be done several times on the same two images taken at times t and $t + dt$, or on images taken at successive times $t, t+dt, t+2dt, \ldots, t+mdt$. From (3.7.4, 7) it becomes apparent that no estimation of motion is possible at image points where the partial derivatives with respect to x and y vanish. Equation (3.7.18) shows that at points with small partial derivatives the numerator becomes small if a is small, and that in this case errors have a significant effect. Therefore, the value of a should be comparable to the error in f_x^2 and f_y^2.

The above two sections have only given the basic properties of optical flow. It was mentioned that severe problems are caused by motion discontinuities occurring, for example, at the boundary of a moving object and a static background. One approach is to modify the smoothness constraint such that larger changes of u and v are tolerated at image positions where corners occur. Another approach is the introduction of discontinuities across line processes. The large amount of computation time caused by the many unknowns in (3.7.18) is a further problem. An approach to its reduction is to use a resolution hierarchy.

3.7.4 Matching Corresponding Image Points

Another approach for estimating velocity components is to isolate an object and trace it in the image. A more basic approach is to select a certain neighborhood or window f_w of image points, say $m \times m$ points at position (i, j, k) and try to find these $m \times m$ points at position $(i + \mu, j + \nu, k + 1)$ in the next image of the sequence. Sine μ, ν are unknown, this requires a matching operation for different values of μ, ν. The problem is similar to image registration (see Sect. 2.2.8) and template matching (see Sect. 4.2.4). The most advisable approach is to use a resolution hierarchy in order to reduce the computational effort. However, an important difference is that in template matching or image registration only one matching position is determined. In motion estimation a set of matches is to be determined such that this set yields a vector field which is optimal in some sense. Therefore, it is appropriate to give a short account of this approach.

Let the image be partitioned into blocks B_{mn} of size $2^k \times 2^k$, for example, $k = 4$ or $k = 5$ in an 512×512 image. It is assumed that the velocity is constant within a block. The velocity can be determined if the displacement $(u, v)_{mn}$ of block B_{mn} from the image at time k to the image at time $k + 1$ is known. Given a displacement (u, v) the *displaced frame difference* is defined by

$$E_{f,mn}(u, v) = \sum_{i,j \in B_{mn}} |f_{ijk} - f_{i+u,j+v,k+1}| \,. \tag{3.7.19}$$

In order to avoid large displacements at homogeneous regions a constraint

$$E_{c,mn}(u, v) = a(|u| + |v|)\,, \quad a > 0 \tag{3.7.20}$$

is introduced. A matching displacement $(u, v)_{M1,mn}$ is one minimizing the combined criterion

$$E_{1,mn} = E_{f,mn}(u, v) + \alpha E_{c,mn}(u, v)\,, \quad \alpha > 0\,. \tag{3.7.21}$$

This is determined by evaluating (3.7.21) for displacements in a certain search interval. The second term avoids large errors at fairly homogeneous regions. In this way matching displacements are determined for every block in the image.

The next step is to smooth the displacements which were determined independently in the previous step. A smoothness criterion

$$\begin{aligned} E_{s,mn}(u, v) = \;& |u_{mn} - u_{m+1,n}| + |v_{mn} - v_{m+1,n}| \\ & + |u_{mn} - u_{m,n+1}| + |v_{mn} - v_{m,n+1}| \\ & + |u_{mn} - u_{m-1,n}| + |v_{mn} - v_{m-1,n}| \\ & + |u_{mn} - u_{m,n-1}| + |v_{mn} - v_{m,n-1}| \end{aligned} \tag{3.7.22}$$

is introduced and a new combined matching criterion is defined by

$$E_{2,mn}(u, v) = E_{f,mn}(u, v) + \beta E_{s,mn}(u, v)\,, \quad \beta > 0\,. \tag{3.7.23}$$

Using the displacements of the first step as initial values a new matching displacement $(u, v)_{M2,mn}$ is determined which minimizes $E_{2,mn}$. Having done this for all blocks the process is iterated for a block if the displacement in one of its neighboring blocks has been changed. The process stops if no more changes occur. This results in a local minimum of the sum of $E_{2,mn}$ over all blocks.

The last step is a refinement of the displacements. A block is split into four blocks of equal size. The initial displacement per block is the displacement of the larger block. The process of computing smoothed displacements is repeated as above. The block size is reduced step by step until a desired resolution is achieved.

Two results obtained by this approach are shown in Fig. 3.31.

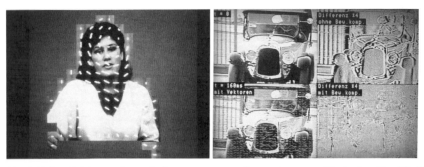

Fig. 3.31. The left image shows displacement vectors of a moving person (block size 16 × 16). The upper and lower left parts of the right image show a moving car at $t = 0$ and $t = 160$ ms, respectively (block size 8 × 8); displacement vectors are shown in the lower left panel. The upper and lower right parts of the right image show the difference image without and with motion compensation, respectively. Results according to W. Tengler, SIEMENS AG, Zentralbereich Forschung und Technik, München

3.8 Depth

3.8.1 Approaches to Depth Recovery

It was mentioned in Sect. 2.2.10 that depth (or the distance of an object point to an image point in the image plane of a camera) is lost when projecting a three-dimensional object onto the image plane of a camera. The recovery of this depth requires special methods. The main possibilities for determination of depth are:

1. Evaluation of two (or more) *stereo images*. This approach will be treated in more detail below.
2. Recovering depth from focusing of the image using (2.2.49). Automatic focusing of an image can be achieved, for example, by maximization of the high frequencies of an image.
3. Illumination of the scene by *patterned light*. The patterned light consists of a line or grid pattern having known geometrical properties, for example, a set of horizontal and vertical lines of known spacing. From the known spacing of the line pattern in the illumination and the spacing of lines in the image it is possible to compute the depth of a line in the image if in addition the camera properties are known.
4. Measuring depth using a laser range finder. An object point is illuminated by a laser sight source and the time of flight of the emitted and reflected light is measured. From this time the distance of the point is obtained directly.

Methods 1 and 2 are *passive* in the sense that they do not require special illumination of a scene; methods 3 and 4 are *active* and do require special illumination. Active methods are computationally simpler, but need specialized equipment and so far are restricted to situations where only one sensory device is looking at a scene. For example, two robots or vehicles illuminating the same scene with patterned light in the same spectral channel would interfere with each other. The stereo method is computationally sophisticated, but works with standard cameras and is not impaired if several stereo cameras are looking at the same scene. In the following we will only treat depth recovery from two stereo images and leave other methods to the references.

3.8.2 Stereo Images

The principle of stereo imaging is to take two (or more) images of the same object from different positions. One object point P will be recorded as two *corresponding image points* P_L and P_R in the two stereo images. If the imaging geometry is known, the depth of P can be recovered from the position of the two corresponding points. The main problem in automatic depth computation using stereo images is the determination of corresponding points because there are many candidates for correspondences and it may also occur that a certain object point is visible only in one image.

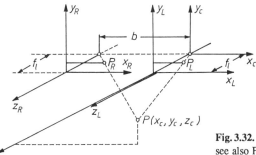

Fig. 3.32. Imaging geometry in binocular stereo, see also Figs. 2.11, 12

The imaging geometry used here for binocular stereo is shown in Fig. 3.32. The optical axes of the cameras are assumed to be parallel. The main parameter in this case is the *stereo basis* b. The two images are the left and right stereo image, respectively. In addition we assume that the relevant camera parameters are known, for example, from camera calibration. We also assume that the images are normalized such that the stereo basis b is parallel to the horizontal image coordinate. Therefore, a displacement between two corresponding image points is only a difference in x values. Using the notation in Fig. 3.32 the relation between three-dimensional camera coordinates (x_c, y_c, z_c) of the point P and the two-dimensional image coordinates (x_L, y_L) and (x_R, y_R) in the left and right stereo images, respectively, are

$$\left| \begin{matrix} x_L \\ y_L \end{matrix} \right| = \frac{f_l}{z_c} \left| \begin{matrix} x_c \\ y_c \end{matrix} \right| \quad \text{and} \quad \left| \begin{matrix} x_R \\ y_R \end{matrix} \right| = \frac{f_l}{z_c} \left| \begin{matrix} x_c + b \\ y_c \end{matrix} \right| . \tag{3.8.1}$$

The *disparity*, that is the displacement between two corresponding image points, is

$$d = x_R - x_L = f_l b / z_c .\tag{3.8.2}$$

It is useful to define the *disparity coefficient*

$$u = f_l / z ,\tag{3.8.3}$$

which is a function of the position in the image although this is not stated explicitly in the equation. If d can be determined from the two images of a point P, the three-dimensional coordinates of P are

$$\begin{vmatrix} x_c \\ y_c \\ z_c \end{vmatrix} = \frac{1}{u} \begin{vmatrix} x_L \\ y_L \\ f_l \end{vmatrix} .\tag{3.8.4}$$

The problem now is to obtain the disparity d of two corresponding image points. Apparently there is a close analogy between depth and motion computation in the sense that both require the identification of corresponding points (or structures) in two (or more) different images.

It is useful to measure the disparity d in units of pixel distance. If M and x_B are the width of the image in pixels and in [cm], respectively, the disparity in pixels is

$$d' = dM/x_B$$
$$= \gamma M b / z_c , \quad \text{with} \quad \gamma = f_l / x_B .\tag{3.8.5}$$

If depth ranges from $z_{\min}$ to $z_{\max}$, disparity d' ranges from

$$d'_{\min} = \gamma M b / z_{\max} \leq d' \leq \gamma M b / z_{\min} = d'_{\max} .\tag{3.8.6}$$

3.8.3 Optical Flow

It is almost straightforward to adapt the optical flow approach to the stereo problem. If we introduce formally

$$f_L(x,y) = f(x,y,b) \quad \text{and} \quad f_R(x,y) = f(x,y,0)\tag{3.8.7}$$

we have an "image sequence" consisting of two images separated by b (instead of Δt). The two equations (3.7.8,9) reduce to

$$E_s^2 = \left(\frac{\partial u}{\partial x}\right)^2 + \left(\frac{\partial u}{\partial y}\right)^2 ,\tag{3.8.8}$$

$$E_f = f_x u + f_b .\tag{3.8.9}$$

The constraint (3.8.8) is slightly modified in order to allow a different weighting of u_x and u_y

$$E_s^2 = u_x^2 + \beta^2 u_y^2 , \quad \beta^2 \leq 1$$

$$= (\nabla u)_t \mathbf{W}(\nabla u) \ . \tag{3.8.10}$$

In the ideal case $E_f = 0$ the *direct solution*

$$u^* = -f_b/f_x \tag{3.8.11}$$

is possible if $f_x \neq 0$. The problem of computing f_b is treated below; formally f_b replaces f_t in the computation of motion.

If $f_x \approx 0$, u^* from (3.8.11) becomes very inaccurate. Therefore, a variational problem in analogy to (3.7.10) is introduced by specializing this equation to

$$E = \iint \left[\alpha^2 E_s^2 + E_f^2 \right] dx \, dy \tag{3.8.12}$$

$$= \iint \left[\alpha^2 \left(u_x^2 + \beta^2 u_y^2 \right) + (f_x u + f_b)^2 \right] dx \, dy$$

$$= \iint F(u, u_x, u_y) dx \, dy \ . \tag{3.8.13}$$

From Theorem 3.7.1 in Sect. 3.7.2 the condition

$$f_x^2 u + f_x f_b = \alpha^2 \left(u_{xx} + \beta^2 u_{yy} \right) \tag{3.8.14}$$

results. An iterative solution of (3.8.14) is obtained by starting with an arbitrary u_0 and computing improved values of u from

$$u_{N+1} = \frac{\alpha^2 \bar{u}_N - f_x f_b}{f_x^2 + \alpha^2} = \frac{\alpha^2 \bar{u}_N + f_x^2 u^*}{f_x^2 + \alpha^2} \ . \tag{3.8.15}$$

The right hand expression in this equation results from (3.8.11). The term $\bar{u}$, which of course is a function of pixel location (i, j), is introduced from the relation

$$u_{xx;ij} + \beta^2 u_{yy;ij} = \bar{u}_{ij} - u_{ij} \ . \tag{3.8.16}$$

It remains to compute discrete approximations of $\bar{u}, f_x$, and f_b. A discrete version of $\bar{u}$ is

$$\bar{u}_{ij} = (4 + 8\beta^2)^{-1} \left[2u_{i-1,j} + 2u_{i+1,j} + \beta^2 \left(u_{i-1,j-1} + 2u_{i,j-1} + u_{i+1,j-1} \right. \right.$$

$$\left. \left. + u_{i-1,j+1} + 2u_{i,j+1} + u_{i+1,j+1} \right) \right] \ . \tag{3.8.17}$$

A problem arises in the computation of suitable values of f_x and f_b – or of a suitable initial value of u – if the disparity becomes large. Usually the range $z_{\min}, z_{\max}$ of the depth values will be known from knowledge about the task domain. Then from (3.8.6) $d_{\min}$ and $d_{\max}$ are known. A *local correspondence criterion* is introduced which is based on the two requirements that in corresponding image points the intensities should be similar and the partial derivatives with respect to x should be similar. The unknown disparity, measured in integer valued pixels, is denoted by s. A useful local correspondence criterion is

$$L_{ij}(s) = \left[\left(f_{L;ij} - f_{R;i+s,j}\right)^2 + \left(f_{L;i+1,j} - f_{R;i+s+1,j}\right)^2\right]$$
$$+ \delta \left[\left(f_{xL;ij} - f_{xR;i+s,j}\right)^2 + \left(f_{xL;i+1,j} + f_{xR;i+s+1,j}\right)^2\right]. \quad (3.8.18)$$

The minimum of $L_{ij}(s)$ is determined by searching over $s \in (d_{min}, d_{max})$. s^* is the value of s which minimizes L, and it is possible to use it as an initial value in (3.8.15). However, s^* is an integer and u need not be an integer. A refined value u^* of u can be computed by fitting a parabola through the three points $(k, L_{ij}(k); k = s^* - 1, s^*, s^* + 1)$ and determining the point u_{ij}^* where this parabola has its minimum. This point is

$$u_{ij}^* = s^* + \frac{L_{ij}(s^* - 1) - L_{ij}(s^* + 1)}{4a}, \quad (3.8.19)$$

$$a = \frac{L_{ij}(s^* - 1) - 2L_{ij}(s^*) + L_{ij}(s^* + 1)}{2}. \quad (3.8.20)$$

An estimate of f_x is

$$f_{x;ij} = \frac{f_{x;R;i+s^*,j} + f_{x;L;i,j}}{2} \quad (3.8.21)$$

and an estimate of $f_{x;L;ij}$ can be computed either by

$$f_{x;L;ij} = f_{L;i+1,j} - f_{L;ij} \quad (3.8.22)$$

or by the Sobel operator (3.3.2).

It was mentioned that the iteration (3.8.15) may be started with an arbitrary u_0, for example, $u_0 = 0$. A much better initial value is obtained from the rule

IF ($f_x^2 > \Theta$), THEN ($u_0 = u^*$ from (3.8.19)), ELSE (u_0 is linear
interpolation of u^* from the nearest left and right
neighbor with $f_x^2 > \Theta$). (3.8.23)

By this approach one obtains depth at every pixel, not only at the corners of an image. However, the problems with depth discontinuities are the same as those mentioned for motion discontinuities. An example is shown in Fig. 3.33.

It was mentioned in Sect. 3.7.3 that a modified smoothness constraint may be introduced. This can also be done in (3.8.8). A modification of (3.8.8) is

$$E_s = u_x^2 \left(f_y^2 + a^2\right) + u_y^2 \left(f_x^2 + a^2\right). \quad (3.8.24)$$

It is an approximation of a constraint allowing discontinuities in the direction of the image gradient and enforcing continuity in the direction normal to the gradient or tangential to lines of constant image intensity. The effect of this constraint is also shown in Fig. 3.33.

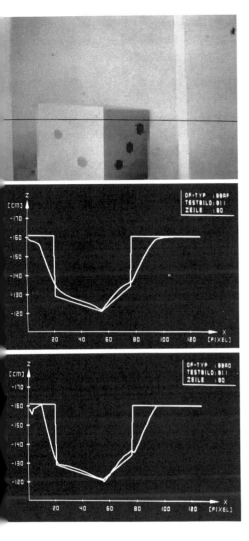

Fig. 3.33. From top to bottom: the right image of a pair of stereo images; correct and computed depth along the line shown in the original image; correct and computed depth using a modified continuity constraint [3.10]

3.8.4 Area Matching

Corresponding image points can also be found in analogy to Sect. 3.7.4. After selecting one image point P_L in the left stereo image, the corresponding image point P_R in the right stereo image is found by matching a certain neighborhood of P_L with neighborhoods of points on the epipolar line in the right image. If normalized stereo images are used, the search for a match is restricted to image points in the same scan line in the right image. The point in the right image giving the best match is assumed to be the corresponding image point. The matching criterion may be the correlation or the sum of absolute differences of image points. The search for a match is local, if P_L and P_R are determined without regard of neighboring point pairs. However, the example in Sect. 3.7.4 shows that global criteria can also be incorporated in the pixel matching approaches. Since here the similarity of pixel values is determined, this approach assumes constant intensities of corresponding pixels. In addition, there must be sufficient structure of the gray value function within the neighborhood used for matching.

3.8.5 Line Matching

Instead of matching neighborhoods of pixels one may also determine certain segmentation objects in the images and try to find corresponding segmentation objects. Often short line segments or longer lines are determined first and a matching of the line segments or lines is attempted. Depending on the matching criterion a local or a global match can be achieved. As an example we give an outline of an algorithm using a so-called minimal differential disparity criterion. The general idea is to find a set of matching line pairs such that the disparities of those line pairs in a certain neighborhood are similar; this is a continuity criterion for depth values.

Let the stereo images be normalized. It is assumed that a set of lines has been extracted from the two stereo images together with a measure of line contrast and line orientation. The set of lines in the left and right images is $A = \{a_i\}$ and $B = \{b_j\}$, respectively. For each line a_i in the left image a window $w(a_i)$ is defined such that it contains all lines from the right image which may correspond to a_i. The same is done for lines in the right image. The window is a parallelogram having a_i as one side and a width determined by the maximal disparity in the image. A predicate $H(i,j)$ indicates whether two lines a_i and b_j are potential matches. H is true if the lines "overlap", denoted by $a_i \leftrightarrow b_j$, and if they have similar contrast and orientation. Two lines overlap if they have a certain number of common horizontal scan lines. The disparity d_{ij} of the two lines is defined to be the average disparity in the common scan lines. Now define the terms

$$S(a_i|H) = \{b_j | b_j \text{ is in } w(a_i) \wedge H(i,j) \text{ is true}\},$$
$$S(a_i|\overline{H}) = \{b_j | b_j \text{ is in } w(a_i) \wedge H(i,j) \text{ is false}\}, \quad (3.8.25)$$
$$\text{card}(a_i) = \text{number of lines in } S(a_i|H) \cup S(a_i|\overline{H}).$$

In analogy to (3.8.25) the terms $S(b_j|H), S(b_j|\overline{H}), \text{card}(b_j)$ are defined.

For each potential match between lines a_i, b_j the following criterion $\varepsilon(i,j)$ is a measure of the difference of the disparities in the neighborhood of a_i, b_j. For each line pair with $H(i,j) = $ true the criterion is defined by

$$\varepsilon(i,j) = \text{card}(b_j)^{-1} \left[\sum_{a_k \text{ in } w(b_j)} \min_{\{b_l \text{ verifies } C1(a_k)\}} r_{ijkl} |d_{ij} - d_{kl}| \right]$$

$$+ \text{card}(a_i)^{-1} \left[\sum_{b_l \text{ in } w(a_i)} \min_{\{a_k \text{ verifies } C2(b_l)\}} r_{ijkl} |d_{ij} - d_{kl}| \right]$$

$$r_{ijkl} = \min \{\text{overlap } (a_i b_j), \text{overlap } (a_k, b_l)\}, \qquad (3.8.26)$$

where overlap (a_i, b_j) is the number of common scan lines. The conditions "verifies $C1$" and "verifies $C2$" are defined below. Having the criterion ε the notion of a preferred match can be defined. A match (a_i, b_j) is preferred if

$$\begin{aligned}&\varepsilon(i,j) < \varepsilon(i,l) \text{ for all } b_l \in S(a_i|H) \text{ and } b_l \leftrightarrow b_j \text{ AND}\\&\varepsilon(i,j) < \varepsilon(k,j) \text{ for all } a_k \in S(b_j|H) \text{ and } a_k \leftrightarrow a_i\,.\end{aligned} \qquad (3.8.27)$$

A line may have more than one preferred match in order to account for a match between a line and two fragments of the corresponding line in the other image. If (a_i, b_j) is a preferred match, then b_j is in the set $Q(a_i)$ of preferred matches, and a_i is in $Q(b_j)$. Initially, all sets Q are empty. The criterion "b_l verifies $C1(a_k)$" is defined by the two conditions

IF $(Q(a_k) \neq \phi)$, THEN (b_l is in $Q(a_k)$), ELSE (b_l is in $S(a_k|H)$),

EITHER $(b_l \neq b_j)$, OR (a_i and a_k do not overlap). (3.8.28)

The criterion "a_k verifies $C2(b_l)$" is defined in analogy to (3.8.28).

First the terms in (3.8.25) are computed. Then for all pairs (a_i, b_j) the difference (3.8.26) is evaluated. Since the sets Q are empty initially, the "min" expression in (3.8.26) is computed for all lines in the window w meeting the second condition in (3.8.28). After this first step a set of preferred matches can be computed by (3.8.27). The difference (3.8.26) is evaluated again, but now the "min" expression is computed for a restricted set according to (3.8.28). Another set of preferred matches is computed and (3.8.26) is evaluated for the third time. Experience indicates that three iterations for the computation of differences are sufficient.

Once the set of preferred matches of line pairs are obtained, then the average disparity and depth of the corresponding lines can be computed. This approach gives depth estimates only at contour lines in the image, that is a *sparse depth map*. If depth estimates are also desired at intermediate points, that is a *dense depth map*, an interpolation of depth values or a combination of line matching with area matching or optical flow is necessary.

3.9 Shape from Shading

3.9.1 Illumination Models

If an object having completely known surface properties (position, orientation, spectral reflectance, absorption, diffusion, ...) is illuminated by a light source (or several light sources) having known properties (position, spectral distribution of energy, focusing of the light beam, ...), the image recorded by a camera having known properties can be computed. The image is determined by the triple {surface, light source, camera}, where camera may be replaced by some viewer. If in turn the image and the properties of the illumination and the camera are known, some properties of the object can be computed. In image analysis an important property is the *surface* of the object, for example, represented by the equation

$$z = g(x, y) \tag{3.9.1}$$

of the surface or represented by the surface normal $n(x_n, y_n, z_n = g(x_n, y_n))$.

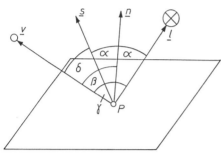

Fig. 3.34. Directions and angles used to characterize surface properties. P is some surface point, V an observation point, l is the direction of the light source, n is the surface normal, s is the direction of specular reflection, and v is the direction of the viewer

The methods of *shape from shading* attempt to recover the object surface $z = g(x, y)$ or the normal n from the recorded gray-level image $f(x, y)$. This is an inverse problem requiring a surface model, an illumination model, and a camera model. Figure 3.34 shows the relevant vectors and angles of this triple; they are the vector n normal to the surface, the vector l pointing towards the light source, and the vector v pointing towards the viewer or camera as well as the angle α between n and l, β between n and v, and γ between l and v. Apparently, n, l, and v need not be in the same plane. In addition s is the direction of specular reflection. The angle between s and n is equal to α, and s, n, and l are in the same plane.

A simple surface model is the *Lambertian surface* or a *diffuse radiator*. It assumes that the luminous intensity I_V in an observation point V depends on the intensity I_P illuminating a surface point P by

$$I_V = I_P r_d \cos \alpha = I_P r_d n \cdot l \; . \tag{3.9.2}$$

In this equation r_d is a factor summarizing the dependency of I_V on the surface reflectivity and the distance to the surface point P. The main advantage of this model is its simplicity and the fact that in restricted scenes it often is a reasonable approximation. Apparently, shape from shading is only feasible if sufficiently realistic surface and illumination models are available. Similarly in computer graphics only in this case will one get realistic images. Therefore, more involved models have been developed taking into account points such as background intensity, specular reflection, spectral dependency of surface properties, shadow, multiple light sources, spectral properties of the illumination, or directional dependency of surface properties. A more realistic model is

$$I_V = I_U + I_P r_d \cos \alpha + I_P r_s \cos^n \delta . \tag{3.9.3}$$

The first term corresponds to a constant background intensity, the second is the diffuse radiation (3.9.2), the third term is an approximation to specular reflection. The angle δ is between s and v as shown in Fig. 3.34. The term $\cos^n \delta$ models the fact that the intensity due to specular reflection is high if $s \approx v$ or $\delta \approx 0$, and decreases rapidly if δ increases. The constant n is dependent on the material, e.g. 10–100 for nickel, and values common in computer graphics are around $n = 10$. As an additional complication the factor r_s is dependent on α. There are other models of observed intensity, for example, models using the wavelength-dependent bidirectional reflectivity. However, the determination of shape from shading uses (3.9.2) in most cases, and (3.9.3) only in some cases. The term I_V is recorded as the observed image $f(x,y)$.

3.9.2 Surface Orientation

A surface can be characterized by (3.9.1) or some other type of equation, but it can also be characterized by its orientation, that is, the direction of its surface normal n at every point P of the surface. A surface point P having coordinates $(x, y, z = g(x,y))$ has a normal vector $n = (x_n, y_n, z_n)_t$. In the following we assume $|n| = 1$, that is n is a unit vector. Therefore, two parameters uniquely define n. From analytic geometry there is the relation

$$n = \begin{vmatrix} -\partial g(x,y)/\partial x \\ -\partial g(x,y)/\partial y \\ 1 \end{vmatrix} = \begin{vmatrix} -p(x,y) \\ -q(x,y) \\ 1 \end{vmatrix} . \tag{3.9.4}$$

An alternative is to give the *slant angle* σ_n and the *tilt angle* τ_n of the surface normal. These angles are shown in Fig. 3.35. Between slant and tilt and the parameters p and q there are the relations

$$\tau_n = \arctan(q/p) \text{ if } p \neq 0 , \quad \tau_n = \pm 90° \text{ if } p = 0 ,$$
$$\sigma_n = \pm \arccos(1 + p^2 + q^2)^{-1/2} . \tag{3.9.5}$$

Between the coordinates of n and slant and tilt there are the relations

$$z_n = \cos \sigma_n ,$$
$$x_n = \pm \left[\left(1 - \cos^2 \sigma_n\right) / \left(1 + \tan^2 \tau_n\right) \right]^{1/2} ,$$
$$y_n = \pm \left(1 - \cos^2 \sigma_n\right)^{-1/2} , \quad \text{if } x_n = 0 , \tag{3.9.6}$$
$$y_n = x_n \tan \tau_n , \quad \text{if } x_n \neq 0 .$$

From these coordinates one can compute p and q by

$$p = \frac{-x_n}{z_n} , \quad q = \frac{-y_n}{z_n} . \tag{3.9.7}$$

Two coordinate systems have become common in shape-from-shading analysis, the *viewer system* (x_v, y_v, z_v) and the *light source system* (x_l, y_l, z_l) shown in

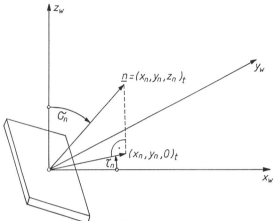

Fig. 3.35. The definition of the angles slant σ and tilt τ

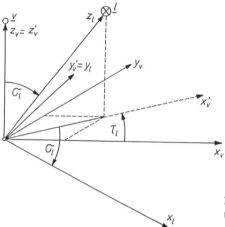

Fig. 3.36. The light source system (x_l, y_l, z_l) and the viewer system (x_v, y_v, z_v)

Fig. 3.36. The z_v axis of the viewer system points towards the viewer v. The z_l axis of the light source system points towards the light source l whose location must be known in order to use this system. In the viewer system the vector l has slant σ_l and tilt τ_l. The light source system is obtained by two rotations. The first rotates the x_v, y_v axis around the fixed z_v axis in a counter-clockwise direction by τ_l and gives the axes $x'_v, y'_v = y_l$, and $z'_v = z_v$. The second rotates the z'_v, x'_v axes around the fixed y_l axis by σ_l such that z_l points towards l. The coordinate transformation is given by

$$\begin{vmatrix} x_l \\ y_l \\ z_l \end{vmatrix} = \begin{vmatrix} \cos\sigma_l \cos\tau_l & \cos\sigma_l \sin\tau_l & -\sin\sigma_l \\ -\sin\tau_l & \cos\tau_l & 0 \\ \sin\sigma_l \cos\tau_l & \sin\sigma_l \sin\tau_l & \cos\sigma_l \end{vmatrix} \begin{vmatrix} x_v \\ y_v \\ z_v \end{vmatrix}, \qquad (3.9.8)$$

$$\begin{vmatrix} x_v \\ y_v \\ z_v \end{vmatrix} = \begin{vmatrix} \cos\sigma_l \cos\tau_l & -\sin\tau_l & \sin\sigma_l \cos\tau_l \\ \cos\sigma_l \sin\tau_l & \cos\tau_l & \sin\sigma_l \sin\tau_l \\ -\sin\sigma_l & 0 & \cos\sigma_l \end{vmatrix} \begin{vmatrix} x_l \\ y_l \\ z_l \end{vmatrix}. \qquad (3.9.9)$$

If the surface normal n has been determined in every pixel and a surface element orthogonal to n has been attributed to every normal n, this usually does not give a smooth surface due to errors in n. In order to obtain a smooth surface a smoothness criterion is defined similar to the approaches used to compute motion or depth from optical flow. One criterion is the smoothness constraint

$$E_s = \iint \left(p_x^2 + p_y^2 + q_x^2 + q_y^2\right) dx\, dy ,\qquad(3.9.10)$$

where p_x (and so on) denotes the partial derivative of p with respect to x (and so on). Its disadvantage is that it does not guarantee a unique surface. In this case the *criterion of integrability* must hold

$$\frac{\partial^2 g(x,y)}{\partial x\, \partial y} = \frac{\partial^2 g(x,y)}{\partial y\, \partial x}, \quad p_y = q_x .\qquad(3.9.11)$$

If P and Q are two points on the surface, if a is some curve on the surface from P to Q, and if (3.9.11) holds, then the integral of $g(x,y)$ from P to Q along a is independent of the curve a and depends only on P and Q.

3.9.3 Variational Approach Using the Reflectance Map

The *reflectance map* $R(p,q)$ is defined to give the dependence of object radiance on surface orientation. It can be shown that for a Lambertian surface illuminated by one distant light source, R is given by

$$R(p,q) = (1 + p_s p + q_s q)\left[(1+p^2+q^2)(1+p_s^2+q_s^2)\right]^{-1/2} ,\qquad(3.9.12)$$

where p_s, q_s is the direction of the source and $I_P r_d = 1$. In general, $R(p,q)$ depends on the material of the surface and on the illumination.

The recorded image irradiance $f(x,y)$ is proportional to radiance in the object point corresponding to the image point (x,y). Assuming proper normalization of $R(p,q)$ the basic relation

$$f(x,y) = R(p(x,y), q(x,y))\qquad(3.9.13)$$

holds. The problem then is to find orientations p, q such that (3.9.11, 13) hold. This can be formulated as a variational problem similar to (3.7.10). We replace (3.9.13) by the criterion

$$E_R = \iint [f(x,y) - R(p,q)] dx\, dy\qquad(3.9.14)$$

and introduce the smoothness constraint

$$E_s = \iint \left(p_y^2 - q_x^2\right) dx\, dy .\qquad(3.9.15)$$

With a weighting factor α the criteria are combined to

$$E = \iint (E_R + \alpha E_s) dx\, dy \ . \tag{3.9.16}$$

A variant of (3.9.16) is to use a stereographic projection of the gradients instead of p, q directly.

3.9.4 Local Shading Analysis

Another approach to the computation of surface orientation depends on the assumption that a small neighborhood of a surface point may be approximated by a sphere and that the surface is Lambertian. In a small neighborhood of a surface point P one may assume that I_P, r_d and l in (3.9.2) are constant. Therefore, one can compute the first and second derivatives of I_V with respect to x, y, and so on. Denoting I_V and its derivatives by $f, f_x, f_y, f_{xx}, f_{xy}, f_{yy}$ there are six equations relating observable quantities to properties of the surface. They are the slant and tilt of the surface normal n and of the light source l and the product $I_P r_d$ in (3.9.2). If the surface is approximated by a sphere in a neighborhood of P, the derivatives of n depend on the slant and tilt of the surface and on the radius R of this sphere, which is an additional parameter. So there are six parameters and six equations. Solving them for slant and tilt of the surface normal yields

$$\tan \tau_n = \frac{f_{yy} - f_{xx} + \sqrt{(f_{xx} - f_{yy})^2 + 4f_{xx}^2}}{2f_{xy}} ,$$

$$\cos^2 \sigma_n = \frac{(f_{xx} + f_{yy}) \tan \tau_n - f_{xy} \sec^2 \tau_n}{(f_{xx} + f_{yy}) \tan \tau_n + f_{xy} \sec^2 \tau_n} . \tag{3.9.17}$$

The advantage of (3.9.17) is that the direction of the light source need not be known, but the computation of second derivatives of the observed gray-level function $f(x, y)$ is influenced by noise and the assumption of a locally spherical surface may be too restrictive.

If the direction of the light source is assumed to be known and the surface is locally approximated by a sphere at some point P, the tilt of the surface normal in P can be computed easily in the light source system. Let f_x, f_y be the partial derivatives of f in the viewer system. They are transformed to the light source system by

$$\begin{vmatrix} f'_x \\ f'_y \end{vmatrix} = \begin{vmatrix} \cos \sigma_l \cos \tau_l & \cos \sigma_l \sin \tau_l \\ -\sin \tau_l & \cos \tau_l \end{vmatrix} \begin{vmatrix} f_x \\ f_y \end{vmatrix} . \tag{3.9.18}$$

The surface tilt at P with respect to the light source system is then

$$\tau'_n = \arctan (f'_y / f'_x) . \tag{3.9.19}$$

The computation of the slant angle requires an additional assumption. Let Q and R be points on the surface lying on each side of P such that the image gradient has the same direction as the line connecting P, Q, and R. The image intensity in P, Q, and R is denoted by f_P, f_Q, and f_R, respectively. The slant of the surface normal in P, Q, and R in the light source system is denoted by $\sigma'_P, \sigma'_Q,$ and σ'_R, respectively. It is required that

$$\sigma'_P - \sigma'_R = \Delta_1 = \sigma'_Q - \sigma'_P = \Delta_2 = \Delta \approx 0 \,. \tag{3.9.20}$$

In this case

$$\sigma'_P \approx \arctan\left[\frac{1}{2}|f_R - f_Q|(f_P(2f_P - f_R - f_Q))^{-1/2}\right] \,. \tag{3.9.21}$$

In order to ensure (3.9.20), the slant is only evaluated if

$$\frac{1}{a} \leq \frac{|f_R - f_P|}{|f_Q - f_P|} \leq a \,, \tag{3.9.22}$$

where a is between 2 and 10. If the viewer and the light source system coincide and the slant of a sphere is computed, the error in the slant angle is low at points P whose normal is close to the direction of the light source; if the angle between the normal and the light source increases, the error increases.

The actual computation of slant and tilt is done for four images obtained from the original image, a 3×3, 5×5, and 7×7 median filtered image. The tilt and slant values computed for the four images are smoothed by median filtering, and may be used as initial values for an iterative improvement of the values as described below.

3.9.5 Iterative Improvement

Using the slant (3.9.21) one can compute from (3.9.2) an "image"

$$\alpha(x,y) = \frac{f(x,y)}{\cos \sigma'_n} \,. \tag{3.9.23}$$

In this equation α summarizes the term $I_P r_d$ in (3.9.2) which should be a constant for an object having a uniform color and being homogeneously illuminated. Due to errors in the slant, α depends on the position (x,y). The idea of an iterative improvement of the slant is to compute a "slant" in the $\alpha(x,y)$ image, use this to improve α, and iterate this process a few times. The resulting value of α is then used to compute an improved value of the slant by (3.9.23). If the iteration index is m, this process is as follows:

1. Compute $\alpha(x,y)_0$ by (3.9.23) and smooth it by a 5×5 median filter.
2. Compute the $3 \times 3, 5 \times 5, 7 \times 7$ median filter of the α-image.
3. Compute σ_{m+1} by (3.9.21) ensuring (3.9.22), but replace f by α_m, using the four images.
4. Compute the mean value of slant and tilt obtained for the four images.
5. Compute α_{m+1} by (3.9.23) replacing σ'_n by σ_m; smooth the new α-image.
6. Iterate steps 2 and 3 a few times until $m = m'$.
7. Compute the final value of the slant from (3.9.23), that is,

$$\sigma'_n = \arccos \frac{f(x,y)}{\alpha(x,y)_{m'}} \,. \tag{3.9.24}$$

3.9.6 Depth Recovery

There are at least two approaches to depth recovery from the tilt and slant or from p and q, which both exploit the criterion of integrability (3.9.11). The first approach is based on a series expansion of the surface $z(x, y)$, the second on the independence of the line integral from the actual path. In the first approach consider a surface where the criterion of integrability holds. If it is represented by basis functions $\Phi_{\mu\nu}(x, y)$, we have the relations

$$z(x, y) = \sum_\mu \sum_\nu Z_{\mu\nu} \Phi_{\mu\nu}(x, y) , \tag{3.9.25}$$

$$p(x, y) = \sum \sum Z_{\mu\nu} \Phi_{x,\mu\nu}(x, y) ,$$

$$q(x, y) = \sum \sum Z_{\mu\nu} \Phi_{y,\mu\nu}(x, y) . \tag{3.9.26}$$

The values of p and q computed by shape-from-shading are denoted by $p_{\text{old}}(x, y)$ and $q_{\text{old}}(x, y)$ and may be represented using the same basis functions

$$p_{\text{old}}(x, y) = \sum \sum P_{\mu\nu} \Phi_{x,\mu\nu}(x, y) ,$$

$$q_{\text{old}}(x, y) = \sum \sum Q_{\mu\nu} \Phi_{y,\mu\nu}(x, y) . \tag{3.9.27}$$

Since the criterion of integrability may not hold for the $p_{\text{old}}, q_{\text{old}}$, the coefficients of the expansion are different in general. The goal is to find new values of p and q, denoted by $p_{\text{new}}, q_{\text{new}}$, such that the criterion of integrability holds and the new p, q values are a good approximation of the old ones. This is summarized in the equations

$$p_{\text{new}}(x, y) = \sum \sum Z_{\mu\nu} \Phi_{x,\mu\nu}(x, y) ,$$

$$q_{\text{new}}(x, y) = \sum \sum Z_{\mu\nu} \Phi_{y,\mu\nu}(x, y) , \tag{3.9.28}$$

$$E = \iint \left(|p_{\text{new}} - p_{\text{old}}|^2 + |q_{\text{new}} - q_{\text{old}}|^2 \right) dx\, dy = \min . \tag{3.9.29}$$

Theorem 3.9.1. The coefficients $Z_{\mu\nu}$ minimizing (3.9.29) are given by

$$Z_{\mu\nu} = \frac{R_{\mu\nu} P_{\mu\nu} + S_{\mu\nu} Q_{\mu\nu}}{R_{\mu\nu} + S_{\mu\nu}} , \tag{3.9.30}$$

$$R_{\mu\nu} = \iint |\Phi_{x,\mu\nu}(x, y)|^2 dx\, dy , \quad S_{\mu\nu} = \iint |\Phi_{y,\mu\nu}(x, y)|^2 dx\, dy . \tag{3.9.31}$$

Proof. See [3.11]. □

The functions $p_{\text{old}}, q_{\text{old}}$ are given as discrete functions and we look for a discrete function z_{jk}. The values of p and q are obtained from the central differences

$$p_{jk} = \frac{z_{j+1,k} - z_{j-1,k}}{2}, \quad q_{jk} = \frac{z_{j,k+1} - z_{j,k-1}}{2}. \qquad (3.9.32)$$

They can be transformed by the DFT or FFT defined in Sect. 2.3.3 to yield

$$\overline{P}_{\mu\nu} = \text{DFT}\{p_{jk}\}, \quad \overline{Q}_{\mu\nu} = \text{DFT}\{q_{jk}\}. \qquad (3.9.33)$$

In the discrete case the goal is the same as defined in (3.9.29) and there is an equivalent to Theorem 3.9.1.

Theorem 3.9.2. *In the discrete case the coefficients $Z_{\mu\nu}$ giving a function z_{jk} minimizing (3.9.29) if p and q are computed by (3.9.32) are defined by*

$$Z_{\mu\nu} = \frac{-\text{i}\sin(2\pi\mu/M)\overline{P}_{\mu\nu} - \text{i}\sin(2\pi\nu/M)\overline{Q}_{\mu\nu}}{\sin^2(2\pi\mu M) + \sin^2(2\pi\nu/M)}, \qquad (3.9.34)$$

where $\text{i} = \sqrt{-1}$ *and M^2 is the size of the image.*

Proof. See [3.10]. □

From $Z_{\mu\nu}$ one can obtain z_{jk} from an inverse DFT or FFT and also $p_{\text{new}}, q_{\text{new}}$ from (3.9.28) or (3.9.32). The surface z is determined up to a constant scaling factor since from shading alone one cannot distinguish between a large distant object and a small nearby one. The coefficient Z_{00} is undefined from (3.9.34) since the denominator becomes zero. It determines the average z value which can be fixed arbitrarily. An example of two surfaces recovered by the approach described in the last sections is shown in Fig. 3.37. The main steps in the recovery of the surface are computation of slant and tilt in the light source system and iterative improvement of the values, computation of p and q in the viewer system, and computation of the surface using the criterion of integrability.

The second approach to the incorporation of an integrability criterion is based on the fact mentioned above that in this case the integral from a point P to another point Q on the surface is independent of the line of integration. The surface integral is approximated by a sum over the discrete values of the surface z_{jk} and partial derivatives are approximated by differences. Some point z_{jk} (or better some points) on the surface must be known or its z-coordinate must be fixed to some arbitrary value. Given z_{jk} the other values of z on the surface can be computed from the known values of surface orientation. Evaluating the integral from z_{jk} to a point (l,m) along different lines will give different values of z at (l,m) and the arithmetic mean is assumed to be a good approximation to the true value of z_{lm}. This process is iterated for all points on the surface until a stable set of z-values is obtained.

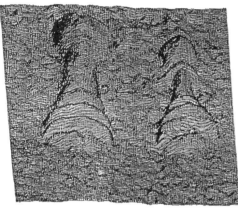

Fig. 3.37. A gray-level image and the surface recovered by shape from shading, [3.12]. The light source is (almost) in the direction of the viewer

3.10 Segmentation of Speech

3.10.1 Subword Units

In automatic speech recognition (ASR) the aim of *speech segmentation* is the same as of image segmentation, i.e., to find borders separating (speech) units which are different in some sense, or to find regions containing (speech) units which are homogeneous in some sense. Since speech is a one-dimensional pattern, these borders are just points on the time axis whereas in image processing the borders (contours) may be complicated two-dimensional curves. Similarly, the regions are just intervals of the time axis whereas in image processing they may have a complicated shape. From Sect. 3.1 it is clear that speech segmentation is nevertheless a difficult problem.

From linguistic and phonetic research a good deal of knowledge is available about transcription of languages to phonetic representations, number and type of phonemes present in a language, possibilities for replacement of phonemes due to individual pronunciation, and so on. In ASR units of speech are often represented in a hierarchical manner, proceeding from rough to fine representation of speech details (or from fairly simple and reliable automatic extraction to more complicated and erroneous extraction). An example is indicated in Fig. 3.38.

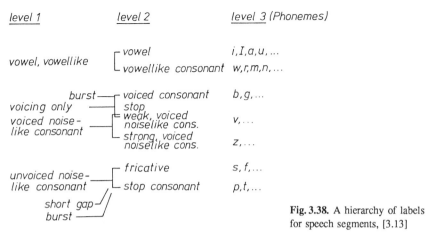

Fig. 3.38. A hierarchy of labels for speech segments, [3.13]

So far there is no general agreement about the best units for segmentation of speech. It seems however that at least for large vocabularies (about 1000 or more words) it is useful to first segment speech into *subword units*; word recognition then is performed on the basis of these subword units. Common subword units are the phoneme, the allophone, the phone, the diphone, the demisyllable, and the syllable. The *phoneme* is a theoretical construct which represents the smallest unit distinguishing two words of different meaning. For example, /d/ and /t/ are different phonemes because they are necessary to distinguish between "down" and "town". Depending on the language there are about 20–60 phonemes. An *allphone* denotes a certain realization of a phoneme in a certain context. For example, the phoneme /t/ may have a voiced or unvoiced, aspirated or unaspirated realization, each resulting in several different allophones of /t/. The number of allophones is fairly large, on the order of 1000. More useful for the purpose of ASR is the *phone*, which is often defined as "the smallest auditorily distinguishable unit". Nevertheless such a unit has different acoustic realizations, for example, due to different pitch, intonation, context, or time duration. In automatic recognition the different acoustic realizations of one phone must be grouped into the same class. Although phones are acoustically a continuum, the number of units distinguishable by a phonetician is about 200, the IPA set is about 100, and a rough transcription requires about 40.

A *syllable* consists of a nucleus which usually is a vowel (an exception being the syllabic nasals) and an initial and a final consonant cluster. Syllable borders in ASR are often obtained from the function of loudness over time. The nucleus corresponds to a local maximum, the minima to the left and right are the syllable borders. An estimate of the combinatorial number of syllables in German is about 500,000. However, this depends very much on the language since, for example, Japanese only has about 100 syllables. A syllable consists of two *demisyllables*. The first consists of the initial consonant cluster and the first half of the vowel, the second consists of the second half of the vowel and the final consonant cluster. According to the definition of the demisyllable they can be grouped into initial and final demisyllables. An estimate for German is that there are about 800 initial and 2500 final demisyllables. From this follows the *double-demisyllable* which covers the center of one vowel to the center of the next following vowel. The combinatorial number of double-demisyllables for German is almost 2,000,000. Of course, in a certain task domain requiring a vocabulary of about 1000 words this number would be significantly less. Finally, the *diphone* is usually defined as the region between the middle of a phonetic segment and the middle of the next phonetic segment; the number of diphones is about 1000.

The above discussion omits many of the details from linguistic and phonetic research, but it shows that a variety of subword units are available, that there may be significant differences between languages, and that there seems to be no ideal subword unit. An ideal subword unit would consist of a small number of classes and allow easy and reliable automatic recognition. The "small" units (e.g. phones) have only a few classes, but are hard to recognize automatically because of the dependence of their acoustic realization on the context. The "large" units (e.g. demisyllables) absorb a good deal of the context dependence, but require a large number of classes. No unit seems to offer significant advantages with respect to speaker-independent recognition. In any case the problem of segmentation is to find the unit boundaries and the class name of the unit within a boundary. In order to cope with errors it is common practice to label a certain speech segment with several alternative class names and/or to provide alternative segment boundaries in the speech signal.

3.10.2 Measurements

The characterization of speech relies to some extent on studies of human speech perception which indicate that humans use some kind of frequency analysis. Some of these results are presented in Sect. 1.8. Therefore, methods which use spectral properties predominate. Among these are methods which pass the speech signal through a bank of bandpass filters and sample the filter outputs every 10 ms or so, which calculate a discrete Fourier transform of a windowed speech portion at regular intervals, which use linear predictive modeling of speech spectra (Sect. 2.5), or cepstral coefficients. In addition, the use of autocorrelation coefficients and time-domain measurements are possible, and it seems that performance is comparable to frequency-domain methods. This indicates that a variety of parameters are available for measuring changes or homogeneity of speech characteristics. Some examples are given below. The term "frame" or "window" is used here as in Sect. 2.5.2.

Energy measurements are used to distinguish periods of silence from speech and to aid in distinguishing fricatives. If a spectral representation is available with coefficients F_ν (taken from a DFT of f_n) or α_ν (taken from a LP model), then energy in the spectral band ($f_1 = \nu_1 f_s/M$; $f_2 = \nu_2 f_s/M$) is

$$E(\nu_1, \nu_2) = \sum_{\nu=\nu_1}^{\nu_2} |\alpha_\nu|^2 . \tag{3.10.1}$$

The notation $F_\nu, \alpha_\nu, f_s, M$ is as introduced in Sect. 2.5.2. In the full spectral band energy may also be computed from N windowed sample values as stated in Sect. 2.2.3.

Frequency, amplitude, and bandwidth of the first two, three, or four *formants* are important for vowel classification. One possibility is to pick the first peaks of the model spectrum in (2.5.21) and use peak frequency as formant frequency as indicated in Fig. 2.21. A further problem is the tracking of formant frequencies since they may change from frame to frame. In one particular algorithm formant tracking is done only in vowel-like segments of speech and starts at the middle of a vowel. Because of the high energy in the middle it is assumed that formants can be found reliably; a forward and backward branch is made from there. Starting from initial estimates of the first four formant frequencies spectral peaks are attributed to the formants with the closest frequency estimate. These peaks then become new estimates for the next frame. Provisions are made to remove peaks attributed to two formants, to deal with peaks not assigned to a formant, and to smooth formant trajectories. Another possibility is to compute the poles of $G(z)$ in (2.5.13) which are the roots of $A(z)$ in (2.5.2). For iterative solutions initial values of roots may be obtained from the peaks of the model spectrum. If

$$z_\nu = z_{\nu r} + i z_{\nu i} \tag{3.10.2}$$

is a complex root of $A(z)$, the frequency f_ν and half bandwidth b_ν of the corresponding pole are

$$f_\nu = \frac{f_s}{2} \arctan \frac{z_{\nu i}}{z_{\nu r}}, \quad b_\nu = \frac{f_s}{4\pi} \log \left(z_{\nu r}^2 + z_{\nu i}^2 \right). \tag{3.10.3}$$

An example for finding the first formant from the frequency and bandwidth of the poles is as follows. If $f_1 < 900\,\text{Hz}$ this is the first formant; however, if $f_2 < 950\,\text{Hz}$, and $f_2 - f_1 < 150\,\text{Hz}$, and $4(f_2 - f_1) < b_1 - b_2$, then f_2 is taken as the first formant (assuming poles to be ordered such that $f_1 < f_2 < \ldots$). If there are no poles below 900 Hz with small bandwidths, no first formant is assumed to be present.

Decisions between voiced and unvoiced speech may be based on extraction of the fundamental frequency or the pitch period, which is the period of the glottal pulses. A direct approach is the use of the autocorrelation function. Since the autocorrelation of a periodic function is itself periodic, the pitch period should result in peaks of the autocorrelation. The problem is that these peaks may be heavily influenced by formants. Therefore, several other methods have been developed. A short outline of the so-called simplified inverse filter tracking algorithm will be given. It is based on (2.5.1) which shows that the output of the inverse system would be a sequence of impulses if $A(z)$ were the true system function and not an approximation as in (2.5.2). Since only the pitch period is desired, a simplified filter $A(z)$ which removes the fine structure of the spectrum is used. The algorithm has the following steps:

1. Prefilter the speech signal f_j with a low-pass filter with cutoff at about 0.8 kHz. Take every fifth sample from the filter output to reduce the effective

sampling frequency to 2 kHz, assuming an initial frequency of 10 kHz. The results are the sample values f'_j.
2. Calculate an inverse filter $A(z)$ as described in Sect. 2.5 with $m = 4$, $N = 64$.
3. Pass the samples f'_j through $A(z)$ and call the result y_j where

$$y_j = f'_j + \sum_{\mu=1}^{4} a_\mu f'_{j-\mu} \quad j = 0, 1, \ldots, 63 . \quad (3.10.4)$$

4. Compute the autocorrelation sequence r_j of y_j from (2.5.7). Divide r_j by r_0 to obtain normalized values r'_j.
5. Let r'_p be the peak of the normalized values r'_j. If $r'_p < 0.4$, assume unvoiced speech; otherwise assume voiced speech with pitch period $p/2$ ms.

An interpolation scheme may be used to improve pitch resolution. A variable threshold in step 5 is useful for reduction of errors. For small p a larger number of pitch periods is included in the analysis; this should result in a larger autocorrelation peak and consequently in a larger threshold.

Another example is an algorithm which determines pitch period only in voiced segments of speech. This requires that a voiced/unvoiced decision is made first. The speech wave is sampled at 10 kHz, a 12.8 ms frame is weighted by a Hamming window, and a 512 point FFT is computed in this window. Let the Fourier coefficients be F_ν. As a measure of energy the term

$$E = a \cdot \log \left(\sum_{\nu=\nu 1}^{\nu 2} |F_\nu|^2 \right) - b \quad (3.10.5)$$

is computed, where the constants a and b are chosen to scale E to some desired interval and $\nu 1$ and $\nu 2$ are chosen to cover the frequency band 300–2300 Hz. The pitch period is determined only if $E > \Theta$. The speech signal is band-pass filtered in the interval 300–1100 Hz if telephone bandwidth speech is used and it is downsampled 1:4 to reduce the sampling rate to 2500 Hz. In the corresponding speech segments the FFT is computed every 12.8 ms on intervals of 38.4 ms duration. This interval includes two pitch periods even at a pitch frequency of 55 Hz. The peaks in the FFT correspond to pitch frequency and its harmonics. Let $\nu_1, \nu_2, \ldots$ be the frequency of the largest, second-largest and so on, peak in the FFT and let $F_1, F_2, \ldots$ be the height of those peaks. The first pitch frequency estimate is $p_1 = |\nu_1 - \nu_2|$. If $F_3 < \Theta'$, p_1 is the final result. However, if $p_1 > 550$ Hz and $F_3 < \Theta'$, ν_1 is the final result. Otherwise, ν_3 is considered. This yields two new pitch estimates p_2 and p_3. The values p_1, p_2 and p_3 are stored in an accumulator of pitch estimates. If $F_4 > \Theta'$, the peak at ν_4 is considered. The distances between the four selected peaks give three new pitch estimates which are entered into the accumulator. This process of computing additional estimates is repeated until either $F_j < \Theta'$ or a certain pitch estimate occurs eight times in the accumulator. The final pitch estimate then is the frequency occurring most often in the accumulator. The pitch estimates are smoothed by a 3-point median

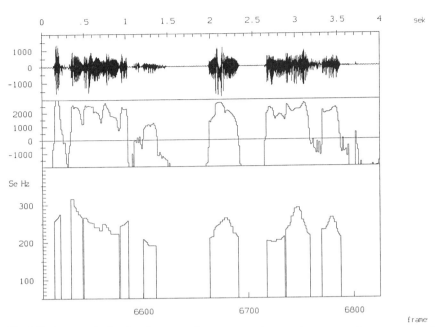

Fig. 3.39. An example showing, from top to bottom, a speech wave, the measure of energy computed by (3.10.5), and the estimate of pitch frequency

filter followed by a 5-point median filter. A result of this algorithm is shown in Fig. 3.39.

The determination of the boundary and class of a subword unit is usually based on a *parametric representation* of the speech signal or on the extraction of a *feature vector;* some examples were mentioned at the beginning of this section. According to experimental evidence a very useful set of parameters or features are the spectral and cepstral coefficients based on an approximation of the mel frequency scale. This approximates the frequency groups mentioned in Sect. 1.8 by triangular filters; usually between 10 and 40 filters are used, typically about 24. An example of a set of coefficients, called mel-frequency coefficients (MFCs), is the following. The speech signal is sampled at 10 kHz, nonoverlapping frames with 128 sample values are used, and the sample values are weighted by a Hamming window. The frame is expanded to 512 values by filling in zeros. Let the result be $f_j, j = 0, 1, \ldots, 511$. The squared absolute values of a DFT are computed by

$$F_\mu = \left| M^{-1} \sum_{j=0}^{M-1} f_j \exp\left(-2\pi i \mu j / M\right) \right|^2 , \quad M = 512 , \quad \mu = 0, 1, \ldots, 255 . \tag{3.10.6}$$

Using a set of 32 triangular filters shown in Fig. 3.40 the 32 MFC denoted by c_k are computed by

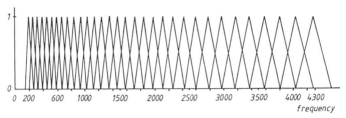

Fig. 3.40. Triangular filters approximating a mel-frequency scale

$$c_k = \sum_{\mu=0}^{M'-1} \log\left[t_{k\mu}F_\mu\right], \quad M' = 256, \quad k = 0, 1, \ldots, 31 \qquad (3.10.7)$$

where $t_{k\mu}, k = 0, 1, \ldots, 31$ are the coefficients of the 32 triangular filters. From the MFCs a set of mel-cepstral coefficients (MCCs) denoted by γ_l can be computed by

$$\gamma_l = \sum_{k=0}^{M''-1} c_k \cos\left[l(2k+1)\pi/(2M'')\right], \quad M'' = 32, \quad l = 1, \ldots, M''. \quad (3.10.8)$$

The MCCs γ_l provide a very efficient set of parameters for the representation of speech frames.

3.10.3 Segmentation

There are two main approaches to speech segmentation which are roughly comparable to boundary and region extraction in image processing. In both of them the computation of speech parameters is repeated, say, every 10 ms. This gives, from a time sequence of speech frames, a time sequence of parameters. In the first approach parameters are traced in time, and a boundary of a speech segment is assumed if significant parameter changes occur. In the second approach each frame of speech is classified and compared to its neighboring frame. All similar (or homogeneous) neighboring frames are merged to obtain one larger segment. The problem of assigning class labels is solved in mainly two ways. The first is to obtain a classified sample ω according to (1.3.1). This requires hand segmentation and hand classification of a sample of speech into the desired subword units. The second employs some type of unsupervised learning or clustering to automatically obtain a set of classes. Some examples are given to illustrate these approaches.

A distinct example of the parameter tracing approach is the following. The outputs of a 16-channel analog filter bank in the range from 0.2 to 6.4 kHz are sampled every 5 ms and are a feature vector c. This is compared to a set of prototype vectors $c_{\kappa\nu}, \kappa = 1, \ldots, k, \nu = 1, \ldots, k_\kappa$, representing phoneme classes

$\Omega_\kappa, \kappa = 1, \ldots, k$. If c is an n-dimensional vector, assume $n > k_\kappa$. The k_κ prototype vectors of class Ω_κ span a subspace of R^n, and c is projected onto this subspace. This allows one to write

$$c = c_\kappa^{(p)} + c_\kappa^{(n)}, \qquad (3.10.9)$$

where $c_\kappa^{(p)}$ is the projection of c onto the subspace spanned by $c_{\kappa\nu}$ and $c_\kappa^{(n)}$ is normal to $c_\kappa^{(p)}$. Since $c_\kappa^{(n)}$ is the distance of c from the subspace, the segment borders are assumed at a time (or frame number) where a local maximum of

$$\left|c_{\min}^{(n)}\right| = \min_\kappa \left|c_\kappa^{(n)}\right| \qquad (3.10.10)$$

occurs, whereas classification is performed at the minima (in time) of $|c_{\min}^{(n)}|$. Classification is according to the rule

$$\text{if } \left|c_\lambda^{(p)}\right| = \max_\kappa \left|c_\kappa^{(p)}\right|, \text{ then } c \in \Omega_\lambda. \qquad (3.10.11)$$

It should be noted that the indexing of the frame numbers was omitted in (3.10.9–11) to avoid notational complexity.

There are numerous examples of the classification and segmentation approach. For example, frames corresponding to 4 average pitch periods have been used. The 12 linear prediction coefficients of a frame are used as components of a feature vector, plus an additional component which is the log ratio of prediction error to signal variance. The feature vector is classified by statistical decision methods into one of k phonemic classes. Several frames with the same class name give a segment corresponding to a phoneme. In a somewhat different technique a rough segmentation is first made with four broad classes, which are vowel-like, volume dip within vowel-like sound, fricative-like sound, and STOP including a following voice burst or aspiration if present. Decisions are based on a linear prediction model spectrum. Silence is assumed if the energy in the band from 0.3 to 5 kHz is not more than 3 dB above background noise. Having removed silence segments the remaining portions are segmented into sonorant and nonsonorant speech segments. If the ratio of the energy in the 3.7 to 5 kHz range is very high, the frame is labeled sonorant even if the pitch detector indicates unvoiced speech. If the pitch detector indicates voicing and the aforementioned energy ratio is not very low, the frame is also labeled sonorant. Detection of bursts, aspiration, and volume dip are also based on energy measurements.

Due to various ambiguities segmentation and labeling are not ideal. To cope with these problems – which are common to all areas of pattern analysis – a segmentation lattice may be used to represent alternative segmentations, and several labeling alternatives may be provided to increase the probability of including the correct one. Some kind of smoothing is carried out in most approaches to eliminate discontinuities. For instance, if one isolated unvoiced frame occurs between voiced frames, it seems reasonable to relabel it to voiced, too. This is the same principle as used in image processing when a white point surrounded by black points is also assigned the gray-value black. In this way a fairly, but not completely, correct string of phonemes is passed to further stages of analysis which will have to be able to resolve ambiguities and correct errors to a certain extent. In essence, the labeling process is a classification of simple patterns (Sect. 1.2 and 4.2), which in this case are usually a parametric representation of 10 ms frames of speech.

An example of a complete module for the transformation of a speech signal (resulting from continuous German speech) into a string of phones will be outlined next. Nonoverlapping frames of 12.8 ms duration are obtained from the signal. The module consists of speech prefiltering and sampling, energy normalization, feature extraction on the frames, classification of frames into phone components, and segmentation (that is in this case combination of similar frames) into phones. The result of classification into phone components is a string of elements

$$v_c = \{(\text{phone component, a posteriori probability})_i, \quad i = 1, \ldots, 5\}. \quad (3.10.12)$$

In addition loudness and pitch are computed, but were omitted in (3.10.12). A set of $k = 49$ phone components is distinguished. In many cases, for example, in the case of vowels, a phone component coincides with a phone. In some cases a phone consists of more than one phone component, for example, the /t/ consists of the phone components /stop of t/, /burst of t/, /aspiration of t/. The phone components are grouped into the four broad classes "voiced non-fricative", "voiced fricative", "unvoiced", and "stop". The result of segmentation is a string of elements

$$v_p = \{\text{begin, end, (phone, score)}_i, \quad i = 1, \ldots, 5\}. \quad (3.10.13)$$

Again loudness and pitch were omitted. A set of 36 phones is distinguished with 15 vowels, 10 consonants, 4 nasals, 6 plosives, and as an additional symbol, the pause. The main steps of acoustic-phonetic segmentation are summarized below:

1. Band-pass filtering of the speech signal (0.3–3.4 Hz, 48 dB/octave), 10 kHz sampling frequency, 12 bit quantization.
2. Formation of nonoverlapping frames of 12.8 ms duration weighted by a Hamming window (2.2.9).
3. Energy normalization according to (2.2.12).
4. Computation of 10 mel-cepstral coefficients MCCs according to (3.10.8).
5. The feature vector consists of the 10 MCCs plus a measure of loudness which is $10 \log (r_0/(E_U/E_I))$, see (2.2.12) and (2.5.7).
6. Classification into the four broad classes using a Bayes classifier for normally distributed feature vectors, see (4.2.19).
7. Classification into phone components using (4.2.19). Training of the classifier is by means of a hand segmented and labeled sample of speech. The result is (3.10.12).

3.10 Segmentation of Speech   147

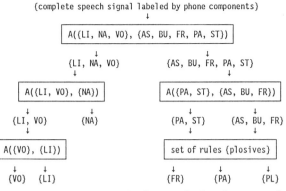

{complete speech signal labeled by phone components}
↓
A({LI, NA, VO}, {AS, BU, FR, PA, ST})
↓   ↓
{LI, NA, VO}   {AS, BU, FR, PA, ST}
↓   ↓
A({LI, VO}, {NA})   A({PA, ST}, {AS, BU, FR})
↓   ↓   ↓   ↓
{LI, VO} {NA}   {PA, ST}   {AS, BU, FR}
↓   ↓
A({VO}, {LI})   set of rules {plosives}
↓   ↓   ↓   ↓   ↓
{VO} {LI}   {FR}   {PA}   {PL}

Fig. 3.41. Segmentation hierarchy for acoustic-phonetic segmentation. The symbol $A(\{a\}, \{b\})$ denotes a finite automaton distinguishing the sets $\{a\}$ and $\{b\}$. AS: set of 4 aspiration types; BU: set of 6 burst types; FR: set of 10 fricatives; LI: set of 3 liquids; NA: set of 4 nasals; PA: pause; PL: set of 6 plosives; ST: set of 6 stop types; VO: set of 15 vowels

schoensten							
T X S S S I S Z T Q E E E E N N N N Z S Z Z S S S S T - T T G M M M V M								
D S X X X X S V G Z Y Y A E E Q M M M M M C Z S S Z Z Z Z - T - - - N N N Z N								
S C Z I Z I P F Z V Y L Q A A Y N E N T N S T X T T S T T S T D B T N N N L G								
A Z T T C Z D T B P E E E Q I I N N N L L T T T T D T T T S G G K G L G T -								
P T C L T I X Z G - E Y N L Q L E I E D T T X V C T X S T Z D B D P I N N C T								

SH 66	ER 29	N 53	Z 48	T 61	M 63
XI 23	QH 16	M 35	S 47	G 15	NE 20
Z 5	E 13	NG 6	XI 2	B 10	N 11
V 3	YH 8	NE 4	SH 0	D 6	NG 4
F 1	AR 7	ER 0	V 0	K 4	L 0

Fig. 3.42. The figure shows, from top to bottom, a speech wave, five alternatives of the phone components, and up to five phone alternatives per segment together with a measure of reliability

8. Hierarchical segmentation by means of segmentation automata and segmentation rules. The segmentation hierarchy is shown in Fig. 3.41. The result is (3.10.13). An example is shown in Fig. 3.42.

3.10.4 Vector Quantization

The last section gave approaches to the labeling of speech by perceptually meaningful subword units. An alternative is provided by *vector quantization* which attributes to a speech frame a code word obtained from a formal optimization procedure. In Sect. 2.1.2 and Fig. 2.3 the problem was considered of coding the values of a scalar continuous variable by a finite set of discrete values. Vector quantization can be viewed as the direct generalization of this problem, that is, coding values of a vector valued variable having continuous real components by a finite set of prototype vectors. In our case the vector valued variable is a feature vector $c \in R^n$ representing a frame of speech. This feature vector can be the set of sample values in the frame or one of the parametric representations of a speech frame mentioned above. Let there be L prototype vectors $a_l \in R^n, l = 1, \ldots, L$. Each prototype vector represents a certain volume V_l in R^n. Every vector $c \in V_l$ is coded by the corresponding prototype a_l or equivalently by the corresponding index l. The set of prototype vectors is also called the *code book*. In analogy to (2.1.9) the quality of a code book can be measured by the mean square error

$$\varepsilon = \sum_{l=1}^{L} \int_{V_l} |c - a_l|^2 p(c) dc \ . \tag{3.10.14}$$

There are no generalizations of Theorem 2.1.3 in the n-dimensional case because due to the arbitrary shape of V_l no differentiation is possible as in the one-dimensional case (2.1.9). However, given a set of prototype vectors a_l it can be shown that minimization of ε requires that a vector c be attributed to the volume V_l according to the rule

$$\text{if } |c - a_l|^2 \leq |c - a_k|^2 \quad \text{for all} \quad k \neq l, \quad \text{then} \quad c \in V_l \ . \tag{3.10.15}$$

A vector quantizer using the same code book but different volumes than those defined by (3.10.15) cannot have a smaller error. On the other hand, given a partition of R^n into volumes V_l it can be shown that the best prototype vectors are defined by

$$a_l = E\{c | c \in V_l\}, \quad l = 1, \ldots, L \ . \tag{3.10.16}$$

In numerical computations (3.10.16) is approximated by

$$a_l = (N_l)^{-1} \sum_{c \in V_l} c \ , \tag{3.10.17}$$

where N_l is the number of vectors c observed in V_l.

These results are the basis for the iterative *LGB-algorithm* for the computation of a locally optimal code book. This algorithm in turn is basically a modification of a clustering algorithm. It works in the steps:

1. Initialize by providing a sample of training vectors $\omega = \{c^r, r = 1, \ldots, N\}$; an initial code book $A^{(0)}$ consisting of L prototype vectors $a_l, l = 1, \ldots, L$; and a threshold Θ of the relative error.

2. For a given code book $A^{(m)} = \{a_l^{(m)}, l = 1, \ldots, L\}$ compute the code book index l of every vector c by (3.10.15). Compute $\varepsilon^{(m)}$ from (3.10.14).
3. If the relative error $(\varepsilon^{(m-1)} - \varepsilon^{(m)})/\varepsilon^{(m)} < \Theta$, then END algorithm.
4. Compute new prototypes from (3.10.17) for $l = 1, \ldots, L$. These will give the new code book in the next iteration. Repeat from step 2 with $m = m+1$.

There are some alternatives for providing the initial code book in step 1. A useful approach is to start with the mean vector $a_l^{(0)}, L = 1$, of the sample ω and split it into two vectors $(1 \pm \delta)a_l^{(0)}, L = 2, \delta \ll 1$, apply the LBG algorithm to the two initial code book vectors, split the resulting two vectors into four vectors, apply the LBG algorithm to the four vectors, and so on, until the desired number of code book vectors is obtained or a prespecified error in (3.10.14) is achieved.

The vectors of the code book can be used to label speech frames by (3.10.15). The index l of the corresponding volume V_l can be used formally in the same manner as the class index of some other subword unit of the type discussed above.

3.11 Bibliographical Remarks

An account of segmentation objects in image understanding is given in [3.14], and of subword units for speech in [3.15].

A survey of different threshold selection techniques is given in [3.16] and some examples of the application of thresholding in [3.17–23]. The filtered histogram was described in [3.24]. Optimization of threshold selection is treated in [3.25] and an iterative technique in [3.26]. Examples of contour following in binarized images are given in [3.18, 27, 28]. Operations on binary images like noise cleaning, shrinking, and skeletonization are described in [3.1, 29–34]. An algorithm for thresholding under nonuniform illumination was developed in [3.35].

There is an extensive literature on detection of gray level changes or of edges, and surveys are [3.36, 37]. The examples of difference operators in Sect. 3.3.1 are from [3.38–44]. The matched filter is well known from statistical communication theory, for example, [3.45, 46]. The filters mentioned were developed in [3.4, 47–50]. The Canny operator is described in [3.5], details of a realization in [3.6]. The zero crossings of the Laplacian of Gaussian filters were introduced in [3.51, 52]. Hueckel's operator is developed in [3.53, 54] and the plane fitting approach in [3.55]. The example of a statistical approach to edge detection is from [3.56]; other examples are [3.57–59]. Several iterative schemes for edge enhancement are presented in [3.60–63]; the example discussed here is [3.64]. Extensions of edge detection techniques to color images are given, for example, in [3.65].

Computation of lines from gray level changes is treated, for example, in [3.33, 40, 66–71]. The Hough transform is developed in [3.72, 73], and generalizations and extensions in [3.74]. Dynamic programming for contour following is treated in [3.75–79]. The characterization of contours is covered in many articles;

some examples of the scalar transform techniques are [3.80–86] and examples concerning the space domain techniques are [3.87–92].

A survey of region growing techniques is given in [3.93]. The merging techniques described in the text are from [3.94–96], the splitting techniques from [3.97, 98], and the split-and-merge technique from [3.99]. Additional material on split-and-merge is found in [3.100, 101]. An expert system combining the results of line and region finding was described in [3.102]. An approach not covered in the text is the hidden Markov mesh [3.103, 104].

A collection of texture tables is given in [3.8]. Human perception of textures was studied in detail by [3.105, 106]. Surveys are [3.107, 108]. Some approaches to texture modeling are [3.109–111]. The examples of numerical characterization of textures are from [3.112–116], and the syntactical approach from [3.117].

Overviews on motion estimation are given in [3.118, 119]. The treatment of optical flow given here follows the development in [3.120]. A comparison of different approaches may be found in [3.121]. The line processes for motion discontinuities are treated in [3.122]. Estimation of three-dimensional motion parameters is treated in [3.123, 124]. The block matching approach is described, for example, in [3.125, 126]. An overview of the use of resolution hierarchies or multigrid methods (not only for optical flow) is given in [3.127].

Overviews on depth computation are given in [3.128–130]. The patterned light method is treated, for example, in [3.131]. The optical flow for stereo images was introduced in [3.132], area matching is described in [3.133, 134], the example of line matching is from [3.135], and other examples of stereo algorithms are [3.136–141].

Shape from shading was introduced by [3.142]. The variational approach to shape from shading is developed in [3.143], the use of second order derivatives in [3.144], the light source system in [3.145]. The integrability criterion based on an expansion of the surface is from [3.11]. In addition there are approaches for computing shape from specularity [3.146], but this was not covered in the text. A discussion of illumination models is found in [3.147].

A discussion of subword units is given in [3.15]. A detailed treatment of phonology is given, for example in [3.148]. Examples and experiments with different parametric representations are presented in [3.149–154]. Examples for formant detection and tracking are given in [3.149, 155, 156]. A detailed account of pitch detection is given by [3.157]. The algorithms outlined here are according to [3.158–160]. The cepstrum for pitch detection was investigated by [3.161]. Time domain measurements are used, for example, in [3.152, 162]. One point which was omitted in Sect. 3.10 is suprasegmental features like stress, intonation, and melody, which are summarized by prosodics. References to this topic are [3.160, 163–166]. The segmentation by tracking approach is developed in [3.167]. Examples of acoustic-phonetic segmentation using different subword units are given in [3.168–177]. Additional material can also be found in part I of [3.178]. The acoustic-phonetic module outlined in Sect. 3.10.3 is developed in [3.15]. The concept of vector quantization is treated, for example, in [3.179, 180], the LBG algorithm is from [3.181].

4. Classification

It became apparent in Chap. 3 that it will often be necessary to classify (or label, or recognize) parts of a complex pattern. This was mentioned explicitly, for instance, in Sect. 3.6 where differently textured parts of an image had to be classified, or in Sect. 3.10 where single frames of connected speech had to be classified (labeled). As indicated in (1.3.2), classification by numerical methods amounts to a mapping of vectors onto integers. Another equivalent viewpoint is that the feature space is partitioned into parts corresponding to pattern classes. This task is fairly well understood and has a sound theoretical basis.

There are other more general problems of classification which occur if a model of a class has to be compared (matched) to an observation of some pattern. Typical examples are the recognition of words in speech or the recognition of three-dimensional objects from certain views of the object. This area still has several open problems.

The above problems are among the central problems of pattern recognition because they require the transition from a *numerical to a symbolic* representation, a problem which is, so far, not solvable with 100% accuracy. The transition between two representation domains is not yet completely understood and solved. Common names for methods achieving this transition are *classification, recognition* or *matching*. Regardless of their complexity, all these problems have some similar properties: There is a set of *classes* of objects, events, situations, or facts which are represented by appropriate internal *models* of a system. There are *observations* of certain individual *patterns*. The observations are subject to various transformations, distortions, and degradations. The general problem of classification is to identify the class given the observation, or to map different observations having the same meaning to the same symbolic description.

In this chapter the following topics are considered:

1. Statement of the problem – outline of the problem considered here.
2. Classification of feature vectors – the "classical" approach.
3. Classification of symbol strings – the linguistic approach.
4. Recognition of words – specialized techniques for speech recognition.
5. Recognition of objects – specialized techniques for classifying objects.
6. Neural nets – massively parallel, adaptive structures.

4.1 Statement of the Problem

The early approach to classification originates from a fairly simple model for the observation. It is assumed that there is a set of ideal *prototypes* $s_\kappa(x), \kappa = 1, \ldots, k$ of *classes* Ω_κ which is given in some reference coordinate system x. The available *observation* consists of this prototype corrupted by additive noise $n_\kappa^r(x)$ to yield the rth *pattern* from class Ω_κ

$$f_\kappa^r(x) = s_\kappa(x) + n_\kappa^r(x) . \qquad (4.1.1)$$

This assumption was motivated by the approach taken by communication and detection theory. If this idea did not fit to the real world, it was postulated that by suitable preprocessing, feature extraction, normalization, enhancement and so on, the actual observation was transformed to another one which could be approximated with sufficient accuracy by the above equation. So the original observation or pattern was transformed to a *feature vector*

$$f_\kappa^r(x) \to c_\kappa^r . \qquad (4.1.2)$$

The advantage of this approach is that it can use the by now well established field of numerical classification and *statistical decision theory* and that in fact there are real world problems which can be solved this way. Some basic results are given below.

However, experience showed that this approach has its limitations and that there are important problems requiring different methods. As an *example* we consider the following situation where we start with a more general definition of an observation:

1. There are some (possibly unobservable) *ideal objects* (or events, situations, facts, and so on) $o_{\kappa\lambda}(u_1, u_2, \ldots, u_s)$, where $\kappa = 1, \ldots, k$ is the number of classes of objects, and $\lambda = 1, \ldots, l$ is the number of ideals per class. The ideal objects may be, for example, three-dimensional objects moving at different velocities or an event such as the grasping of an object.

2. It is possible to make *observations* $f^r(x_1, x_2, \ldots, x_n)$, where $r = 1, \ldots, N$ is the number of observations. An observation may contain one, several, or none of the objects, and there may be additional information not corresponding to a known object. The observations of visual objects may be represented by gray level images, color images, or image sequences.

3. Before an observation, the ideal object may undergo various transformations and distortions, for example:

3.1. *Coordinate transformation* T_r yielding

$$(x_1, x_2, \ldots, x_n) = T_r\{(u_1, u_2, \ldots, u_s)\} . \qquad (4.1.3)$$

3.2. *Linear distortion* by $G_r(x_1, x_2, \ldots, x_n)$ yielding

$$F_1 = o_{\kappa\lambda}(T_r(u_1, u_2, \ldots, u_s)) \otimes G_r(x_1, x_2, \ldots, x_n) , \qquad (4.1.4)$$

where $\otimes$ means the convolution operation.

3.3. *Occlusion* represented by a mask $V_r(x_1, x_2, \ldots, x_n)$ yielding

$$F_2 = F_1 V_r \ . \tag{4.1.5}$$

3.4. *Additive noise* $N_r(x_1, x_2, \ldots, x_n)$ yielding

$$F_3 = F_2 + N_r \ . \tag{4.1.6}$$

3.5. *Shadow* $S_r(x_1, x_2, \ldots, x_n; f^r)$ yielding

$$F_4 = F_3 + S_r \ . \tag{4.1.7}$$

4. This combines to the *observation* or *pattern*

$$f^r = o_{\kappa\lambda}(T_r\{u\}) \otimes G_r \cdot V_r + S_r + N_r \ . \tag{4.1.8}$$

Apparently, other sequences of transformations to obtain an observation from an object are possible. In word recognition, for example, other transformations are relevant. The above should be considered as an illustrative example demonstrating that the simple case (4.1.1) may not be sufficient in general. In any case, the sequence of transformations amounts to certain unknown parameters which for convenience of notation are summarized here in the vector a. So we can combine all changes of the ideal objects in one transformation to yield

$$f^r = T\{o_{\kappa\lambda}, a^r\} \ . \tag{4.1.9}$$

The general *classification* or *matching problem* may now be stated as:

Problem 1. Take a pattern f^r according to (4.1.8, 9) resulting from an ideal object $o_{\kappa\lambda}$ and compute a *distance*

$$d^r_{\kappa\lambda} = d(f^r, o_{\kappa\lambda}) \ . \tag{4.1.10}$$

Problem 2. Compute a *model* $M_\kappa(o_{\kappa\lambda}, \{I^r(f^r), a^r\}, r = 1, \ldots, N)$ of the ideal objects $o_{\kappa\lambda}$ and an *initial representation* or description $I^r(f^r)$ of the observation f^r. The initial description is not to be confused with the symbolic description B^r introduced in Sect. 1.3; B^r has to meet the needs of a certain task domain, whereas I^r as introduced in (3.1.3) is just the first step towards B^r. It should be noted that initital descriptions are usually ambiguous and erroneous.

Problem 3. Compute a *distance*

$$d^r_\kappa = d(I^r(f^r), M_\kappa) \ . \tag{4.1.11}$$

This distance should be *invariant* or *robust* with respect to variations in the unknown parameter a^r and with respect to errors in the initial description. Different classification algorithms use different models for the observation, different models for object classes, different initial descriptions, and different distance measures to achieve classification. So far, no unifying theory is available.

Two well-known special cases of initial descriptions are the feature vector $c^r \in R^n$ and the symbol string $v^r \in V_T^*$, where V_T is a finite set of terminal symbols or simple constituents of a pattern. A more involved type of initial description is introduced in Sect. 3.1. Short accounts of the classification of feature vectors and symbol strings are given in Sects. 4.2 and 4.3, respectively.

4.2 Classification of Feature Vectors

The initial representation I^r is a vector c^r of real numbers, the model M_κ is a suitable parametric function allowing one to partition the space R^n into k regions belonging to the k classes Ω_κ. In addition, nonparametric techniques are available, some of which use the whole sample ω as a model, some of which use a parametric representation, too.

4.2.1 Statistical Classification

In *statistical classification* the properties of the k classes $\Omega_\kappa, \kappa = 1,\ldots,k$ are characterized by k *conditional densities* $p(c|\Omega_\kappa), \kappa = 1,\ldots,k$. Thus a value c_κ^r of a *feature vector* extracted from a pattern f_κ^r is a value of a random variable c_κ. It is assumed that a parametric family

$$\tilde{p} = \{p(c|a) \mid a \in R_a\} \tag{4.2.1}$$

of densities is known, where a is a parameter vector, R_a the parameter space, and $p(c|a)$ is completely specified except for a. It is further assumed that

$$p(c|\Omega_\kappa) \in \tilde{p}, \quad \kappa = 1,\ldots,k \ . \tag{4.2.2}$$

The determination of conditional densities is then reduced to the estimation of the parameter vectors $a_\kappa, \kappa = 1,\ldots,k$.

Furthermore, the following additional assumptions are made:

1) The pattern classes occur with a certain known *a priori probability* p_κ.
2) The *cost* of classifying a pattern $f_\kappa^r \in \Omega_\kappa$ as belonging to Ω_λ is $r_{\kappa\lambda}$, $\kappa, \lambda = 0, 1, \ldots, k$, $\kappa \neq 0$. As discussed in Sect. 1.2, Ω_0 is the *reject class*. It is reasonable to require

$$0 \leq r_{\kappa\kappa} \leq r_{\kappa 0} \leq r_{\kappa\lambda}, \quad \kappa \neq \lambda \ . \tag{4.2.3}$$

3) The *decision rule* $\delta(\Omega_\kappa|c)$ gives the probability of deciding for class Ω_κ if the vector c was observed. We require

$$\sum_{\kappa=0}^{k} \delta(\Omega_\kappa|c) = 1 . \qquad (4.2.4)$$

With these assumptions the *expected costs* or the *risk* of classifying patterns by using δ is

$$V(\delta) = \sum_{\kappa=1}^{k} p_\kappa \sum_{\lambda=0}^{k} r_{\kappa\lambda} \int_{R_c} p(c|\Omega_\kappa)\delta(\Omega_\lambda|c) dc . \qquad (4.2.5)$$

The *optimal decision rule* δ^* is now defined by

$$V(\delta^*) = \min_{\{\delta\}} V(\delta) , \qquad (4.2.6)$$

that is δ^* minimizes the risk. It is given by the following result.

Theorem 4.2.1. The optimal decision rule δ^* in (4.2.6) is defined as follows: Compute the *test variables*

$$u_\lambda(c) = \sum_{\kappa=1}^{k} (r_{\kappa\lambda} - r_{\kappa 0}) p_\kappa p(c|\Omega_\kappa) , \quad \lambda = 0, 1, \ldots, k \qquad (4.2.7)$$

IF $u_\kappa(c) = \min_\lambda \{u_\lambda(c)\}$,

THEN $\delta^*(\Omega_\kappa|c) = 1$; $\delta^*(\Omega_\lambda|c) = 0$ for $\lambda \neq \kappa$, $\lambda = 0, 1, \ldots, k$.(4.2.8)

Proof. See, for instance, [4.1–5]. □

The decision rule in (4.2.8) is nonrandomized, that is it selects precisely one class, although initially a randomized rule was not excluded. The structure of this classifier is shown in Fig. 4.1.

For the special cost function

$$r_{\kappa\kappa} = r_c ,$$
$$r_{\kappa 0} = r_0, \quad \kappa, \lambda = 1, \ldots, k \qquad (4.2.9)$$
$$r_{\kappa\lambda} = r_e, \quad \lambda \neq \kappa,$$

the risk reduces to

$$V(\delta) = r_c p_c + r_0 p_0 + r_e p_e , \qquad (4.2.10)$$

where p_c, p_0, p_e are the probabilities of a correct decision, a reject, and an erroneous decision, respectively. Of course, (4.2.7) still applies, but it may be specialized as shown in the following result.

156   4. Classification

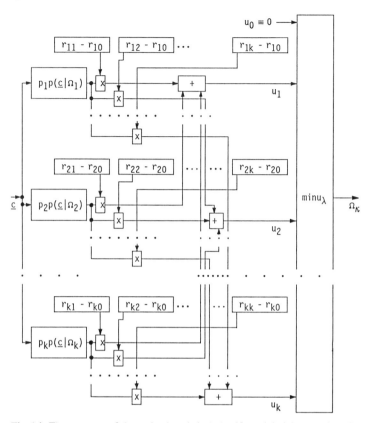

Fig. 4.1. The structure of the optimal statistical classifier minimizing the risk of classification

Theorem 4.2.2. The optimal decision rule δ^* for the cost function (4.2.9) is identical to a rule which minimizes p_e for a fixed p_0. With the test variables

$$u_\lambda(c) = p_\kappa p(c|\Omega_\lambda) \tag{4.2.11}$$

the rule is

1. compute $\quad u_\kappa(c) = \max_\lambda \{u_\lambda(c)\} \tag{4.2.12}$

2. IF $\quad u_\kappa(c) \geq \beta \sum_{\lambda=1}^{k} u_\lambda(c)$,

   THEN $\quad \delta^*(\Omega_\kappa|c) = 1$, $\quad \delta^*(\Omega_\lambda|c) = 0 \quad$ for $\quad \lambda \neq \kappa \tag{4.2.13}$

   ELSE $\quad \delta^*(\Omega_0|c) = 1$, $\quad \delta^*(\Omega_\lambda|c) = 0 \quad$ for $\quad \lambda = 1,\ldots,k$. $\tag{4.2.14}$

The relation between β, r_c, r_0, r_e is

$$\beta = \frac{r_e - r_0}{r_e - r_c} > 0 . \tag{4.2.15}$$

Proof. See, for instance, [4.1, 4.5]. □

Another specialization of the decision rule is achieved if a decision is forced, that is no reject possibility is allowed, and the cost function

$$r_c = 0 , \quad r_e = 1 \tag{4.2.16}$$

is used. In this case the risk reduces to the *error probability*

$$V(\delta) = p_e . \tag{4.2.17}$$

By (4.2.10) together with (4.2.15) and by (4.2.17) it is evident that proper choice of the costs in (4.2.3) also allows control of error and reject probabilities.

Theorem 4.2.3. The optimal decision rule δ^* for the cost function (4.2.16) is identical to a rule which minimizes p_e (no rejects are allowed). With the test variables from (4.2.11) it is given by the rule

IF $u_\kappa(c) = \max_\lambda \{u_\lambda(u)\}$,

THEN $\delta^*(\Omega_\kappa|c) = 1$, and $\delta^*(\Omega_\lambda|c) = 0$ for $\lambda \neq \kappa$. (4.2.18)

Proof. See, for instance, [4.1, 4.5]. □

With (4.2.11) it is seen that (4.2.18) is equivalent to computing the a posteriori probabilities

$$p(\Omega_\lambda|c) = \frac{p_\lambda p(c|\Omega_\lambda)}{\sum_{\lambda=1}^{k} p_\lambda p(c|\Omega_\lambda)} \tag{4.2.19}$$

and assigning c to the class Ω_λ with maximal a posteriori probability.

In order to obtain explicit formulas for the test variables in (4.2.7, 11, 13, 19) it is necessary to choose $\tilde{p}$ in (4.2.1), the most common choices being the multivariate normal densities

$$\begin{aligned} p(c|a) &= p(c|\mu, K) \\ &= (|2\pi K|)^{-1/2} \exp\left[-(c-\mu)_t K^{-1}(c-\mu)/2\right] \end{aligned} \tag{4.2.20}$$

or the statistical independence assumption

$$p(c|a) = \prod_{\nu=1}^{n} p(c_\nu|a) . \tag{4.2.21}$$

In the first case one has to estimate conditional *mean vectors* μ_κ and *covariance matrices* K_κ from classified samples $\omega \supset \omega_\kappa$ in (1.3.1); in the second case one has to estimate one-dimensional component densities $p(c_\nu|a_\kappa), \nu = 1,\ldots,n; \kappa = 1,\ldots,k$. For instance, with (4.2.11, 20) one obtains for the classifier minimizing p_e the test variables

$$u_\lambda(c) = -(c - \mu_\lambda)_t K_\lambda^{-1}(c - \mu_\lambda) + \gamma_\lambda,$$

$$\gamma_\lambda = 2 \ln \left[p_\lambda(|2\pi K_\lambda|)^{-1/2}\right]. \qquad (4.2.22)$$

In the special case that K_λ is a diagonal matrix

$$K_\lambda = \text{diag}\left(k_{\lambda 1}, k_{\lambda 2}, \ldots, k_{\lambda n}\right)$$

the weighted Euclidian distance criterion

$$u_\lambda(c) = -\sum_{\nu=1}^{n} \frac{(c_\nu - \mu_{\lambda\nu})^2}{k_{\lambda\nu}} + \gamma_\lambda \qquad (4.2.23)$$

results. If all covariance matrices K_λ are equal to K, the linear criterion

$$u_\lambda(c) = 2c_t\left(K^{-1}\mu_\lambda\right) + \gamma_\lambda - \mu_{\lambda t}K^{-1}\mu_\lambda \qquad (4.2.24)$$

is obtained. Classifiers using a distance such as (4.2.23) or (4.2.22), with γ_λ omitted and possibly K_λ^{-1} set to the identity matrix, are very popular because of their simplicity.

The main problem with statistical classifiers is the proper choice of a parametric family $\tilde{p}$ of densities in (4.2.1).

4.2.2 Distribution-Free Classification

This approach does not require assumptions about conditional densities and, therefore, is called *distribution free*. Properties of the classes are now modeled by a parametric family

$$\tilde{d} = \{d(c, a) | a \in R_a\} \qquad (4.2.25)$$

of discriminant functions $d(c, a)$ which have to be chosen in advance. For k classes we specify a vector

$$d = (d_1(c, a), \ldots, d_k(c, a))_t; \quad d_\lambda(c, a) \in \tilde{d} \qquad (4.2.26)$$

and represent $d_\lambda(c, a)$ with m independent functions $\varphi_\nu(c), \nu = 1, \ldots, m$ by

$$d_\lambda(c, a) = \sum_{\nu=1}^{m} a_{\lambda\nu}\varphi_\nu(c). \qquad (4.2.27)$$

With the notation

$$A = (a_1, \ldots, a_k), \quad \varphi(c) = (\varphi_1(c), \ldots, \varphi_m(c))_t \qquad (4.2.28)$$

the vector d may be written compactly as

$$d = d(c) = A_t \varphi(c) .\qquad(4.2.29)$$

A decision is made by the rule

$$\text{IF}\quad d_\kappa(c, a) = \max_\lambda \{d_\lambda(c, a)\}, \quad \text{THEN}\quad c \in \Omega_\kappa .\qquad(4.2.30)$$

It remains to specify A in (4.2.28) such that reliable decisions are achieved. We now assume an ideal discriminant function δ_λ to be given, which allows error-free classification. An example are functions defined by

$$\text{IF}\quad c \in \Omega_\kappa, \quad \text{THEN}\quad \delta_\kappa(c) = 1, \quad \text{ELSE}\quad \delta_\kappa = 0 .\qquad(4.2.31)$$

If a classified sample ω of patterns is given, the values of $\delta_\kappa(c)$ are known for patterns $f_\kappa^r \in \omega$. Now we wish to approximate δ_κ by d_κ in the best possible way. In this case reliable classifications by (4.2.30) may be expected.

It is obvious that discriminant functions other than (4.2.27) could be chosen, for instance, piecewise linear functions. Ideal functions δ_κ can also be chosen which are different from (4.2.31). Furthermore, an error criterion for the approximation of δ_κ by d_κ and a computational procedure for minimizing the error have to be selected. These remarks show that many different solutions are possible, some of which are mentioned in the references. We only give one particular example here which allows a closed-form solution.

Measuring the difference between d and δ by the expected value of the squared error and minimizing with respect to A results in

$$T(A) = \min_{\{A\}} E\{(\delta - d)^2\} = \min_{\{A\}} E\{(\delta - A_t\varphi(c))^2\} .\qquad(4.2.32)$$

Differentiating T with respect to the elements of A and equating the result to zero gives

$$\frac{\partial T}{\partial A} = 0 = E\{\varphi\varphi_t\}A - E\{\varphi\delta_t\} .\qquad(4.2.33)$$

If the matrix $E\{\varphi\varphi_t\}$ is nonsingular, the closed-form solution for A is

$$A = E\{\varphi\varphi_t\}^{-1} E\{\varphi\delta_t\} .\qquad(4.2.34)$$

Estimates of the matrices in (4.2.34) can be computed from the sample ω by

$$E\{\varphi\varphi_t\} \approx N^{-1} \sum_{r=1}^{N} \varphi(c^r)\varphi_t(c^r) ,$$

$$E\{\varphi\delta_t\} \approx N^{-1} \sum_{r=1}^{N} \varphi(c^r)\delta_t(c^r) \qquad(4.2.35)$$

since the value of δ follows from (4.2.31).

An example of functions $\varphi_\nu(c)$ is

$$\varphi_1(c) = 1, \quad \varphi_{\nu+1}(c) = c_\nu, \quad \nu = 1, \ldots, n. \tag{4.2.36}$$

In this case the functions $d_\lambda(c)$ in (4.2.27) are hyperplanes. A generalization would be to include quadratic terms

$$\varphi_{\nu+n+1}(c) = c_i c_j, \quad i = 1, \ldots, n, \quad j = 1, \ldots, i \\ \nu = 1, \ldots, n(n+1)/2. \tag{4.2.37}$$

Apparently, further generalization to higher order terms (or polynomials of order p) is limited only by the allowable number m of terms in (4.2.27). The structure of this classifier is shown in Fig. 4.2.

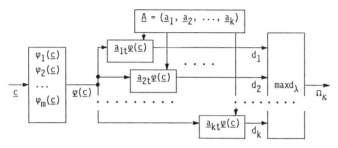

Fig. 4.2. The structure of a distribution free classifier, called a polynomial classifier if $\varphi(c)$ is a polynomial of order p

In the limiting case of unrestricted d a relation to Sect. 4.1 is possible.

Theorem 4.2.4. For unrestricted functions $d(c)$ the optimal function δ^* minimizing $E\{(\delta - d)^2\}$ is given by the conditional expectation (regression function)

$$\delta^* = E\{\delta | c\}. \tag{4.2.38}$$

Together with (4.2.31) the rule (4.2.30) then is identical to the classifier minimizing p_e by (4.2.18).

Proof. See, for instance, [4.3, 6]. □

4.2.3 Nonparametric Classification

A nonparametric decision rule requires no assumptions about a parametric family of densities as in (4.2.1) or discriminant functions as in (4.2.25). For instance, (4.2.22) is optimal only for normally distributed features, but a nonparametric rule should hold for different families of densities, too.

An important example is the *nearest-neighbor rule*. Assume that it is possible to measure the distance $d(c, c^r)$ between a newly observed pattern c and a pattern $c^r \in \omega$, and let $d(c, c^r)$ be a metric, for instance, the Euclidian distance. The pattern c is classified according to the rule

IF $d(c, c^s) = \min_r \{d(c, c^r)\}$ and $c^s \in \Omega_\kappa$, THEN $c \in \Omega_\kappa$. (4.2.39)

This means that a new pattern is assigned to the class of its nearest neighbor in the sample. The performance of the rule (4.2.39) can be stated as follows.

Theorem 4.2.5. Let p_{en} be the error probability of the rule (4.2.39) and p_e the corresponding error probability of the optimal classifier in (4.2.18). Under very general conditions p_{en} is bounded by

$$p_e \leq p_{en} \leq p_e \left(2 - p_e \frac{k}{(k-1)} \right) \qquad (4.2.40)$$

if the sample size N tends to infinity. These bounds are as tight as possible.

Proof. See, for instance, [4.7, 8]. □

Some modifications of this rule are mentioned briefly. The m-nearest-neighbor rule determines the m nearest neighbors of a pattern and assigns it to class Ω_κ if the majority of these m patterns belongs to Ω_κ. A reject criterion may be introduced by requiring that all of the m patterns belong to Ω_κ or that at least m' out of the m belong to Ω_κ. A problem with the application of (4.2.39) is that the whole sample ω of patterns has to be stored. There are several suggestions for reducing the sample size.

4.2.4 Template Matching

It is often the case that one part of an image f has to be compared with another image (or template or signal) s. For instance, s may represent a known object and it should be decided whether the image contains the object and at what location. The template s may also be any segmentation object like a contour element or a homogeneous region. However, due to noise and distortions one cannot expect to find an exact replica of s in f. It is then necessary to find a part of f which is most similar to s (see also Sect. 2.2.8).

Assume that, as in Sect. 2.1, f is an M^2 image and let s be a M_s^2 template with $M > M_s$. With the notation of Fig. 4.3 the difference between f and s may be measured by

$$\varepsilon_{jk} = \sum_{\mu=1}^{M_s} \sum_{\nu=1}^{M_s} |f_{j+\mu, k+\nu} - s_{\mu\nu}| \qquad (4.2.41)$$

or also by

$$\varepsilon_{jk} = \sum_{\mu=1}^{M_s} \sum_{\nu=1}^{M_s} \left(f_{j+\mu, k+\nu} - s_{\mu\nu} \right)^2. \qquad (4.2.42)$$

The last equation is evaluated to obtain

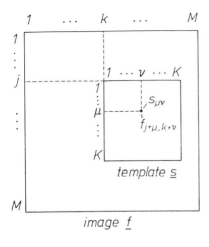

Fig. 4.3. The difference ε_{jk} between image f and template s is evaluated for the upper left corner of s located at position (j,k) of f

$$\varepsilon_{jk} = \sum\sum \left(f^2_{j+\mu,k+\nu} + s^2_{\mu\nu} - 2f_{j+\mu,k+\nu}s_{\mu\nu}\right). \tag{4.2.43}$$

Since the smallest possible value of ε_{jk} is zero in case of a perfect match between f and s at position (j,k) and since the first and second terms in (4.2.43) are positive, the *cross correlation*

$$R'_{jk} = \sum_\mu \sum_\nu f_{j+\mu,k+\nu} s_{\mu\nu} \tag{4.2.44}$$

will become large in case of a match. But R' will also be large if f or s are large without a match. Therefore, it is useful to take the normalized cross correlation

$$R_{jk} = R'_{jk} \Big/ \left(\sqrt{\sum_\mu\sum_\nu f^2_{j+\mu,k+\nu}} \sqrt{\sum_\mu\sum_\nu s^2_{\mu\nu}}\right). \tag{4.2.45}$$

If $f_{j+\mu,k+\nu} = \alpha s_{\mu\nu}$ holds for some value (j,k), some constant α, and $\mu,\nu = 1,\ldots,s$, then R_{jk} takes its maximal value equal to one. Other normalizations are possible, e.g. (3.6.1 or 12).

Since the object may be located anywhere inside the image f, R_{jk} in (4.2.45) has to be evaluated for all possible positions (j,k) of the template. This makes computations very costly although the fast Fourier transform may be applied to find R'_{jk} in (4.2.44). If s may also be rotated, this would require additional computations. Template matching is a special method for the classification of patterns, that is computation of a distance (4.1.10), and may be done in the frequency domain using matched filters (Sect. 3.3.2), or by optical methods.

Reduction of computational expenditure is an important problem for application of template matching. This can be achieved by using a resolution hierarchy and sequential decision methods, see also Sect. 2.2.8. Let $f^{(l-1)}$ be the image at level $(l-1)$ and $f^{(l)}$ the image at level l with resolution reduced by a factor of 2 in each dimension. The reduced-resolution image is obtained by low-pass filtering as outlined in Sect. 2.3.7. Thus at the lowest resolution or highest level l_0 of the hierarchy the image $f^{(l_0)}$ contains only $(M/2^{l_0})^2$ sample points. The same is done with the template s.

Similar to (2.2.46) a nonrepeating sequence of n points $(\mu,\nu), 1 < n < (M_s/2^l)^2$ is taken which may be random or data dependent. With (4.2.41) the error $\varepsilon^{(l)}_{n,jk}$ at the lth level computed for n points is

$$\varepsilon^{(l)}_{n,jk} = \sum_{\mu,\nu} \left| f^{(l)}_{j+\mu,k+\nu} - s^{(l)}_{\mu,\nu} \right| . \qquad (4.2.46)$$

A set of points (j,k) where a match might occur is defined by

$$Q^{(l)} = \left\{ (j,k) \big| \varepsilon^{(l)}_{n,jk} < \Theta^{(l)}_n , \quad n = 1,\ldots,(M_s/2^l)^2 \right\} , \qquad (4.2.47)$$

where $\Theta^{(l)}_n$ is a threshold to be defined below. At the next higher resolution level $l-1$, except $l-1=0$, only test locations $(2j, 2k)$ with $(j,k) \in Q^{(l)}$ are considered. This test continues until one of the following two conditions occur.

1. At some level l there is only one element (j,k) in $Q^{(l)}$. If $l=0$, this point (j,k) is the location of the match. Otherwise a local search of the four neighbors is done until $l=0$.
2. At $l=0$ more than one element (j,k) is in $Q^{(0)}$. Then the point (j,k) with smallest error is taken as the most likely location of a match.

The threshold in (4.2.47) is computed from

$$\Theta^{(l)}_n = 2^{(l_0-l)/2} \left(n + g_l n^{1/2} \right) \gamma_{l_0} . \qquad (4.2.48)$$

The factor g_l is related to the probability p_l of a match. For instance, for $g_l = 1$ one has $p_l = 0.841$, and for $g_l = 5, p_l = 0.999$. The factor γ_{l_0} is the averaged error between f and s at the true match position. A constant threshold $\Theta^{(l)}$ at level l is obtained if $n = (M_s/2^l)^2$. Otherwise, tables for $\Theta^{(l)}_n$ are given in the references.

4.2.5 Relaxation Labeling

Relaxation labeling is a procedure to find a consistent labeling of a set of elements (or patterns, or segmentation objects). Whereas the above classification techniques treated every element or pattern independent of others, the goal here is to take into account the neighboring patterns. In this sense relaxation labeling is a technique for context-dependent classification. This problem also may be treated as a generalization of (4.2.19). Initially each pattern may have several possible labels which are obtained, for example, by the approaches described in Sect. 4.2.1–4. For a particular pair of patterns not every pair of labels will be meaningful. By a suitable algorithm meaningful labels are emphasized and meaningless ones are reduced. Two such relaxation schemes are discussed in the following. They are a discrete relaxation process and a nonlinear probabilistic one. A variety of applications, such as edge detection (see also Sect. 3.3.5), noise cleaning, template matching, and region identification have been reported in the literature.

Suppose a set $V_T = \{s_1, \ldots, s_n\}$ of segmentation objects was extracted from pattern f, for example, regions, contours, or both extracted from an image. The set of possible labels $L = \{l_1, \ldots, l_k\}$ is known from a priori knowledge about the task domain Ω. These labels may be considered as class names and obtained from independent classification of each element. The set L_i of labels which is possible for a particular segmentation object $s_i \in V_T$ is $L \supseteq L_i$ and is determined only from evidence about s_i. For example, a label "lake" is not compatible with a region s_i of color "green", but a label "grass" is compatible. For each pair $s_i, s_j, i \neq j$ of segmentation objects some pairs of labels will be impossible. Let $L_{ij}, L_i \times L_j \supseteq L_{ij}$, be the set of pairs of labels which is compatible with s_i, s_j. For instance, a pair of labels "car, sky" would not be possible for a "yellow region surrounded by a blue region", but a pair "boat, lake" is compatible. A labeling $L' = \{L_1, \ldots, L_n\}$ assigns a set L_i of labels to each segmentation object s_i. A *consistent labeling* has the property

$$(l_i \times L_j) \cap L_{ij} \neq \phi, \quad i,j = 1, \ldots, n \quad \text{and} \quad l_i \in L_i . \tag{4.2.49}$$

This definition requires that for each pair of segmentation objects s_i, s_j and each label $l_i \in L_i$ there must be a label $l_j \in L_j$ which is compatible with l_i, that is, $(l_i, l_j) \in L_{ij}$. Such a consistent labeling is obtained by the following algorithm:

1. Start with an initial labeling L'^0.
2. Obtain L'^{N+1} from L'^N by removing from each $L_i^N \in L'^N$ all the labels l_i violating (4.2.49) for some index j.
3. Iterate step 2 until no more changes occur.

If no more changes occur for some N, then L'^N obviously is a consistent labeling, and since all sets are finite, the algorithm stops for a finite N. However, it is not guaranteed that this final labeling will be unambiguous in the sense that each L_i^N contains one and only one label l_i. By a search procedure it possible to find unambiguous labelings provided they exist. The algorithm is discrete in the sense that a pair of labels either is compatible with a pair of segmentation objects, or it is not. This may be reasonable or not, depending on the application. There are cases where one would like to account for varying degrees of compatibility, not just for compatibility or incompatibility. This is achieved by assigning weights to labels and pairs of labels.

In addition to the set L_i of labels assigned to each segmentation object s_i, a set of probabilities $p_i(l_\mu)$ is assigned to each label $l_\mu \in L_i$ such that

$$0 \leq p_i(l_\mu) \leq 1 \quad \text{and} \quad \sum_\mu p_i(l_\mu) = 1 . \tag{4.2.50}$$

The compatibility of label l_μ for segmentation object s_i and of l_ν for s_j is determined by the *compatibility coefficient* $r_{ij}(l_\mu, l_\nu)$ between -1 and $+1$. A negative value indicates that the pair (l_μ, l_ν) for segmentation objects s_i, s_j seldom occurs, a positive value indicates that it occurs often, and a value close to zero means that labels l_μ and l_ν are fairly independent for these segmentation objects. The compatibilities may thus be interpreted as correlations. The Nth iteration of a nonlinear probabilistic relaxation process is defined by the equations

$$\beta_{iN}(l_\mu) = \sum_j \alpha_{ij} \sum_\nu r_{ij}(l_\mu, l_\nu) p_{jN}(l_\nu) , \quad \sum_j \alpha_{ij} = 1 , \qquad (4.2.51)$$

$$p_{i,N+1}(l_\mu) = p_{iN}(l_\mu)[1 + \beta_{iN}(l_\mu)] / \sum_\mu p_{iN}(l_\mu)[1 + \beta_{iN}(l_\mu)] . \qquad (4.2.52)$$

The basic operation in (4.2.52) is to iteratively adjust the probabilities or weights of each label l_μ assigned to each segmentation object s_i. The denominator only achieves normalization according to (4.2.50), and the two essential contributions to the new probability are the old probability $p_{iN}(l_\mu)$ and the term $\beta_{iN}(l_\mu)$. From (4.2.51) it follows that this term has a significant positive value if other segmentation objects with highly probable levels are highly compatible with label l_μ of segmentation objects s_i, and this term has a significant negative value if these other segmentation objects are highly incompatible. In this way segmentation objects with highly probable labels influence the labels of another segmentation object. On the other hand, segmentation objects with low probability labels, that is $p_{jN}(l_\nu) \approx 0$, have little influence regardless of their compatibility. Experience indicates that a relaxation process of this type converges within a few iterations. Linear relaxation is also possible, but it can be shown that it converges to a set of probabilities which is independent of prior assignments.

4.2.6 Learning

In order to design one of the classifiers from Sect. 4.2.1, 2 it is necessary to estimate the unknown parameters occurring in (4.2.1, 25). This is usually referred to as *learning* or *training*. Several texts given in the references treat this subject. A prerequisite for learning is that a sample of patterns is known or observable as indicated in (1.3.1) of Sect. 1.3.

If the sample is classified, the learning is called *supervised*. For instance, in connection with (4.2.22) or (4.2.29) it amounts to estimating μ_κ, K_κ or evaluating (4.2.35), respectively. This may be done directly, iteratively, or by stochastic approximation. An analysis of (4.2.19) shows that the estimated parameters may be used as if they were the true parameters.

If the sample is unclassified, the more complicated case of *unsupervised learning* results. It may be treated by methods for parameter estimation (identification of mixtures) or by cluster analysis, but the former tend to be computationally more involved than the latter. Computational complexity is reduced by decision-directed schemes.

Examples of training procedures for neural nets are given in Sect. 4.6.

4.2.7 Additional Remarks

There are types of classifiers which do not quite fit into Sects. 4.2.1–3. One such type are sequential classifiers. The aim here is to use as small a number of features as possible. The basic approach is to classify a pattern at first with a small number of features and a high reject threshold. In the case of a rejection, classification is tried again with more features and possibly a reduced reject threshold. The last step is iterated until a decision is made or a prespecified number of trials has been carried out.

Another type is hierarchical classifiers or decision trees. With these classifiers some broad classes are distinguished first. In the next step each of the broad classes is further subdivided. Subdivision continues until a final classification is obtained. Systematic design of these classifiers is still unsolved in general, but several special methods have been developed.

In certain applications simplified classifiers seem to be sufficient. For instance, in the classification of multispectral images rectangular regions are attributed to classes, thereby reducing classification to some threshold tests.

4.3 Classification of Symbol Strings

The initial representation $\varGamma^r$ in the classification of symbol strings or in *syntactical classification* is the finite string v^r of terminal symbols, where the model M_κ is some formal grammar G_κ representing pattern class Ω_κ. It is not the intent here to give an extensive account of this approach since it is more suited to classification of simple patterns than to analysis and understanding of complex patterns. So we only summarize some basic properties. A *formal grammar* G is defined by a quadruplet

$$G = (V_N, V_T, R, S), \qquad (4.3.1)$$

where V_N is a finite set of *non-terminal symbols*, V_T is a finite set of *terminal symbols*, R is a finite set of *rules* or *productions*, and $S \in V_T$ is a *start symbol*. The sets V_N and V_T are assumed to be disjoint. Every rule $r_i \in R$ has the form

$$r_i : \beta_i \to \gamma_i, \quad i = 1, \ldots, l,$$

$$\beta_i \in (V_N \cup V_T)^* V_N (V_N \cup V_T)^*. \qquad (4.3.2)$$

A string $\beta_{\nu+1} \in (V_N \cup V_T)^*$ is said to be *directly derivable* from $\beta_\nu \in (V_N \cup V_T)^*$ if there are substrings $\beta_i, \beta_j, \beta_k, \gamma_j$ such that

$$\beta_{\nu+1} = \beta_i \gamma_j \beta_k, \quad \beta_\nu = \beta_i \beta_j \beta_k, \quad \beta_j \to \gamma_j \in R. \qquad (4.3.3)$$

This is abbreviated by $\beta_\nu \to \beta_{\nu+1}$. A string β_ν is said to be *derivable* from a string β_μ if there are strings $\beta_i, i = 1, \ldots, n$, such that $\beta_\mu = \beta_1 \to \ldots \to \beta_n = \beta_\nu$. This is abbreviated by $\beta_\mu \xrightarrow{*} \beta_\nu$.

Starting with the symbol S and using rules in R, other symbol strings may be derived from the initial symbol S using (4.3.3). The set of terminal symbol strings which can be derived from S is called the *language* $L(G)$ defined by the grammar G

$$L(G) = \{v | S \overset{*}{\to} v, v \in V_T^*\} \ . \tag{4.3.4}$$

It can be shown that it is undecidable in general whether a certain terminal symbol string is an element of $L(G)$. However, in certain restricted classes of grammars it is decidable whether some terminal string v is in $L(G)$ or not. Apparently, for the purposes of pattern classification only these restricted classes of grammars are of interest. In this case *recognition algorithms* are known which can be used to classify a string v.

Classification using this formalism requires the following steps:

1. Select a suitable set V_T of terminal symbols, for example, straight and curved line segments of a pattern.
2. Determine k formal grammars representing the k classes.
3. Select a proper recognition algorithm.
4. Transform a new pattern f into a symbol string v.
5. Determine the grammars G_κ with $v \in L(G_\kappa)$. If there is exactly one such grammar, then assume $f \in \Omega_\kappa$, otherwise reject f.

Additional steps are required if errors in the terminal string v have to be tolerated.

These few remarks have to suffice here for the problem of syntactic classification. Additional material is given in the references.

4.4 Recognition of Words

4.4.1 Word Models

There is no uniform structure for the initial representation I^r and no standard model M_κ in the problem of word recognition. Early approaches to recognition of *isolated words* used the feature vector representation of words and word models. The speech wave was segmented into a fixed number of intervals (this amounts to a linear normalization of the time axis), a set of features (e.g. coefficients of a discrete Fourier transformation) was extracted per interval, and a word template or prototype was constructed by averaging feature vectors from different utterances of the same word. A new word was classified by computing the distance between the feature vector of the word and the prototype. This approach is sufficient if the vocabulary is small, if there are only one or a few speakers, and if words are pronounced carefully.

A significant improvement was achieved by the introduction of nonlinear time normalization. This is important because the duration of different phones in a word usually varies by an amount depending on this phone. For example, a vowel may be lengthened considerably, whereas a plosive can vary only within smaller limits. A certain utterance then is segmented into a varying number of intervals of constant duration, typically about 10 ms or 1 cs (one centisecond). Each interval is represented by a feature vector, e.g. coefficients of the discrete Fourier transformation, linear prediction, autocorrelation, or mel cepstral coefficients. A prototype or word model is formed by clustering or simply averaging different utterances of the same word. A new word is recognized by comparing it to all prototypes and computing a distance measure using nonlinear time warping. This approach is discussed shortly in Sect. 4.4.6.

Another very important approach to word recognition is to model a word by a stochastic automaton; the *hidden Markov models* are the most widely used types of automata in word recognition. This allows one to compute the probability that a certain automaton (word model) caused the observation of a certain speech wave. An account of this approach is given in Sect. 4.4.2.

It is well known that words may be segmented into perceptually meaningful *subword units*, for example, phones, diphones, demi-syllables, syllables, or code words of vector quantization as discussed in Sect. 3.10. In particular in the context of continuous speech recognition it has become common practice to first segment a speech wave into subword units and perform recognition using these units instead of the numerical coefficient vectors. The advantages of this approach are that a significant data reduction is achieved, that the problem of speaker independence may be handled on the level of subword units, and that dictionaries for word recognition may be derived automatically from the standard pronunciation of words. The disadvantages are that the choice of subword units is not obvious, that segmentation of the speech wave and recognition of the subword units is erroneous, and that speaker-independent recognition of those units is by no means a simple problem.

In any case the set of words known to the system is represented in a *lexicon* or *dictionary* which is organized such that recognition is facilitated. An example of a lexicon suitable for word recognition is presented below, a more general lexicon is defined in Sect. 7.6.3. The lexicon is often organized as a tree where each node represents some acoustical unit or state of a word; the descendants of a node are possible continuations; a path from the root to a leaf, that is a node with no successors, represents a complete word; in addition there may be specially labeled word nodes which are not leaves. From a word node there is a pointer to the corresponding word in the lexicon. Different organizations of the tree result from the choice of the acoustical unit and the incorporation of phonological rules. As a first example the dictionary shown in Fig. 4.4 is discussed briefly. Starting from the word level, *words* are represented by *syllables* in a syllable-word tree, that is the nodes of the tree are syllables. Part of such a tree is given in Fig. 4.4a. The syllables in turn are represented by up to three *syllable parts* (onset, vowel, coda) in a sylpart-syllable tree shown in Fig. 4.4b. Finally, the three parts onset, vowel, and coda of syllables are represented by *labeled segments* of the input speech utterance in a segment-onset, a segment-vowel, and a segment-coda tree,

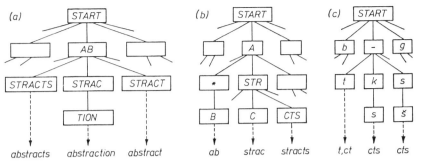

Fig. 4.4a–c. Parts of the tree structures used to store words, syllables, and sylparts of a dictionary. Note that in (**b**) vowels come first because they are searched for first. Lower case letters in (**c**) represent phonemes, otherwise phonetic representation was avoided

respectively. Part of the last tree uses acoustic evidence from the input directly.

An advantage of using syllables is that only few syllable parts are needed for a large dictionary. The knowledge contained in the last tree is obtained from hand segmentation of a training set of utterances into syllable parts. For the other two trees it is obtained from a pronunciation dictionary which supplies base pronunciations. These are processed to account for variations in pronunciation.

4.4.2 Word Recognition Using Hidden Markov Models

Hidden Markov models (HMMs) are currently among the most widely used approaches to word recognition, both in *isolated word recognition* and recognition of words in *continuous speech*. An HMM is a special model for the generation of a *stochastic process* by employing two different mechanisms. The first mechanism uses a finite set of state symbols

$$S = \{S_1, S_2, \ldots, S_I\} \tag{4.4.1}$$

to generate a *state sequence* $s = [s_n]$ of length T. At discrete time t_n the model is in state $s_n \in S$. At time t_{n+1} the model makes a random transition to state $s_{n+1} \in S$. Which state is selected next is determined by the *state transition probability matrix*

$$A = [a_{ij}] = P(s_{n+1} = S_j | s_n = S_i), \quad i, j = 1, 2, \ldots, I. \tag{4.4.2}$$

State transition probabilities are assumed to be independent of time and previous states except for one predecessor state. The first state s_1 is selected at random according to the *initial state probability*

$$\boldsymbol{\pi} = [\pi_i] = P(s_1 = S_i), \quad i = 1, 2, \ldots, I. \tag{4.4.3}$$

However, the states are not observable, hence the term *hidden* Markov model. What is observable is a sequence of *outputs* or *observations* $o = [o_i]$ which are taken from a finite set of *output symbols* or *observation symbols*

$$O = \{O_1, O_2, \ldots, O_L\} \ . \tag{4.4.4}$$

The second mechanism generates an observation $o_i \in O$ if the model is in state s_i by selecting an output symbol according to the *output probability*

$$\begin{aligned} B &= [b_{il}] = P(O_l \text{ emitted in state } S_i \,|\, s_n = S_I) \ , \\ i &= 1, \ldots, I \ , \quad l = 1, \ldots, L \ . \end{aligned} \tag{4.4.5}$$

A hidden Markov model (HMM) is defined by the triple

$$\text{HMM} = (\pi, A, B) \ . \tag{4.4.6}$$

It generates a member

$$o = (o_1, o_2, \ldots, o_T) \tag{4.4.7}$$

of the stochastic process or an *observation sequence* as follows. At first, a state symbol S_i is selected as initial state s_1 with probability π_i. Then a transition to state symbol S_j is made with probability a_{ij} and an output symbol O_l is emitted with probability b_{il}; this gives the new state s_2 at time t_2. State transition and emission of an output symbol is repeated until a sequence o of length T is generated.

An HMM can be used for word recognition. Select as set S an appropriate set of subword units, for example, phone segments. Provide as a word model an HMM having M states $s_i \in S$ where M corresponds to the typical number of subword units in a word, for instance, the number of phones in the standard pronunciation of a word. Segment the speech signal of a new utterance such that each segment corresponds to a subword unit. Consider the segmented speech wave as the member o of a stochastic process. Determine the HMM which most probably generated this sequence (isolated word recognition) or a certain part of the sequence (continuous speech recognition). Classify the speech wave or the interval as being the word modeled by the most probable HMM. In the case of speech recognition the HMM can be restricted to the so-called left-to-right model. This does not include state transitions which go backward in time. An example of an HMM modeling a word is given in Fig. 4.6 in Sect. 4.4.5.

In view of this approach the following questions arise:

1. If an observation sequence o and an HMM are given, what is the probability $P(o|\text{HMM})$ of generating o by HMM?
2. If an observation sequence o and an HMM are given, what is the most probable state sequence of HMM?
3. If several utterances of a word are given, what are the values for A, B, π maximizing $P(o|\text{HMM})$?

We will discuss these questions in the next section.

4.4.3 Basic Computational Algorithm for HMM

The straightforward approach to computation of $P(o|\text{HMM})$ is to determine a state sequence s having length T, to compute $P(o, s|\text{HMM})$ for this state sequence, to repeat the computation for every state sequence of length T, and to sum up all those values. Since there are very many state sequences (e.g. for $I = 10$ states and an observation of length $T = 20$ there are $I^T = 10^{20}$ state sequences), this is not feasible for actual computations. A more efficient approach is the so-called *forward-backward algorithm* outlined next. It exploits the fact that the probability $P(o|\text{HMM})$ can be computed from the first state s_1 forward, or from the last state s_T backwards. The *forward procedure* for computation of $P(o|\text{HMM})$ is the following:

1. Define the *forward variable*

$$\alpha_{ni} = P(o_1, \ldots, o_n, s_n = S_i | \text{HMM}) . \qquad (4.4.8)$$

It is the probability of observing the first n symbols in o and being in state S_i at time t_n, given a particular HMM.

2. Initialize α by

$$\alpha_{1i} = \pi_i b_{il} , \quad \text{where } l \text{ is chosen such that} \quad o_1 = O_l , \quad i = 1, \ldots, I . \quad (4.4.9)$$

3. Recursively compute successive values of α by

$$\alpha_{n+1,j} = \left(\sum_{i=1}^{I} \alpha_{ni} a_{ij} \right) b_{jl} , \quad n = 1, \ldots, T-1 , \quad j = 1, \ldots, I , \qquad (4.4.10)$$

and l is chosen such that $o_{n+1} = O_l$.

4. Obtain the final probability from

$$P(o|\text{HMM}) = \sum_{i=1}^{I} \alpha_{Ti} . \qquad (4.4.11)$$

This forward procedure is illustrated in Fig. 4.5. It depends on two points. First, the observation of a certain output symbol is the product of the probability of being in state S_i times the probability of emitting symbol O_l in this state. Second, the probability of reaching state S_j from predecessor S_i is the product of the probability of being in state S_i times the transition probability a_{ij}; the total probability of reaching S_j is the sum over all possible predecessor states.

In an analogous manner one may start at s_T and define the *backward procedure*:

1. Define the *backward variable*

$$\beta_{ni} = P(o_{n+1}, \ldots, o_T | s_n = S_i, \text{HMM}) . \qquad (4.4.12)$$

This is the probability of observing the last symbols in o beginning from the symbol at $n + 1$, if the state at time t_n is S_i and the model is HMM.

2. Initialize β by

172   4. Classification

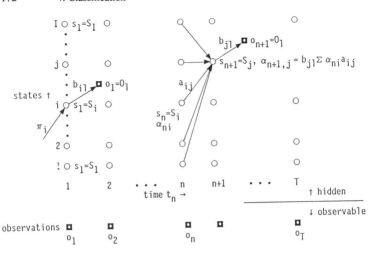

Fig. 4.5. Illustration of the forward procedure

$$\beta_{Ti} = 1, \quad i = 1, \ldots, I. \tag{4.4.13}$$

3. Recursively compute successive values of β by

$$\beta_{ni} = \sum_{j=1}^{I} a_{ij} b_{jl} \beta_{n+1,j}, \quad n = T-1, \ldots, 1; \quad i = 1, \ldots, I, \tag{4.4.14}$$

such that $o_{n+1} = O_l$.

4. Obtain

$$P(o|\text{HMM}) = \sum_{i=1}^{I} \alpha_{ni} \beta_{ni}. \tag{4.4.15}$$

Usually the sequence β_{ni} is not used for computing $P(o|\text{HMM})$, but only for training.

The above algorithms are the basis also for answering the last question from Sect. 4.4.2. The best state sequence s given an observation sequence s is usually computed under the assumption that $P(s, o|\text{HMM})$ is to be maximized, that is the single best state sequence is computed. This can be done by the *Viterbi algorithm*.

1. Initialize the variable δ and the backpointer Φ by

$$\delta_{1i} = \pi_i b_{il}, \quad \text{such that} \quad o_1 = O_l, \quad i = 1, \ldots, I, \tag{4.4.16}$$

$$\Phi_{1i} = 0. \tag{4.4.17}$$

2. Recursively compute successive values

$$\delta_{nj} = \max_{1 \leq i \leq I} \{\delta_{n-1,i} a_{ij}\} b_{jl}, \quad \text{such that} \quad o_n = O_l, \quad n = 2, \ldots, T$$
$$j = 1, \ldots, I, \tag{4.4.18}$$

$$\Phi_{nj} = \underset{1\leq i\leq I}{\operatorname{argmax}}\{\delta_{n-1,i}a_{ij}\}, \quad n,j \text{ as above.} \tag{4.4.19}$$

3. Terminate with

$$P^* = \max_{\{s\}} P(s,o|\text{HMM}) = \max_{1\leq i\leq I}\{\delta_{Ti}\} \tag{4.4.20}$$

$$s_T^* = \underset{1\leq i\leq I}{\operatorname{argmax}}\{\delta_{Ti}\} \tag{4.4.21}$$

4. Obtain the optimal state sequence by tracing the backpointers

$$s_n^* = \Phi_{n+1,k}, \quad \text{such that} \quad s_{n+1}^* = S_k, \quad n = T-1,\ldots,1. \tag{4.4.22}$$

The third question concerns the computation of those parameter values of an HMM which most probably caused the observation of a certain sequence o. This can be done by the so called *Baum-Welch algorithm* for reestimating or updating the parameters of an HMM.

1. Assume that an initial HMM $= (\pi, A, B)$ is given; compute by steps 2–5 an updated model HMM$' = (\pi', A', B')$.

2. Define p_{nij} as the probability of being in state S_i at time n and in S_j at $n+1$

$$p_{nij} = P(s_n = S_i, s_{n+1} = S_j | o, \text{HMM}). \tag{4.4.23}$$

Since
$P(s_n = S_i, s_{n+1} = S_j | o, \text{HMM}) = P(s_n = S_i, s_{n+1} = S_j, o|\text{HMM})/P(o|\text{HMM})$,
and using (4.4.8, 12) we obtain

$$p_{nij} = \frac{\alpha_{ni}a_{ij}b_{j,l}\beta_{n+1,j}}{P(o|\text{HMM})}. \tag{4.4.24}$$

3. Define the variable γ_{ni} as the probability of being in state S_i at time t_n, given HMM and o. From the definition of p_{nij} above one gets

$$\gamma_{ni} = \sum_{j=1}^{I} p_{nij}. \tag{4.4.25}$$

4. Observe that the expected number (over time) of transitions from S_i to some other state is the sum of (4.4.25) over n, and the expected number of transitions from S_i to S_j is the sum of (4.4.24) over n.

5. Compute updated parameter values from

π_i' = expected number of times in S_i at time t_1

$$= \gamma_{1i}, \tag{4.4.26}$$

$$a_{ij}' = \frac{\text{exp. number of trans. from } S_i \text{ to } S_j}{\text{exp. number of trans. from } S_i}$$

$$= \frac{\sum_{n=1}^{T-1} p_{nij}}{\sum_{n=1}^{T-1} \gamma_{ni}}, \qquad (4.4.27)$$

$$b'_{il} = \frac{\text{exp. number of times in } S_i \text{ and obs. } O_l}{\text{exp. number of times in } S_i}$$

$$= \frac{\sum_{n=1}^{T} (\gamma_{ni} | o_n = O_l)}{\sum_{n=1}^{T} \gamma_{ni}}. \qquad (4.4.28)$$

6. If a set $O = \{o^1, \ldots, o^N\}$ of observation sequences is used, the expected number of events in (4.4.26–28) is obtained by adding the corresponding numbers in the individual observation sequences.

7. Repeat the updating until the parameters do not change any more.

Theorem 4.4.1. Using the above algorithm either gives HMM = HMM$'$ (that is the estimation is finished) or results in new parameters yielding

$$P(o|\text{HMM}') > P(o|\text{HMM}), \qquad (4.4.29)$$

that is one is more likely to observe o from HMM$'$ than to observe it from HMM.

Proof. See, for example, [4.9, 10]. □

The model parameters can be computed using a set of training data. In particular, for the above mentioned left-to-right model, multiple observation sequences are essential because for a single observation sequence the number of events will be very small.

The above two sections present only the most basic properties of HMM. There are several useful extensions when applying them to problems of speech recognition. Some of them are mentioned briefly leaving the details to the references.

One extension concerns the use of continuous emission probability densities instead of the discrete probabilities introduced in (4.4.5). This is useful if observations are not from a finite set of symbols, but from a continuous set of values. For example, if speech is segmented into fixed intervals of about 10 ms duration and every interval is represented by a parameter vector, the observations are elements of R^n. In this case the emission probability can be represented by a normal density or, more generally, by a mixture of normal densities. It has been shown that a reestimation procedure is also possible in this case.

Another extension is the inclusion of state duration probability densities into the HMM. Again, reestimation formulas can be derived for this case.

Some precautions are necessary when actually computing parameter values in an HMM. The first is that in the numerical calculations of the above algorithms the dynamic range of numbers, for example α_{ni} in (4.4.10), is usually very large. Therefore, it is necessary to scale the coefficients by multiplication by a scaling factor which depends on the index n but not on i. In addition it may be useful to impose certain constraints on the updated parameter values, for example, that a value should not fall below a certain minimum value.

The above algorithm for computing the HMM parameters requires an initial model and inital values of parameters to start the iteration. According to experimental evidence, initial values for π and A may be chosen at random or as uniform values, whereas reasonable initial values are useful for B. These may be obtained from hand-segmentation and labeling of a training data set.

So far there is no theoretical approach for selecting a reasonable number of states for a model or for finding adequate subword units; this in turn has effects on the type of emission probabilities (continuous or discrete).

4.4.4 Isolated and Connected Words

When using HMM for *isolated word recognition* a separate HMM_κ is constructed per word $\Omega_\kappa, \kappa = 1, \ldots, k$ in the vocabulary. If N_κ training utterances per word are available, the algorithm (4.4.23–28) can be used to estimate the parameters of the corresponding HMM_κ. An utterance of some word is represented by an observation sequence o obtained from segmentation of this word. From Theorem 4.2.3 it is known that classification of o with minimum error probability can be achieved by computing the a posteriori probabilities $P(HMM_\lambda|o)$ for every word Ω_λ and deciding for the word Ω_κ maximizing the a posteriori probability. Because of the definition of conditional probabilities this is equivalent to computing $P(o|HMM_\lambda)P(HMM_\lambda)$, where the last term is the a priori probability of a word Ω_λ. The first term can be obtained from (4.4.11 or 15). The last term can be estimated from a set of training data, the a priori probabilities can be assumed to be equal, or they can just be ignored. In the last two cases only the term $P(o|HMM_\lambda)$ is used. A frequently used alternative is to compute $P(s, o|HMM_\lambda)$ in (4.4.20). This bases the decision on the single state sequence maximizing the probability of observing o instead of summing over all state sequences yielding o.

A set of individual word models is also used in *recognition of connected words*, for example, recognition of strings of spoken digits. The details are left to the references.

4.4.5 Continuous Speech

A basic assumption in isolated word recognition is that word boundaries can be determined fairly easily, for example, by applying an energy threshold to the speech signal. Obviously, there is no such simple method for the determination of word boundaries in continuous speech. This poses the additional problem of finding word boundaries in the speech signal. We mention two approaches to this problem.

One approach to word recognition in continuous speech is the strict top-down approach. Words and their expected position are predicted from linguistic knowledge, for example, from a grammar generating a certain subset of a language.

This reduces the problem to the verification of a word at the predicted position. Verification can be done by computing a word score as described in the last section.

Another approach is to position words in the speech signal. We assume that the speech signal is represented by an observation sequence o having N symbols as above and that every word in the vocabulary is represented by an HMM having a state sequence of length M.

The search for words is split into the two phases of *word hypothesization* and *word verification*. The goal of the first phase is to find quickly a small subset of the lexicon which contains the spoken words with a high probability. No syntactic information is used at this stage in order to concentrate on word recognition in continuous speech. The subset of the lexicon is called the set of *word hypotheses*. The main components of a word hypothesis (WH) are

$$\text{WH} = (word\_number, start\_frame, end\_frame, score). \quad (4.4.30)$$

The word number refers to the entry of the lexicon. The score gives a measure of the quality or reliability of a hypothesis and is described below. Word verification has the goal of modifying the score of a word hypothesis; ideally it will improve the score of a correct hypothesis and debase the score of a false hypothesis.

Two types of approaches to word hypothesization are in use:

1. A lexicon-based (top-down) approach which systematically computes a score for all the words in the lexicon. This is described below.
2. A feature-based (bottom-up) approach which uses phonetic or other features of the utterance to mark certain words in the lexicon.

If the vocabulary is large, it is useful to represent it by a compressed *lexicon tree* (CLT). The CLT is a data structure where the root of the tree is the start of all words in the lexicon, a node represents a certain phone (or some other subword unit), words having the same initial sequence of phones also have a common path in the tree, and the end of a word may be a leaf in the tree or a node labeled by a special symbol. This data structure allows a significant reduction of storage space. It will be seen below that the computation time for generating word hypotheses is also reduced.

Common techniques for word recognition are based on dynamic programming (DP) and on hidden Markov models (HMM). An HMM for a reference word R in the lexicon can be constructed by using the standard pronunciation of a word as listed in a pronunciation dictionary. Every phone is modeled by an HMM having two states and three transitions as shown in Fig. 4.6. The transitions are:

1. SUB – the substitution of a phone of the reference word at a particular position of the speech signal; the phone suggested by phonetic labeling may coincide with the phone of the reference word or not (replacement of one segment of the reference word by one segment of the speech signal, abbreviated by 1-1).

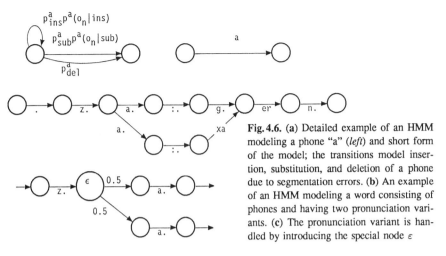

Fig. 4.6. (a) Detailed example of an HMM modeling a phone "a" (*left*) and short form of the model; the transitions model insertion, substitution, and deletion of a phone due to segmentation errors. (b) An example of an HMM modeling a word consisting of phones and having two pronunciation variants. (c) The pronunciation variant is handled by introducing the special node ε

2. DEL – the deletion of a phone occuring in the reference word, but not found in the speech signal (1-0).
3. INS – the insertion of an additional phone in the speech signal which is not present in the reference word (0-1).

Additional transitions may be introduced to account for the peculiarities of phonetic labeling. For example, it may occur that a segment border is not found in the signal which causes a merging of the phones in the two segments into the alternatives of the phone labels in the merged segments. The opposite effect is the introduction of an additional segment border within the segment of one phone which causes the same phone label to occur among the alternatives of the two segments. Finally, a border between two segments may be detected but shifted one unit to the left or right.

Let a reference word R suggested by the lexicon consist of the sequence of M phones $R = (r_1, \ldots, r_i, \ldots, r_M)$, and let the speech signal be represented by a sequence of T segments $Q = (q_1, \ldots, q_j, \ldots q_T)$ which are provided by acoustic-phonetic segmentation. Each segment q_j has the form

$$q_j = ((O_{jk}, w_{jk}); \quad k = 1, \ldots, K) \tag{4.4.31}$$

where O_{jk} is the kth alternative label of the jth segment and w_{jk} is its score or reliability. Let $T(r, o)$ be a matrix of phone confusions which is estimated by the forward-backward algorithm. Then a local similarity between phone r_i and segment q_j is defined by

$$\text{SIM}(r_i, q_j) = \sum_k T(r_i, O_{jk}) w_{jk} . \tag{4.4.32}$$

A rectangular table of nodes is built up as shown in Fig. 4.7. Node (i, j) in row i (state s_i) and column j (segment q_j) is given a cumulative similarity $P(i, j)$ which is computed recursively using the transitions described above and indicated in Fig. 4.6. The recurrence equation to compute $P(i, j)$ is

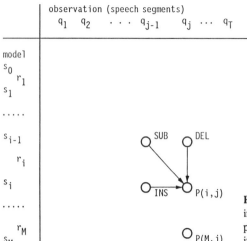

Fig. 4.7. The lattice structure for computing $P(i,j)$; note that an initial sequence of phones is usually shared by several words in the CLT

$$P(i,j) = P(i-1, j-1)\text{SIM}(r_i, q_j)C_S + P(i-1, j)C_D + P(i, j-1)C_I \,. \tag{4.4.33}$$

The term $P(i,j)$ is the cumulative similarity between the string $r_1, \ldots r_i$ of phones in the reference word R and segments of speech up to q_j. Computation proceeds row-wise from left to right, starts with row 1 and stops with row M. So the terms $P(i-1, j-1)$ and so on in (4.4.33) are available from previous computations. Initialization is done with $P(1,j) = 1, j = 1,\ldots,n$, meaning that the reference word may start in any segment; this is necessary because in continuous speech the start segment of a word is unknown. The constants C_S and so on are obtained from initial training. It can be seen that due to the organization of the CLT initial phone sequences shared by several words are evaluated only once.

During the computation of $P(i,j)$ the origin of the maximal contribution to the sum in (4.4.33) is determined and stored in a back pointer. After computation of row M the numbers $P(M,j), j = 1,\ldots,T$, are a measure of certainty that the reference word R ends at segment j. Using $P(M,j)$ word hypotheses are generated as follows:

1. Select segment j_e with maximal $P(M, j_e)$.
2. Select segment j_b by tracing back the back pointers from $P(M, j_e)$.
3. Generate word hypotheses (word, begin, end, score) with word = reference word, begin = j_b, end = j_e, score = $P(M, j_e)$.
4. Exclude segments $S(j_e - \text{mask}), \ldots, S(j_e + \text{mask})$ from further consideration by tagging their $P(M, j)$.
5. Repeat steps 1 to 4 until a specified number of word hypotheses is generated.

The operation in step 4 means that segments in the vicinity of segment j_e are not considered for the generation of additional hypotheses. This is reasonable since the segment next to a large relative maximum of P is likely to be the next relative maximum. The computation of $P(M,j)$ and generation of word hypotheses is repeated for every reference word in the lexicon. The word hypotheses are given a new score B according to the scoring functions described below. Finally, the H hypotheses with maximal values of B are selected as word hypotheses. The number H is chosen proportional to sentence length. Thus generating word hypotheses results on average in H/T best scoring hypotheses per segment of speech in the format of (4.4.30).

As shown above, the initial score $P(M, j)$ is used to generate word hypotheses for every reference word. Then a new score B is computed and among all the hypotheses generated for the whole lexicon the H best scoring according to B are selected. Since P defined in (4.4.33) is a sum of products of small real numbers the tendency of $P(M, j)$ is to become smaller for longer words. Therefore, selection of word hypotheses according to P would give too much preference to shorter words. This can be avoided by using

$$B = \frac{-\log P(M, j_e)}{M}, \qquad (4.4.34)$$

where M is the number of phones in the reference word, or by using

$$B = (-\log P(M, j_e)) - CM. \qquad (4.4.35)$$

The constant C is determined experimentally. Figure 4.8 shows an example of word hypotheses generated by the above procedure.

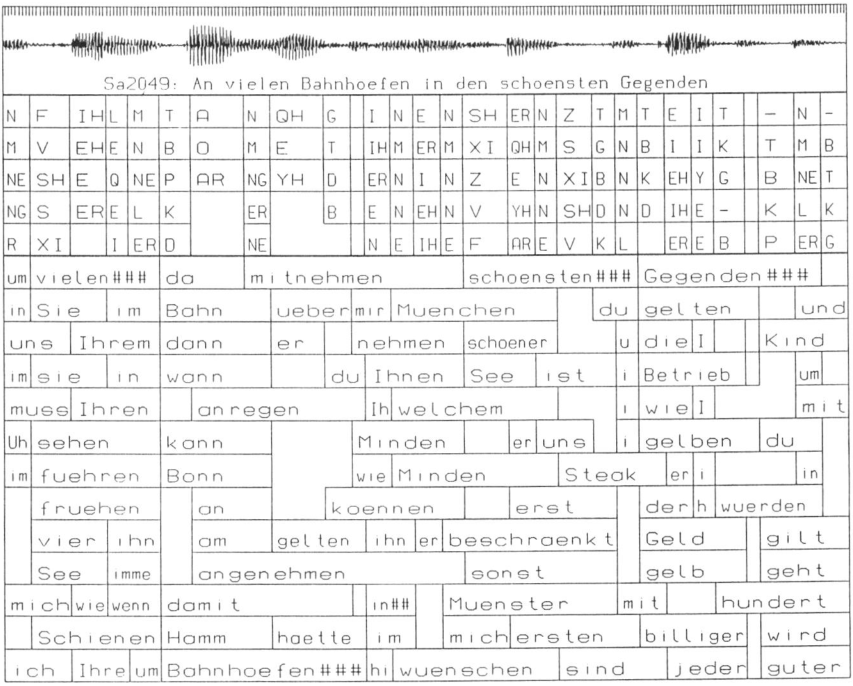

Fig. 4.8. A speech signal, the phone segments, and word hypotheses generated in an utterance. Correct words are marked by a # ; the good scoring words are at the top

The basic idea of the feature-based bottom-up approach is to represent both the words from the dictionary as well as portions of the input speech by subsets of an appropriate feature inventory. There is evidence that the temporal structure of spoken or written words is also sufficiently described in terms of sets instead of series of speech segments, provided that these segments include context to some extent. One possibility is to use the set of all consecutive phone triples as features. Consequently, vocabulary words are characterized by the triples occurring in their standard pronunciation, whereas speech input can be represented by the phone triples contained in the result of the acoustic-phonetic decoding stage. A word is assumed to be part of the spoken input if it has a certain number of features in common with some portion of the utterance. A very efficient calculation can be made if the pronunciation lexicon is stored in an associative manner, that is, each word is retrievable by using its features as keys for access. This feature addressable lexicon can be implemented by means of redundant hash adressing. Details are available from the references.

4.4.6 Word Recognition Using Dynamic Programming

There are quite a few variants of dynamic programming (DP) techniques which have been developed for speech recognition. The basic idea for isolated word recognition was given already in Sect. 2.2.2 in connection with nonlinear time normalization. The speech utterance is represented by a sequence of feature vectors, for example, Fourier or cepstral coefficients computed every 10 or so ms. Every word in the lexicon is represented by a typical sequence of those feature vectors. DP is used to compute a distance between the feature vector of the utterance and the reference words. This technique may also be used for the computation of reference words. Since the time structure of two different utterances of the same word may also be nonlinearly distorted, it is not useful to simply average the different utterances in order to obtain a prototype. Rather, a new utterance of the same reference word is nonlinearly aligned to the presently available word prototype. Then the aligned new utterance is added to the prototype.

DP algorithms have been extended to the problem of connected word recognition, too. An example of such an algorithm is the so-called one-stage algorithm. Its idea is to start the DP distance computation at the beginning of an utterance using every word in the lexicon for comparison. At the end of some word a new distance computation is started, again using all the words in the lexicon. This process is repeated until the end of the utterance. What actually is computed is the best scoring sequence of words spanning the utterance. Let the utterance be represented by a sequence of T feature vectors $C = (c_1, \ldots, c_T)$ and the prototype or template k of some word by a feature vector $\boldsymbol{R}_k = (\boldsymbol{r}_{k1}, \ldots, \boldsymbol{r}_{kj}, \ldots, \boldsymbol{r}_{kM(k)}), k = 1, \ldots, K$, if there are K prototypes. The algorithm works on a grid (i, j, k), where i is the ith feature vector or time frame of the test utterance and j is the jth feature vector of the kth prototype. The index i may be associated with the x-axis of a coordinate system. The y-axis

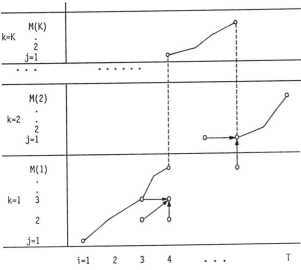

Fig. 4.9. Organization of the one-stage DP algorithm for connected word recognition. An example of an optimal path is given; it can be seen that this may switch arbitrarily between the K prototypes. The two types of predecessors are also shown

is doubly indexed; it has the slowly increasing index k of a prototype and the fast increasing index j per prototype. Thus one prototype is located above the preceding one as shown in Fig. 4.9. It is assumed that $d(i,j,k)$ is a measure of distance between the ith utterance time frame and the jth time frame of the kth prototype. Then the recognition of connected words amounts to finding the path of minimal sum of distances $d(i,j,k)$ through the grid. If this path is within a certain word template, the *continuity constraint*

IF (a grid point $(i,j,k), j > 1$ is given),
THEN (predecessor points are from $\{(i-1,j,k),(i-1,j-1,k),(i,j-1,k)\}$)
(4.4.36)

is assumed as indicated in Fig. 4.9. If the path crosses the border between two templates, the constraint

IF (a grid point $(i,1,k)$ is given),
THEN (predecessor points are from $\{(i-1,1,k),(i-1,M(k),k), k = 1,\ldots,K)$
(4.4.37)

is assumed, that is at a border the path may continue at the beginning of any word template. If additional constraints are available, for example, if there are syntactic constraints, the set of possible predecessors of a word may be further constrained. Using dynamic programming the best path may be found as indicated in the following algorithm:
1. Denote the sum of distances of a path of minimum sum of distances up to a point

(i,j,k) by $D(i,j,k)$ and initialize $D(1,j,k) = \sum_{n=1}^{j} d(1,n,k)$. (4.4.38)

2. FOR $i = 2,\ldots,T$ DO:
   2.1. FOR $k = 1,\ldots,K$ DO:
   $D(i,1,k) = d(i,1,k)$
   $+ \min\{D(i-1,1,k), D(i-1,M(k),k), k = 1,\ldots,k\}$ (4.4.39)
   2.1.1. FOR $j = 2,\ldots,M(k)$ DO:
   $D(i,j,k) = d(i,j,k) + \min\{D(i-1,j,k), D(i-1,j-1,k),$
   $D(i,j-1,k)\}$ (4.4.40)
   END FOR
   END FOR
   END FOR

3. Trace back the optimum path from the end of that prototype having minimum total distance; this is possible using the array $D(i,j,k)$.

This algorithm can be used for connected word recognition if the vocabulary is not too large. Other types of algorithms are mentioned in the references.

4.5 Recognition of Objects

4.5.1 Object Models

There are quite different approaches to modeling of objects for the purpose of automatic recognition since the variability of images in different task domains is very large. We only give some general ideas and treat some special approaches in the following sections. A first general distinction of object models are the two-dimensional and three-dimensional models. The former model properties of images or two-dimensional views, often in image coordinates; the latter model the original three-dimensional object in an object centered coordinate system. By a proper coordinate transformation the object model may be scaled, shifted, and rotated; by a projection from some view point it may be transformed to a 2D image.

Object models may be formed using global features or using local segmentation objects, their attributes and relations. The former case is the feature vector approach to object recognition and classification. Although there are important practical problems where this approach is successful, we will not elaborate this point. In the latter case the model is built by a structural description of the object using segmentation objects. This approach has the advantage that some local properties will remain unchanged even if parts of the object are occluded, whereas global features will change completely if occlusions occur.

It was argued already in Sect. 3.1 that segmentation objects are, for example, vertices, line segments, lines, regions, and volumes; a list of common attributes and relations is given there. In addition it may be useful to allow meaningful parts of an object as components of an object model. For example, it may be useful to model a house by parts such as roof, gable, window, and so on; a window in turn may be modeled by a geometrical figure such as a rectangle, and the rectangle by lines or an appropriate region. Let an object model be denoted by M, its attributes by A, relations by S, and parts by P. Then (initial) segmentation objects (3.1.1) are themselves objects of a primitive type, that is objects having no parts. In accordance with (3.1.2) an object model may now be summarized by a structure

$$M : (O : T, (A : (TA, R \cup AV))^*, (P : M)^*, (S(P) : R)^*,$$

$$CF(A, P, S) : R^n) \,. \tag{4.5.1}$$

The symbols in (4.5.1) are as introduced in (3.1.2) with the addition of O which is an object name. The above notation $M : (t)$ indicates that M is a structure having components t. This is a recursive structure since M may contain parts P which themselves are of type M. The attributes are in general vectors of real numbers or strings of terminal symbols or a combination of them. A structural relation S is defined by a procedure or a test which returns a real number measuring the degree of fulfilment of the relation. In addition, each model is attributed a certainty factor CF, which again is assumed to be defined by a procedure having attributes, parts, and structural relations as allowed arguments. Using the available (image) data, it returns a vector of real numbers measuring the certainty of recognition of the object represented by M. The following sections will give some examples of actual models M constructed for certain applications.

Those models may be on quite different *levels of abstraction* as introduced in Sect. 1.2, for example, on the level of lines and regions (which are examples of segmentation objects of an initial description), on the level of three-dimensional objects, on the level of motions of objects, on the level of interpretations of images and image sequences. An example of an interpretation is the diagnostic interpretation of medical images taking into account objects or organs and their motional behavior (see also Sect. 1.2). If we allow different levels of abstraction we actually have a sequence of models $M_{\kappa l}$ representing the κth class on the lth level of abstraction. *Classification* may thus be viewed as a sequence of mappings starting with the observation, going to an initial description and leading to various levels of abstraction

$$f^r \to I^r \to M_{\kappa 1} \to \ldots \to M_{\kappa l} \to \ldots \to M_{\kappa L} \,,$$

$$\kappa = 1, \ldots, k_l \,, \quad l = 1, \ldots, L \,. \tag{4.5.2}$$

If we use the model in the above described sense, then we are in the same representational framework from I^r to $M_{\kappa l}$ with the only exception being f^r, which for common recording devices is an array of integers.

In this view, we may characterize numerical pattern recognition, syntactical pattern recognition, and the general classification problem introduced in Sect. 4.1 by numerical classification : mapping $R^n \to \Omega_\kappa$,

syntactical classification : mapping $V_T^* \to \Omega_\kappa$, (4.5.3)

general classification : mapping $M \to M$.

Meanwhile there is a wealth of material concerning the properties of R^n and V_T^*, in particular also in its relation to classification. The references provide an overview on this. It would be tremendous progress to have a comparable body of theory concerning M (4.5.1). Such a theory is probably the prerequisite for a transition from heuristic matching algorithms to systematically derived "optimal" classification procedures and the automatic derivation of logical pictures of facts as stated in Sect. 1.1. Statistical decision theory offers an excellent example of this, showing that such optimal procedures can in fact be derived if adequate theoretical background is available. However, until a general theory is available, general classification will rely heavily on heuristics, and the following sections give a few examples.

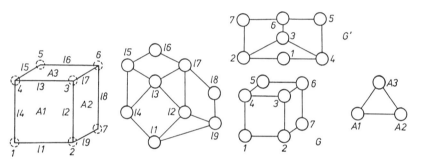

Fig. 4.10. A cube as an example of a simple object and its representation as a graph using lines, corners, and surfaces as nodes of the graph

Models of – or knowledge about – images and objects which might occur in an image may be represented or visualized by graphs or graph-like structures. A node of the graph may correspond to a part or a segmentation object, and an edge to a relation between parts. But this is not mandatory since, for example, one may also have edges corresponding to lines in an image, and nodes corresponding to points of line intersection. Figure 4.10 gives a simple example. The important point is to establish a correspondence between an object or an image and its model which contains sufficient detail, that is includes all information important to the user or the task domain, allows discrimination between models of different objects, and has no irrelevant detail. If one has models of objects it is possible to look for particular instances of these models (see also Sects. 7.1 and 7.4). Furthermore, arrangements of objects may be obtained from an image, and this again may be guided by a model of the images belonging to a particular task domain. Of course, the specification of the models will have varying degrees of accuracy and constraint depending on the application.

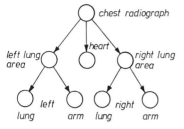

Fig. 4.11. A model of a chest radiograph, [4.11]

Two examples should clarify this. Figure 4.11 shows the graph definition of a chest radiograph. Obviously any instance of a chest radiograph will have a left lung, a right lung, and so on. Thus the model has the same nodes and edges as an instance of it. Only some details of the nodes, like area of the heart or abnormalities of the lung, if present, will vary in individual radiographs. Figure 4.12a shows a model of "a scene with streets, houses, and cars" where it is assumed that at least one street must be present and an arbitrary number of houses, cars and other areas may be present. Also some relations between the objects are specified, for example, a street may intersect a street. This model may be used to analyze a particular scene, say that of Fig. 4.12b, resulting in an instance of the model given in Fig. 4.12c. In this case, the instance of the model is quite different from the model, but only objects and relations allowed by the model occur.

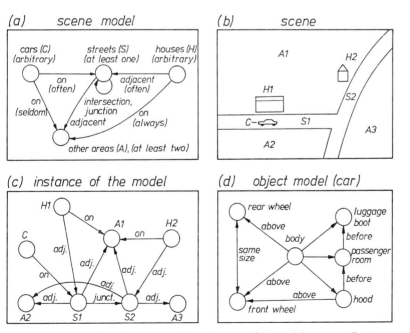

Fig. 4.12a–d. Illustration of a scene model, an instance of the model corresponding to a particular scene, and an object model

Nodes in the object model need, in turn, further definition until primitive nodes are obtained which can be recognized by low-level processing. Usually it is not sufficient to give a node just a symbolic label such as $A1$ or $H2$ in Fig. 4.12c, but rather one would like to include some descriptive information. For the node C in Fig. 4.12c this might be, for instance, an expected range of values for size, color, and velocity, and a pointer to the object model. Quantifiers like "often" or "seldom" may be defined by assignment of probabilities or possibilities (see Sect. 6.4). A possible representation of these models is by the semantic net formalism introduced in Sect. 7.4.

Another example of object models is given in Fig. 4.13. It shows the representation by a list structure of 2D models of industrial parts, one model per stable position of a part. A model M_κ consists of a name of the part and its stable position, and it has a list of simple constituents or segmentation objects. These are divided into a set sufficient for distinguishing this part from the other ones, and a set necessary for assembly purposes. A segmentation object has a type, a list of attributes (ALIST), a list of parameters (PLIST), and an estimated reliability.

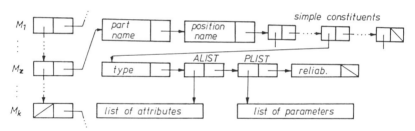

Fig. 4.13. An example of the representation of models of industrial parts [4.12]

This general model of an object, a scene, or something else presents an alternative to specification of structural information by a grammar. A grammar gives explicit rules for the generation and parsing of all allowable patterns of a given task domain Ω. A general model only specifies those properties which any instance of the model must have. The allowable patterns are thus defined implicitly. The model approach seems quite adequate for the scenes with streets, houses, and cars in Fig. 4.12 – at least it would not be easy to find a grammar generating all instances of the general model.

Having modeled objects, and perhaps also parts of objects by graphs and having a particular image it is then necessary to match the objects in the image with the model. As mentioned above it is assumed that at some level primitive nodes (and relations) may be recognized by low-level methods. The general problem then is to find a structural identity or a similarity between an image and a model. Approaches to this problem are given below.

4.5.2 General Recognition Strategies

There are three main approaches to general classification problems and various combinations and modifications of them. A varying subset of the transformations in (4.1.3-7) may be handled. Furthermore, there is a variety of distance measures, models, and descriptions. It is therefore not surprising that there is an ever growing number of different classification or matching algorithms. The three major approaches to classification are:

1. The bottom-up or *data-driven* approach:
   For all segmentation objects of a segmented image:
   Select a segmentation object (for example, a surface having certain attributes) and find those models M having a similar segmentation object.
   Increase a score for those models having this segmentation object (perhaps also decrease a score for those models not having this segmentation object).
   Assume that those objects are present in the image whose models have scores above a threshold.
2. The top-down or *model-driven* approach:
   For all models M_k:
   Select one model and try to find (or to verify) it in the image (for example, by means of the generalized Hough transformation, or by successively selecting primitive parts of the model and trying to find them in the image).
   Compute a score for the match between image and model.
   Assume that those objects are present in the image whose models have scores above a threshold.
3. The *hypothesize-and-test* approach, which is a combination of 1 and 2:
   Determine one or more initial hypotheses ($P_i \in M_\kappa, P_j \in I^r$), that is a model part which may correspond to a certain part of I^r. This may be done on the basis of a few gross attributes of model and image parts.
   Determine an estimate a'' of the unknown parameter vector a or only of some of the components of a. An approach is to try to align P_i and P_j.
   Search for compatible matches (P_k, P_l) having parameter values similar to a''.
   Select one or a few additional matches to compute missing components of a.
   Determine an improved estimate of a.
   Generate from a set of initial matches a complete list of matches between parts of model M_κ and the initial description I^r.
   If the number of matching pairs between the model M_κ and the initial description I^r is sufficiently high, assume that the object corresponding to the model M_κ is contained in the image.
   Remove the parts of the initial description matched by the above process from I^r.
   Repeat from the beginning until there are no more unprocessed parts in I^r.

Of course, the above descriptions of the three approaches to classification can only give a general idea. To find algorithms suited to special problems many details have to be provided. The above algorithms have one point in common: they do not give "one shot" solutions to the problem; instead they require a certain – and in complex problems a very large – amount of searching. This is not uncommon for classification problems, since in the problem stated in (4.1.1) a standard approach is to determine k test variables and search for the maximal (or minimal) value as indicated, for example, in (4.2.7, 8). So some amount of searching seems unavoidable in classification. An extremely important question is what is the minimal amount necessary to solve classification problems of a certain type. As mentioned above, it seems that there are no general solutions to this question so far.

4.5.3 Classification of 2D Images Using 2D Models

A common method for classifying arbitrary two-dimensional shapes is the *generalized Hough transform* which is summarized in the following steps:

1. Let the unknown parameters of the curve be represented by the vector

   $$a = (x_R, s, o) . \quad (4.5.4)$$

   In this equation

   $$x_R = (x_R, y_R) \quad (4.5.5)$$

   are coordinates of a reference point of the curve, s is a scale factor, and o an orientation.
2. Construct the so called *R-table*:
   2.1. Select fixed values of the scale s and orientation o.
   2.2. Choose a reference point x_R of the shape. A reasonable choice is usually the center of gravity.
   2.3. Let $\varphi(x_B)$ be the angle of the gradient direction at the boundary point x_B of the contour and let r be the radius from x_B to x_R.
   2.4. For each boundary point x_B compute $\varphi(x_B)$ and r and store r as function of φ. The result is the *R-table*.
3. If there are values of the parameters s and o other than those selected in step 2.1., then construct R-tables for those values by transforming the original R-table.
4. Initialize an accumulator array $H(x) = 0$, x corresponding to the unknown location parameter. Separately indexed arrays are initialized for all values of s and o.
5. For each boundary point x_I in the image, increment accumulator entries

   $$x = x_I + r \quad (4.5.6)$$

   where r is the entry in the R-table corresponding to gradient direction $\varphi(x_I)$.
6. Smooth the accumulator array.

7. Instances of the shape are expected at local maxima of the accumulator. The unknown parameters a in (4.5.4) follow from the index of the accumulator and the position of the local maximum.

The Hough transform is an element of many classification or matching algorithms. It can be seen that it achieves a solution to the problem of unknown parameters by searching over the set of possible values.

4.5.4 Classification of 2D Images Using 3D Models

Recognition of three-dimensional objects from two-dimensional images in the algorithm described below is accomplished by using the *viewpoint consistency constraint*, which states that all model components, when seen in an image, must be consistent with a projection from a single viewpoint. Starting with an initial guess of the viewpoint a least squares solution to the unknown parameters is obtained by Newton iteration. The algorithm works with the following main steps which may be viewed as an elaboration of the general hypothesize-and-test algorithm outlined in Sect. 4.5.2:

1. Provide a three-dimensional model of the objects by representing them by three-dimensional line segments. Attach a visibility criterion to each segment. This is a simple type of model but sufficient to demonstrate the steps of the algorithm.

2. Provide an initial segmentation of an image by finding line segments; this constitutes a particular example of an initial description I^r of pattern f^r. We omit any details of this processing step.

3. As outlined below, an important step is the computation of unknown viewpoint parameters from some initial values by Newton iteration. These initial values are computed from a few initial matches between image points and model points since most model points will only be visible from a certain range of viewpoints. For example, one may try to match a triple of closely adjacent image lines to a triple of closely adjacent model lines; this point is elaborated in some more detail below under the keyword "perceptual grouping".

4. Assume that the model point $u = (u_1, u_2, u_3)_t$ in three-dimensional model coordinates is translated by a vector $t = (t_1, t_2, t_3)_t$, then rotated by a rotation matrix R into a camera-centered coordinate system $v = (v_1, v_2, v_3)_t$ to give

$$v = R(u - t) . \qquad (4.5.7)$$

The vector v is then projected with a lens of focal length f_l to the two-dimensional image coordinates $x = (x, y)_t$ to yield

$$(x, y) = \left(\frac{f_l v_1}{v_3}, \frac{f_l v_2}{v_3} \right) . \qquad (4.5.8)$$

However, in the following it is more convenient to consider the equivalent equations

$$v = Ru$$

$$(x, y) = \left(\frac{v_1 f_l}{v_3 + T_3} + T_1, \frac{v_2 f_l}{v_3 + T_3} + T_2 \right),$$

$$t = R^{-1} \left(-T_1 \frac{v_3 + T_3}{f_l}, T_2 \frac{v_3 + T_3}{f_l}, -T_3 \right)_t .\qquad(4.5.9)$$

These equations describe the mapping from R^3 to R^2 and the unknown parameters R, T, and f_l. The rotation is viewed as the initial guess R, premultiplied by correction rotation matrices which represent small correction angles around the three orthogonal axes of the camera centered system v. It can be shown that there are simple expressions for the partial derivates of x and y with respect to the unknown seven parameters (three translations, three correction rotations, focal length if this is unknown, too).

5. For each model point corresponding to an image point, the model point is projected onto the image using the current parameter estimates. From the error between the projected image point and the actual image point a correction can be computed by Newton's method. If there are more correspondences than necessary for the computation of the unknown parameters, a least squares solution is obtained. The points are actually determined from perceptual groupings of line segments.

6. It is also possible to use corresponding lines in the image and model instead of corresponding points. An appropriate error measure here is the perpendicular distance of selected model points from the line in the image. This avoids the determination of end points of lines which in image processing is usually fairly prone to error. Also in this case an iterative solution for the unknown parameters is possible. In order to reduce the number of incorrect initial matches only correspondences having at least three line segments are actually used in the algorithm.

7. Extend the initial matches by projecting additional model lines onto the image and evaluate a score of their agreement. Use additional matches to further improve estimates of unknown parameters.

8. Assume that an object represented by some model is present in the image if the final viewpoint estimate is sufficiently overconstrained.

We mentioned above that *perceptual grouping* may be used to select initial matching pairs of model and image components. Such components should be invariant over a large range of viewpoints and should be sufficiently constrained in order to avoid accidental instances. Three important criteria for perceptual grouping are:

1. *Grouping by proximity*: The idea is that two points which are closely adjacent in the model will project to two points which are closely adjacent in the image. However, it is possible that for a certain viewpoint two widely separated model points are projected to closely adjacent image points. The

significance of two closely adjacent image points should vary inversely with the expected number of line endpoints in an image.
2. *Grouping by parallelism*: The idea is analogous to the idea of using proximity.
3. *Grouping by collinearity*: Again, the idea is analogous to the idea of proximity.

These three grouping criteria are reasonably robust with respect to variations of the viewpoint and, therefore, are suited for selecting a few initial matches. The complexity of the search for matching pairs can be reduced significantly by searching only in a neighborhood of a line endpoint which is proportional to line length.

As another example we present in some detail an algorithm for recognizing a 3D polyhedral object in a 2D image. It is assumed that the scale of the object, the orientation of the ground plane on which the object is sitting and the distance from the ground plane to the camera are known. The unknown parameters are the rotation and translation of the object in the ground plane. A point of the 3D object model where two or more edges of the object meet is called a *vertex*. By taking an image of the object a (3D) vertex is projected onto a (2D) *junction* in which some of the vertex edges may be visible and some may be invisible, depending on the viewpoint.

The idea of the classification algorithm is to use a bottom-up process for the estimation of the unknown parameters and a top-down process for their refinement. The bottom-up process compares all pairs of vertex and junction edges, computes an estimate of the parameters, and increments an accumulator array (Hough transform). To cope with processing errors the true value of the rotation angle is assumed to lie within a certain interval of the estimate of that angle. The maximum of the array is used as an initial parameter estimate. This is used by the top-down process to project the vertices of a chosen model and refine the parameter estimates by a least squares analysis.

The geometry of the imaging process is shown in Fig. 4.14. The object has a certain base plane which is parallel to the x_w, y_w-plane of the world coordinate system. For simplicity it is assumed that the object coordinate system of the object model is identical to the world coordinate system. A vertex v_M of the object model is first rotated by an angle α around the z_w-axis; this is represented by a rotation matrix R_o. Next the vertex is translated by $t_o = (x_0, y_0, 0)_t$. This gives the actual object vertex

$$v_o = R_o v_M + t_o . \tag{4.5.10}$$

Then the world coordinate system is transformed to the camera coordinate system by another rotation R_c and translation t_c to give the vertex in camera coordinates

$$v_c = R_c(R_o v_M + t_o) + t_c . \tag{4.5.11}$$

This transforms a vertex of the model, denoted by v_M, to a vertex of the actual object in camera coordinates, denoted by v_c. The transformed v_c is projected

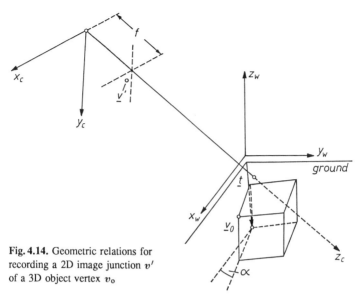

Fig. 4.14. Geometric relations for recording a 2D image junction v' of a 3D object vertex v_o

onto its image $v' = (v_1', v_2', f_l)$ with f_l the focal length of the imaging lens. If a is a constant containing depth information, the relation

$$v_c = av' \qquad (4.5.12)$$

holds, where a can be obtained from (2.2.50) to be $a = z_c/f_l$. It is assumed that R_c and t_c are known or are determined by camera calibration as outlined in Sect. 2.2.10; v_M is known from the model, and v' from the image (assuming that corresponding model and image vertices can be found); so the only unknown parameters are the rotation angle α and the translation t_o.

Next we consider a model edge e_M, $\|e_M\| = 1$, originating at v_M and the point $u_M = v_M + e_M$. This point is transformed to camera coordinates u_c and projected to the point u' in two-dimensional image coordinates. If e', $\|e'\| = 1$ is the image of e_M and k is a constant scaling e' to its actually observed length

$$u' = v' + ke' . \qquad (4.5.13)$$

If d is another constant containing depth information, one obtains from (4.5.12, 13) the relation

$$\begin{aligned} u_c &= d(v' + ke') \\ &= R_c(R_o(v_M + e_M) + t_o) + t_c \\ &= (a+b)v' + ce' . \end{aligned} \qquad (4.5.14)$$

In the last line of (4.5.14) the constants d and k have been substituted by new constants b and c; the constant a is the same as in (4.5.12). Now there is a pair of three equations (4.5.11, 12) relating a model vertex and an image junction and another set of three equations (4.5.14) relating a model edge and an image edge

These six equations contain six unknowns, that is the three position parameters x_o, y_o, and α, and the three imaging parameters a, b, and c.
The next step is to solve the above equations for the unknown parameters. First, insert (4.5.12) into (4.5.11) and solve for t_o. Denoting the three rows of R_c^{-1} by r_{1t}, r_{2t}, and r_{3t}, respectively, yields

$$x_o = a r_{1t} v' - r_{1t} t_c - (v_{M1} \cos \alpha - v_{M2} \sin \alpha) ,\qquad(4.5.15)$$
$$y_o = a r_{2t} v' - r_{2t} t_c - (v_{M1} \sin \alpha + v_{M2} \cos \alpha) ,\qquad(4.5.16)$$
$$0 = a r_{3t} v' - r_{3t} t_c - v_{M3} .\qquad(4.5.17)$$

The last equation can be directly solved for a, but the first two still contain the unknown angle α. Next subtract (4.5.11) from (4.5.14) to get

$$R_c R_o e_M = bv' + ce' ,\qquad(4.5.18)$$
$$R_o e_M = b R_c^{-1} v' + c R_c^{-1} e' .\qquad(4.5.19)$$

Rewriting (4.5.19) gives an equation of the form

$$\begin{vmatrix} \cos\alpha & -\sin\alpha & 0 \\ \sin\alpha & \cos\alpha & 0 \\ 0 & 0 & 1 \end{vmatrix} \begin{vmatrix} e_{M1} \\ e_{M2} \\ e_{M3} \end{vmatrix} = b \begin{vmatrix} w_1 \\ w_2 \\ w_3 \end{vmatrix} + c \begin{vmatrix} h_1 \\ h_2 \\ h_3 \end{vmatrix} .\qquad(4.5.20)$$

From the last row one obtains a relation between b and c. Since $\|e_M\| = 1, \|R_o e_M\| = 1$ and

$$\|bw + ch\| = 1 .\qquad(4.5.21)$$

Using the relation between b and c, both parameters can now be computed. Then the first two rows are solved for $\cos\alpha$. Finally the value of α can be used in (4.5.15, 16). This short outline demonstrates that the three parameters x_o, y_o, and α can be computed in closed form. An overview of a complete algorithm is given in Fig. 4.15.

4.5.5 Classification of 3D Images Using 3D Models

We will not present complete algorithms for this case, but only make a few remarks on this point. Some insight is gained by considering the last algorithm under the premise that a 3D model edge e is not compared to a 2D image edge but to a 3D image edge e', assuming now that 3D information is obtained already in image segmentation. This eliminates the projection part, but leaves the problem of determining the unknown parameters of object location. In general there are three translation and three rotation parameters. In principle, this can be solved by a Hough transform approach using a six-dimensional accumulator H. If some parameters are known in advance, this will reduce the size of the accumulator H. Another possibility is to index the model by position invariant features, for example, the relative angles of edges of a 3D vertex. In polyhedral objects having mostly orthogonal edges this will not give much discriminating power. However, there are other position invariant features such as the distance between two vertices or the area surrounded by a closed contour.

> Order models by importance or ease of recognition, select the first model
> initialize an accumulator array H(rotation, x,y-translation) with zeros
> initialize an empty set M of matching results
>
> BEGIN bottom-up phase:
> FOR each vertex v of the model and each junction v' of the image:
>> FOR each edge of v and each edge of v':
>>> compute rotation angle γ
>>> FOR each value $\gamma + \lambda \Delta \gamma$, $-m \leq \lambda \leq m$:
>>>> compute translation $\underline{t}$, weight w, and error ε
>>>> IF the vertex v is visible under γ, $\underline{t}$, THEN update
>>>> $M = M \cup (\gamma, \underline{t}, w, \varepsilon)$
>>> take the best element $(\gamma, \underline{t}, w, \varepsilon) \in M$ to increment accumulator
>>> $H(\gamma, \underline{t}) = H(\gamma, \underline{t}) + w$
>
> select γ, $\underline{t}$ with maximum weight as current parameter estimate;
> IF multiple occurrences of one object model are possible,
> THEN take all values γ, $\underline{t}$ having weight above a threshold as the
>   set of current parameter estimates.
>
> END bottom-up phase, BEGIN top-down phase
>
> FOR each vertex v of the model and each element in the set of current
> parameter estimates:
>> perform coordinate transformation of model vertex
>> IF the vertex is non-occluded, THEN project it, ELSE try another one
>> FOR each image junction v' which is close to the projected vertex:
>>> compute error $\varepsilon(v, v')$ as minimum angular error between
>>> projected vertex edges and image junction edges; compute weight
>>> as the number of matched image edges
>> match vertex v to junction v' yielding largest weight;
>> IF there are several v' having the same weight,
>> THEN match to the junction having least error $\varepsilon(v, v')$
>
> use the correspondences (v, v') to determine an improved current
> estimate of parameters by a least square procedure
> iterate top-down procedure until termination criterion is met
>
> IF the number of successful matches (v, v') is above a threshold,
> THEN assume that an instance of a model is detected in an image.
> Note that if there are several elements in the set of current parameter
> estimates, multiple detection of an object at different locations
> is possible.
>
> tag the image junctions used in successful matches;
> IF there are more than a prespecified number of untagged junctions,
> THEN start a new bottom-up phase using the next model and the untagged
>   junctions,
> ELSE END algorithm.

Fig. 4.15. An overview of an algorithm for the recognition of objects using 3D object models and 2D image data

4.5.6 Structural Matching

In *structural matching* a pattern is transformed to some symbolic structure and so is the image. Since a graph is a very general symbolic structure we only consider this case in the following. The problem then is to find an identity or a similarity between a model represented by a graph G_1 and an image represented by G_2, or also to look for an object in a larger image. In the first case one has to find an *isomorphism* between G_1 and G_2. Two graphs are isomorphic if there is a one-to-one node mapping between them which preserves a one-to-one correspondence of edges. In the second case one has to find a subgraph in the image which is isomorphic to the model graph. A graph G' is a *subgraph* of a graph G if G' can be obtained from G by removing some nodes from G and all edges adjacent to the removed nodes.

Testing for isomorphism in general is very expensive, but it is facilitated by the fact that nodes and edges of a graph representing an object have certain properties. Let two graphs G and G' have node sets V and V', and edge sets E and E', respectively. If a node $v_i \in V$ is considered, not every node $v'_j \in V'$ is a candidate for correspondence to v_i. If, for example, v_i has the descriptive information or the attributes "red circular region" attached, then only nodes $v'_j \in V'$ which also represent red circular regions may correspond to v_i. A similar argument holds for edges. This makes it possible to reduce the amount of searching necessary to prove or disprove an isomorphism. Obviously, this idea also applies to larger entities. If there are two nodes $v_i, v_j \in V$ related by an edge $e_\mu \in E$ and having the descriptive information that "a red circular region (v_i) is surrounded by (e_μ) a large blue area (v_j)", then only triples of G' with the same descriptive information or labeling can correspond to (v_i, v_j, e_μ).

An algorithm using the above considerations employs the following steps. The first step is set generation. It is based on the principle that if graphs G and G' are isomorphic, then a subset of nodes of G having some property must correspond to that subset of nodes of G' having the same property. Such properties may be attributes of a node, the type of edges of a node, the number of edges of a node (its degree), the least number of edges leading from a node back to itself (the order), and the nodes adjacent to a node (connectivity). If the number of nodes of G having some property is different from the corresponding number of G' (cardinality violation), then G and G' are not isomorphic. With appropriate modifications of the cardinality condition it is also possible to use these principles for testing whether G contains a subgraph isomorphic to G'. The second step is set partitioning. Let V_j and V'_j be corresponding subsets of nodes according to the above criteria and v_i, v'_i corresponding nodes of G and G'. Then it must hold that

$$v'_i \in \bigcap_{\{j | v_i \in V_j\}} V'_j \ . \tag{4.5.22}$$

This relation expresses the obvious fact that if v_i is in several subsets V_j, then v'_i must also be in the corresponding V'_j. Furthermore, if there are two corresponding subsets V_k, V'_k having the same number of nodes, then the elements of both subsets must be isomorphic images or

$$V_k = V'_k \quad \text{and} \quad \overline{V'_k} \supseteq \overline{V_k}, \qquad (4.5.23)$$

where $\overline{V}_k$ is the complement of V_k. So if $V_k = V'_k$ and $v_i \notin V_k$, then $v'_i \in \overline{V'_k}$. In this case a refinement of (4.5.22) to

$$v'_i \in \left\{ \bigcap_{\{j | v_i \in V_j\}} V'_j \right\} \cap \left\{ \bigcap_{\{k | v_i \notin V_k \wedge V_k = V'_k\}} \overline{V'_k} \right\} \qquad (4.5.24)$$

is possible. With (4.5.22 and 24) it is possible to reduce the number of elements in corresponding subsets. Applying set generation and partitioning alternately will result in one of three possibilities. First, there may be pairs of corresponding sets containing one node each in which case an isomorphism is found, or, second, a cardinality violation may occur in which case an isomorphism is disproved, or, third, no more new sets can be generated before the first or second case occurs. The last case may occur if either more than one isomorphism is possible or the properties for set generation are not sufficient to establish an isomorphism. A way out is to systematically try arbitrary assignments of nodes having more than one possible correspondent node.

The above algorithm requires a perfect match between object and model, but there will be numerous cases where such a requirement is too restrictive. Due to processing errors and/or occlusions from other objects some parts of an object in an image may be missing or wrongly recognized. In these cases it is reasonable to look for more flexible approaches to matching. There are three basic approaches to achieve this. The first is to investigate the possible or the most frequent types of errors and occlusions and to provide models of objects containing these errors and occlusions. This would leave the above matching procedures unchanged. The second approach is to develop algorithms which tolerate a certain amount of errors and/or occlusions during matching. This would leave the models unchanged. The third approach is to include possible errors as alternatives in the initial structural description of the image and consider all alternatives during matching. An example is an algorithm which finds the largest substructures in a given image corresponding to the largest substructures of a given object model. If the image and the object model are represented by the graphs G_1 and G_2, respectively, this means that one looks for the largest subgraphs G'_1 and G'_2 of G_1 and G_2, respectively, such that G'_1 and G'_2 are isomorphic. Several authors note that problems of this type may be handled only by hierarchical descriptions, for instance, in the form of h graphs, because otherwise the combinatorial explosion makes the computations unfeasible. A hierarchical description, where the high levels contain a few gross features like size, location, shape, or color, allows one to select only those substructures which match these gross features without considering all the details.

As an example we describe one approach to finding largest substructures. The basic idea of exploiting the descriptive information of nodes and edges, which was already mentioned in connection with (4.5.22–24), is used here, too. Let the graphs G_1 and G_2 have node sets V_1 and V_2, respectively. An *assignment* is a pair (v_1, v_2), $v_1 \in V_1$, $v_2 \in V_2$, where v_1 and v_2 have the same descriptive information. Two assignments (v_1, v_2) and (v_1', v_2') with $v_1, v_1' \in V_1$ and $v_2, v_2' \in V_2$ are *compatible* if, in addition, all relations between v_1 and v_1' also hold for v_2 and v_2', that is, the edges occurring between v_1 and v_1' in G_1 must be the same as those between v_2 and v_2' in G_2. Note that, of course, the special case of no edge between v_1 and v_1', v_2 and v_2' is included. Now define a new graph, the *assignment graph* G_a, by the following procedure. The set of nodes V_a of G_a is the set of assignments. Two nodes in V_a are connected by an undirected edge if the corresponding assignments are compatible. The largest substructures are then given by the *maximal cliques* of G_a. A clique is maximal if no other clique properly includes it, and the cliques of G_a are simply the totally connected subgraphs of G_a. A graph is totally connected if every node has edges to all the other nodes. It should be noted that the use of h graphs for describing patterns and models allows one to start the matching on a high level, where the number of nodes is small, and only to go into details, that is, replace a nonprimitive node by its content, if a reasonably large substructure can be found at the high level.

Finding the cliques of a graph is a well-known problem in cluster analysis. A straightforward algorithm for finding a maximal clique G_a' in G_a is the following, where V_a' is the node set of G_a':

1. Take an arbitrary node $v_{aj} \in V_a$ and set $V_a' = \{v_{aj}\}$.
2. Search in $V_a - V_a'$ for a node v_{ak} which is connected to all the nodes in V_a'. Add v_{ak} to V_a'.
3. Repeat step 2 until no new node v_{ak} can be found. Then V_a' is the node set of a maximal clique G_a'.

Another common structural representation results from the relational data model (see Sect. 5.2). The problem of matching relational representations is treated briefly in Sect. 7.5.

4.5.7 Line Drawings

The analysis of scenes containing polyhedral objects, for example, those shown in Fig. 4.16b, may be reduced to recognizing the straight line contours of such objects. Neglecting shadows and requiring the objects to be trihedral there are four types of lines as indicated in Fig. 4.16. They are boundary lines with two possible directions and interior lines which are either concave or convex. A boundary line is a line separating different objects and its direction is defined such that the object is to the right when walking in this direction. An interior line separates different regions of the same object. Whereas object B on A in Fig. 4.16b is trihedral since there are only vertices where three object surfaces meet, object D on A is not. In a trihedral object there are only four possible vertices as shown in Fig. 4.17a, and in a nontrihedral object, additional vertices, such as those in Fig. 4.17c, may occur. For example, the three lines meeting in the

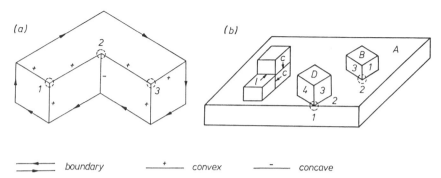

Fig. 4.16a,b. Scenes containing polyhedra and possible line labelings for trihedral objects without shadows

fork-type vertex have $4^3 = 64$ combinatorially possible labelings, but since this vertex corresponds to a three-dimensional object, only the five labelings shown in Fig. 4.17b are physically possible. A similar reduction of combinatorially possible labelings results for the other vertices. Because a line connecting two vertices must have only one label, neighboring vertices mutually constrain possible labels. For example, even without any assumption about the boundary of the object in Fig. 4.16a there is only the unique labeling of vertices 1, 2 and 3 shown in the figure. This labeling is obtained more easily if it is assumed that the direction of the boundary may be determined first. A possible labeling of the lines may be obtained by a discrete relaxation process (see Sect. 4.2.5).

The number of possible labelings and vertices is increased if nontrihedral objects and pseudocontours are allowed. Pseudocontours may result from shadows, which are not shown in Fig. 4.16b, or from cracks, which are the lines marked c in the figure. The concave line l in the figure may be labeled further according to the resulting boundaries if the three objects meeting at l are separated. Finally, the

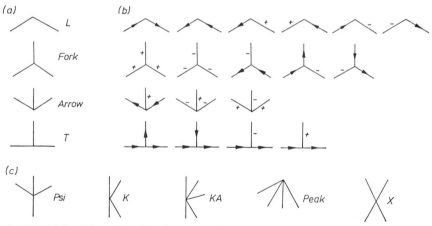

Fig. 4.17. (a) Possible vertices for trihedral objects; (b) phyiscally possible labelings of vertices. Other types of vertices as shown in (c) are possible for nontrihedral objects

consideration of different lighting conditions, such as illuminated, shadowed by another object, and self-shadowed, increases the number of labelings still further. For instance, the lines of an arrow-type vertex have about 10^5 combinatorially possible labelings, but only 70 are physically possible. Again, knowledge about the objects gives a drastic reduction of labelings.

Additional complexity is introduced by allowing curved lines resulting, for example, from cylinders or cones. This case is omitted here. A special problem are "impossible objects", that is line drawings which look like a 3D object, but which cannot be obtained from projecting any 3D object. There are necessary and sufficient conditions for a line drawing to represent a 3D object.

4.6 Neural Nets

4.6.1 Basic Properties

The term *neural net* or *artificial neural network* stands for a computational structure for modeling the properties and behavior of brain structures, in particular, self-adaptation, learning, and parallel processing. Synonyms are *connectionist models, parallel distributed processing models*, and *neuromorphic systems*. They provide densely interconnected, massively parallel structures of simple computational elements.

A neural net consists of a set of nodes and a set of interconnections between nodes. A node contains a computational element taking inputs from incoming interconnections (input links) and providing outputs to outgoing interconnections (output links). A simple, yet typical computational element takes the weighted sum of the input links, subtracts a threshold value, and passes the result through a nonlinearity; additional complexity may be introduced, for example, by temporal integration or delays. The output of the nonlinearity is then passed to output links. The behavior of such a net is determined by the computational elements, in particular the weights, thresholds, and nonlinearity, as well as the available interconnections or the *net topology*. Usually, the net topology and type of nonlinearity are specified in advance and are fixed. The weights and thresholds may be adjusted by a training algorithm to achieve some desired behavior of the net. Three main types of behavior are the *associative memory* (or content-addressable memory), the classification or *recognition* of patterns, and the *vector quantization* of patterns. In the next sections we give some examples of common net topologies.

4.6.2 The Hopfield Net

This type of net can be used as an associative memory or classifier for binary input patterns. It has M nodes containing computational elements with a step nonlinearity. The operation of this net is summarized in the following algorithm:

1. Let the binary input pattern be $f_i, i = 0, 1, \ldots, M - 1$, with $f_i \in \{+1, -1\}$.
   Let there be k prototype patterns $\boldsymbol{f}_\kappa, \kappa = 1, 2, \ldots, k$.
   Let there be M nodes.
   Let the connection between node i and node j have weight w_{ij}.
2. Compute the connection weights according to

$$w_{ij} = \sum_{\kappa=1}^{k-1} f_{\kappa i} f_{\kappa j} \quad \text{if} \quad i \neq j ,$$

$$w_{ij} = 0 \qquad \text{if} \quad i = j , \quad 0 \leq i, j \leq M - 1 . \tag{4.6.1}$$

3. Initialize the net output by

$$h_i(0) = f_i , \quad i = 0, 1, \ldots, M - 1 . \tag{4.6.2}$$

4. Update the output in the mth iteration step by

$$h_j(m+1) = \theta_s \left[\sum_{i=0}^{M-1} w_{ij} h_i(m) \right] , \quad 0 \leq j \leq M - 1 ,$$

$$\theta_s[x] = +1 \quad \text{if} \quad x \geq 0 \quad \text{and} \quad -1 \quad \text{if} \quad x < 0 . \tag{4.6.3}$$

5. Stop if $\boldsymbol{h}(m+1) = \boldsymbol{h}(m)$ and take $\boldsymbol{h} = \boldsymbol{h}(m+1)$ as the output $\boldsymbol{h}$ of the net; otherwise repeat step 4.

It can be shown that the above algorithm converges or stops after a finite number of iterations. If a distorted or truncated version of a prototype pattern is given as an input, the output is considered to be the restored pattern. In this sense the net can be used as an associative memory retrieving a complete output pattern when given an incomplete input pattern. It can be used as a classifier if in addition a distance between the output pattern and the prototype patterns is computed. If the distance is zero or sufficiently small to some prototype pattern $\boldsymbol{f}_\kappa$, then the input is assumed to belong to class Ω_κ, otherwise it is rejected. The number k of classes or prototype patterns should be less than about 0.15 times the number M of nodes in order to guarantee proper retrieval. A stability problem may occur if the prototype patterns have many pixels in common. In this case the output of the net may converge to another prototype if a similar prototype pattern is given as an input. Since the number of weights or interconnections is M^2, this type of net becomes fairly involved even for moderately large input patterns.

4.6.3 The Hamming Net

This net uses a number of weights which grows linearly with the number M of input pixels. It implements the optimal (minimum error) classifier for binary inputs using two subnets, called the input subnet and the output subnet. The optimal classifier for binary patterns which may be corrupted by independent random reversion of bits computes the Hamming distance to each prototype and selects the prototype having minimum Hamming distance. The Hamming net computing this prototype is characterized by the following algorithm.

1. Denote the input and prototype patterns as in step 1 of the algorithm in Sect. 4.6.2. Denote the output by d_κ in association with the decision function in Sect. 4.2.
2. Determine the weights $v_{\kappa i}$ and the offset v_κ of the input subnet by

$$v_{\kappa i} = \frac{f_{\kappa i}}{2}, \quad v_\kappa = \frac{M}{2}, \quad 0 \leq i \leq M-1, \quad 1 \leq \kappa \leq k. \quad (4.6.4)$$

3. Determine the weights $w_{\kappa\lambda}$ in the output subnet by

$$w_{\kappa\lambda} = \begin{cases} 1 & \text{if } \kappa = \lambda \\ -\varepsilon & \text{if } \kappa \neq \lambda \end{cases}$$

$$\varepsilon < 1/k, \quad 1 \leq \kappa, \lambda \leq k. \quad (4.6.5)$$

4. Initialize the output

$$d_\kappa(0) = \theta_t \left[\sum_{i=0}^{M-1} v_{\kappa i} f_i - v_\kappa \right] \quad 1 \leq \kappa \leq k,$$

$$\theta_t[x] = x \quad \text{if} \quad x \geq 0, \quad \text{and} \quad 0 \quad \text{if} \quad x < 0. \quad (4.6.6)$$

5. Compute the output at step $m+1$ from

$$d_\kappa(m+1) = \theta_t \left[d_\kappa(m) - \varepsilon \sum_{\kappa \neq \lambda} d_\lambda(m) \right], \quad 1 \leq \kappa, \lambda \leq k. \quad (4.6.7)$$

6. Stop if $d(m+1) = d(m)$, otherwise repeat step 5.

This net only needs $(M+k)k$ weights. According to experiments after a few iterations (typically about 10) all d_κ are zero, except for one which corresponds to the prototype of the input.

4.6.4 The Multilayer Perceptron

The perceptron is a structure which accepts an input having continuous values and computes an output which may be interpreted as a class name. It consists of one or more layers of computational elements; therefore, the *single layer perceptron* and the two layer, three layer, and in general *multilayer perceptron* (MLP) are distinguished. There is theoretical and experimental evidence that a three-layer perceptron is sufficient in the sense that it can define arbitrary

202   4. Classification

decision regions in R^n. The structure of a three-layer perceptron is shown in Fig. 4.18. It consists of M_0 input elements, two layers containing M_1 and M_2 hidden computational elements, respectively, and an output layer containing M_3 computational elements. The computation in every layer is

$$f_j^{(l+1)} = \Theta\left[\sum_{i=0}^{M_l-1} w_{ji}^{(l+1)} f_i^{(l)} - w_j^{(l+1)}\right], \quad 0 \leq j \leq M_l - 1, \quad l = 0, 1, 2$$

$$= \Theta\left[y_j^{(l+1)}\right]. \tag{4.6.8}$$

The nonlinearity is chosen as

$$\Theta(y) = \frac{1}{1 + \exp[-y]} \tag{4.6.9}$$

having the continuous derivative

$$\frac{d\Theta(y)}{dy} = \Theta(y)[1 - \Theta(y)]. \tag{4.6.10}$$

A training algorithm is available to determine the weights of the multilayer perceptron, the so-called *back-propagation training algorithm*. It minimizes the mean square error between the actual output $f^{(3)}$ in the output layer and a desired or ideal output δ. Since δ must be given, it is a *supervised learning* algorithm. In fact, the training procedure is not limited to a three-layer perceptron, as will become evident. The weights are initialized by small random numbers, and the weight $w_j^{(l+1)}$ is treated as corresponding to an additional component $f_{M_l}^{(l)} = 1$. This way a uniform treatment of weights is possible. Each weight is updated according to the iteration formula

$$w_{ij}^{(l)} \leftarrow w_{ij}^{(l)} - \beta \frac{\partial \varepsilon}{\partial w_{ij}^{(l)}}$$

$$= w_{ij}^{(l)} + \Delta w_{ij}^{(l)}, \tag{4.6.11}$$

where ε is the output error

$$\varepsilon = 0.5 \sum_{n=0}^{M_3-1} \left(\delta_n - f_n^{(3)}\right)^2 \tag{4.6.12}$$

and β is a constant factor, called the gain factor or the learning rate, which is adjusted experimentally. The problem then is to compute the partial derivative of the error in all layers. In the top layer $l = 3$ one obtains

$$\frac{\partial \varepsilon}{\partial w_{ij}^{(3)}} = \frac{\partial \varepsilon}{\partial f_j^{(3)}} \frac{\partial f_j^{(3)}}{\partial y_j^{(3)}} \frac{\partial y_j^{(3)}}{\partial w_{ji}^{(3)}}$$

$$= -\left(\delta_j - f_j^{(3)}\right)\left(1 - f_j^{(3)}\right) f_j^{(3)} f_i^{(2)}$$

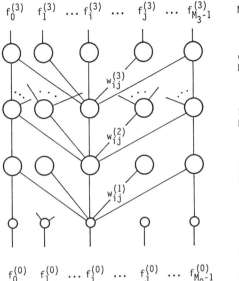

Fig. 4.18. The structure of a three-layer perceptron

$$= -d_j^{(3)} f_i^{(2)} . \tag{4.6.13}$$

This gives the correction term

$$\Delta w_{ji}^{(3)} = \beta d_j^{(3)} f_i^{(2)} . \tag{4.6.14}$$

The next step is to compute the partial derivatives for the hidden (non-output) layers. The result is that the d-terms have the form

$$d_j^{(l-1)} = \sum_{k=0}^{M_l-1} d_k^{(l)} w_{kj}^{(l)} \left(1 - f_j^{(l-1)}\right) f_j^{(l-1)} , \tag{4.6.15}$$

and the correction terms have the form

$$\Delta w_{ij}^{(l)} = \beta d_j^{(l)} f_i^{(l-1)} . \tag{4.6.16a}$$

Taking an input pattern the resulting output is computed first by (4.6.8), then the correction term of the weights is computed for the output layer by (4.6.14) assuming that the desired output δ is known. From the output layer the correction proceeds backward to the input layer by (4.6.15, 16a), hence the name back-propagation algorithm. This is repeated for many patterns until the desired output coincides with the actual output or until the weights stabilize sufficiently. The use of indices to denote individual training patterns or iteration steps has been avoided in the above equations in order to keep the notation simple. A modification is the addition of a *momentum term* in (4.6.16a) to obtain the weight correction in the Nth iteration by

$$\Delta w_{ij,N}^{(l)} = \beta d_{j,N}^{(l)} f_{i,N}^{(l-1)} + \alpha \Delta w_{ij,N-1}^{(l)} \:. \qquad (4.6.16b)$$

The momentum term may speed up the learning process. The actual speed of convergence of the weights is heavily dependent on the values of β and α.

4.6.5 The Feature Map

The *feature map* is an approach to *unsupervised learning* of similarities between feature vectors. It may also be viewed as a *clustering algorithm* or a *vector quantizer* (Sect. 3.10.4). The idea is to map a set of feature vectors to a set of outputs arranged in a two-dimensional grid. Feature vectors belonging to the same class are mapped to the same output, feature vectors belonging to similar classes are mapped to neighboring outputs. However, the classes are not specified from outside (supervised learning), but are obtained unsupervised by a clustering algorithm. The mapping from input to output values is by a structure similar to a single layer perceptron; training again means adjustment of weights. Since there is only one layer the output y_j is computed from the input f_i by

$$y_j = \sum_{i=0}^{M-1} w_{ji} f_i \:, \quad 0 \le j \le M' - 1 \:, \qquad (4.6.17)$$

if we assume M input and M' output values. The goal is to make the weight vector of an output as similar as possible to the input vector, for example, in the sense

$$\varepsilon = \sum_i (f_i - w_{ji})^2 = \min \:. \qquad (4.6.18)$$

Using the same approach as in (4.6.11) results in the equation

$$w_{ji} \leftarrow w_{ji} + \beta (f_i - w_{ji}) \:. \qquad (4.6.19)$$

In order to achieve the above mentioned properties a neighborhood V_j of the output y_j is defined, for example, a square or a hexagonal region of the grid centered on y_j. If a new input $\boldsymbol{f}$ is presented, at first the most similar weight vector $\boldsymbol{w}_j$ is determined by

$$\text{IF} \quad \left(|\boldsymbol{f} - \boldsymbol{w}_j|^2 = \min_{\{\mu\}} |\boldsymbol{f} - \boldsymbol{w}_\mu|^2 \right) \:,$$

THEN (select weight vector $\boldsymbol{w}_j$ and neighborhood V_j) . $\qquad (4.6.20)$

Updating of the weight vectors is done by (4.6.19) for all $\boldsymbol{w}_\mu \in V_j$. The training algorithm starts with a large neighborhood of an output node and reduces its size with increasing number of iterations. Experimental evidence shows that it is useful to start with neighborhoods of 70–300 outputs and a $\beta \approx 0.5$–0.99 and reduce the neighborhood to 3–9 outputs and $\beta \approx 0.01$–0.05 in the last phase of the iteration.

In principle it is possible to use more general functions and criteria than (4.6.17, 20). The main point is the updating of weights in a neighborhood of the selected output and the proper choice of β, V_j and their reduction with time (or number of iterations).

4.7 Bibliographical Remarks

There are quite a few textbooks treating classification of feature vectors in depth and detail, some examples are [4.2–5, 8, 13, 14]. An early example of the application of statistical decision theory is [4.1]. The nearest neighbor rule was introduced in [4.7], a comprehensive text is [4.115], an approach to sample size reduction is [4.16] and an efficient branch-and-bound algorithm for computing nearest neighbors is given in [4.17]. The problem of image template matching is treated in [4.18–20]. Relaxation processes are discussed in [4.21, 22]. The learning of parameters of numerical classifiers is treated, for example, in [4.5, 23, 24]. Some topics not treated in Sect. 4.2 are piecewise linear classifiers [4.25], hierarchical classifiers [4.26–28], sequential classifiers [4.29], and the use of context [4.5, 30].

Classification of symbol strings by syntactical methods (or syntactical classification) was introduced in [4.31] and is treated in detail in textbooks like [4.5, 32–34]. Work in this direction includes, for example, [4.35–40]. In addition, the problem of grammatical inference, that is the automatic construction of a suitable grammar from the observation of a finite set of terminal strings has found considerable interest, e.g. [4.41–45].

The problem of word recognition using hidden Markov models is treated in many publications. A thorough introduction to all aspects of HMM, including an extensive bibliography, is given in [4.46]. Additional material is provided, for example, in [4.10, 47, 48] and in part II of [4.49]. The Viterbi algorithm is presented in [4.50]. Dynamic programming is treated in publications like [4.51–54].

Early approaches to three-dimensional object recognition are [4.55, 56]. This field has found increasing interest recently in publications like [4.57–61]. The generalized Hough transform is treated in [4.62], and approaches to reduce computational complexity are given in [4.63, 64]. The two examples of recognition algorithms are according to [4.65, 66]. The automatic acquisition of object models is treated, for example, in [4.12, 4.67–69]. Algorithms for graph isomorphisms and determination of cliques are given in [4.70–73]. Interpretation of line drawings is treated in [4.55, 74–79].

Early work in the direction of neural nets is covered in [4.80–82] and recent advances in [4.83–85] or the special issue [4.86]. The backward-propagation algorithm is developed in [4.85] and the feature map in [4.87]. Other applications are treated, for example, in [4.88, 89]. A special type of semantic network is encoded by a newal network in [4.90].

5. Data

The result of recording a pattern is a vector or array of sample points. This kind of data has a straightforward representation. The result of preprocessing is usually another vector or array, so that in these cases there would be no reason to talk about data at all. However, there are more complicated cases, too. For instance, the split and merge algorithm of Sect. 3.5.4 made explicit use of a picture tree. This shows that already at this stage of processing more sophisticated structures become necessary, or, at least, useful. Thinking of further stages one might be interested in representing results (for instance, hypotheses) of analysis, in linking results to other results or to pattern sample values, and in deleting results which are not needed any longer. One also might be interested in representing information about structural properties or the task domain. It seems that it will be very convenient, to say the least, if one has the possibility of building general structures, for example, of the type in (3.1.2, 4.5.1, 7.1.1), and of changing them as analysis progresses. Storage of a sample ω of patterns for experimental or archival purposes is another point. It is not useful to store just arrays of points, rather these arrays should be accompanied by additional information facilitating the retrieval of particular patterns. For instance, one might wish to inspect all images containing a certain object or all utterances with at least 10 vowels. Obviously this requires a good deal of data management.

In this chapter some basic techniques of representing and handling data are discussed; they are:

1. *Data Structures* – organization and representation of data.
2. *Data Bases* – organization and manipulation of large amounts of shared data.
3. *Pattern Data* – some examples from pattern analysis.

5.1 Data Structures

5.1.1 Fundamental Structures

Data are a collection of facts and statements represented by numbers or symbols, regardless of whether they are useful or not. Useful data are often termed *information*. A *data structure* may be viewed as exhibiting relations between data, and an *abstract data type* in addition specifies allowed operations on data. This intuitive notion may be replaced by a precise mathematical model, for instance, a directed graph with assignment functions or an algebraic axiomatic definition. In the following an informal treatment of some common data structures is given.

Data used for pattern analysis are represented and manipulated on some kind of digital computer which has some simple types of data "built in". Usually these simple types are numbers like integer or real numbers taking values specified by the computer word length, logical variables taking values true or false, and characters specifying a set of print and control characters (for instance, $A, B, \ldots, Z, 0, 1, \ldots, 9, \%, +, =, \ldots$, backspace, carriage return, end of text, ...). These simple types may then be used to build more complex structures.

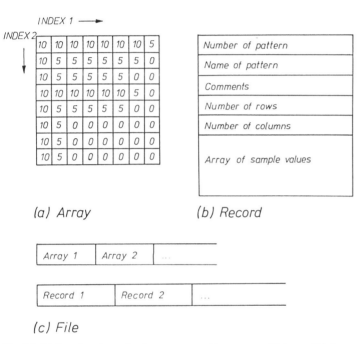

(a) Array (b) Record

(c) File

Fig. 5.1. (a) Sample values of an image are stored in an array of integers; this is convenient for doing numerical computations. (b) The array and additional information are summarized in a record. (c) A sequence of arrays, records, or other data structures is sequentially arranged in a file

A widely used structure is the *array*. It consists of components which are all data of the same type. All components may be accessed arbitrarily by specifying an index which gives the position of the component within the array; access time is independent of the index value if the array is in the working storage. By allowing several independent indices a multidimensional array of data results. For instance, a two-dimensional array whose components are integer numbers is adequate to represent sample values of an image as shown in Fig. 5.1a. The array is a data structure required in any pattern analysis system.

It is convenient to have a data structure containing data of various types, for instance, real numbers, characters, arrays of integers, and so on. This structure is called a *record*. It may be used to summarize different data about a pattern as indicated in Fig. 5.1b. Data in the record are referenced by the name of the respective components, not by an index number. Similar to arrays, random access of components by use of their names is possible.

A data structure allowing only sequential access to its components is a *file*. Its component data may consist of one of the above-defined types, for instance, a sample of patterns as shown in Fig. 5.1c. The access mechanism is viewed as a *pointer* referencing one component. Only the referenced component may be accessed. The pointer may be advanced or backspaced by one component or returned to the first component.

Other structures may be developed and are, in fact, available in some programming languages. The common characteristic of these so-called fundamental structures is that they are specified and attributed to physical storage in a prespecified way (but not necessarily always to the same storage addresses) and only their values may change during the run time of a program.

5.1.2 Advanced Structures

In many cases it is useful to have more complex structures available as well as to have structures which may change dynamically during program execution. These are summarized here as advanced structures although the border is not clear cut. For instance, a tree structure may also be represented by an array as shown below, but this allows no dynamic changes and the representation by pointers seems to be more elegant. Advanced structures should also allow recursive definitions.

A basic structure is the *set* consisting of an unordered collection of elements of some type. Elements may be added or removed from the set during the execution of a program.

If a set of data types, such as arrays or records, is not to be arranged in a file, an alternative is the *linear list* shown in Fig. 5.2. It may be viewed as an ordered set of data types. The first data type is marked by a pointer, and the next is also referenced by a pointer. This allows creation of new list elements at run time, their insertion at arbitrary locations of the list, and the deletion of list elements. Special cases result if elements are always added or deleted at the beginning of a list – called a *stack* operation – or if elements are always added at the end and deleted at the beginning – called a *queue* operation. In addition, the content of elements in a list can be modified, but this operation may also be realized by deleting the old element and inserting the new one. Only the type of the list elements is prespecified, but not their number since elements may be created at the time of program execution. A recursive definition of a possibly empty list with elements of data type D would be that a list is either empty or it is the linkage of a data type D with a list of elements D.

A more general arrangement of data types is the *tree*, shown in Fig. 5.3. A tree is a finite set of *nodes* of data type D such that there is a special node called the

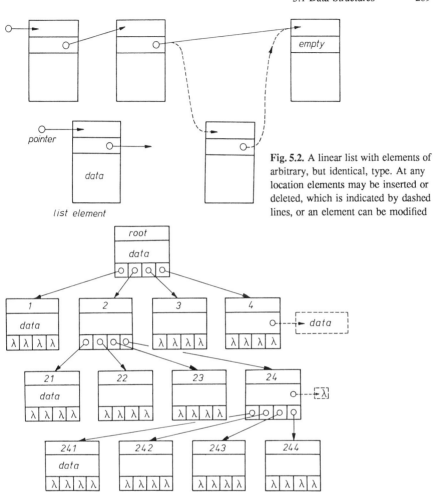

Fig. 5.2. A linear list with elements of arbitrary, but identical, type. At any location elements may be inserted or deleted, which is indicated by dashed lines, or an element can be modified

root (1, 2 (21, 22, 23, 24 (241, 242, 243, 244)), 3, 4)

Fig. 5.3. The tree structure of Fig. 3.22 represented by nodes which contain an identifying key (region number), data about the region, and pointers to subregions. Dashed lines indicate the alternative of pointing from the node to the required data. The symbol λ denotes "empty"

root and the remaining nodes are partitioned into disjoint subsets where each of these are trees called the *subtrees* of the root. Again, this is a recursive definition. The *root* and the *leaf* are two special node types having no predecessor node and no successor node, respectively. It is evident that in image segmentation storage of a segmentation tree is only advisable if nodes do not need to be of exactly the same type. Otherwise nodes "root", 2, and 24 in Fig. 5.3 would contain redundant data. An alternative is to replace the data field of the nodes by pointers to data fields which may be defined appropriately for each level of the tree. Some basic operations on a tree are inspection of all of its nodes, insertion and deletion of nodes, and searching for particular elements in the tree. Special trees are the *binary tree* where each node may have at most two successor nodes, and the *balanced tree* where a path from the root to a leaf always has the same length.

A generalization of a tree is a *graph G*. It consists of a set V of *nodes* or *vertices* and a set $E, V \times V \supset E$, of *edges* or *branches* between nodes. An element $e_{ij} \in E$ is an ordered pair

$$e_{ij} = (\text{node } i, \text{node } j) \tag{5.1.1}$$

indicating that an edge is directed from node i to node j. If the edges are to be undirected, this may be expressed by including an edge

$$e_{j,i} = (\text{node } j, \text{node } i) \, . \tag{5.1.2}$$

Graph G:

(a) $E = \{(1,2),(1,3),(2,3),(3,5),(3,4)\}$
     $V = \{1,2,3,4,5\}$

(c)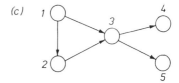

(b) $\underline{M} = \begin{matrix} & j \rightarrow \\ i \downarrow & \begin{bmatrix} 0 & 1 & 1 & 0 & 0 \\ 0 & 0 & 0 & 1 & 0 \\ 0 & 0 & 0 & 1 & 1 \\ 0 & 0 & 0 & 0 & 0 \\ 0 & 0 & 0 & 0 & 0 \end{bmatrix} \end{matrix}$

(e)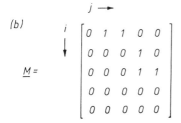

table of nodes

name	description
1	
2	
3	
4	
5	

table of edges

name1	name2
1	2
1	3
2	3
3	4
3	5

(d) $\underline{L} = \{1 + [2,3], 2 + [3] - [1], 3 + [4,5] - [1,2], 4 - [3], 5 - [3]\}$

Fig. 5.4a–e. Some representations of a graph G. (a) List of nodes and edges; (b) Adjacency matrix M; (c) Graphical representation or representation according to Fig. 5.3; (d) Node adjacency list L; (e) Table of nodes including descriptive information such as geometrical properties and table of edges

The corresponding graphs are called *directed* and *undirected graph*, respectively. Two nodes are said to be *adjacent* if there is an edge between them. It is obvious that nodes of a graph may be represented by data types, and branches by pointers from one data type to another. So a graph is a data structure similar to Fig. 5.3. In fact, a tree is a graph without cycles, that is, there is no node N from which a sequence of adjacent nodes leads back to N. An example is given in Fig. 5.4. This also shows that there are different representations of a graph. The *adjacency matrix* M in Fig. 5.4b contains a "1" in element (i, j) if there is a branch from node i to node j; otherwise it contains a "0". The node adjacency list in Fig. 5.4d contains for each node i a list (marked "+") of nodes receiving an edge from node i, and a list (marked "−") of nodes sending an edge to node i. The tables in Fig. 5.4e are similar to the lists in Fig. 5.4a, but allow the inclusion of descriptive information. Some basic operations on graphs are adding, deleting, and relabeling of nodes and edges and traversing a graph (inspecting all of its nodes). A graph may be given by an *explicit definition* which lists all of its nodes and edges, or by an *implicit definition* which states a set of start nodes and a set of transformations allowing one to generate all successor nodes of a given node.

5.1.3 Objects

The term *object* as used in this section is a "software object" and is not to be confused with the "three-dimensional object" (or "physical object") as used in other sections of this book. It is the fundamental entity of object-oriented programming.

It should be noted that there is a complete analogy between program statements and data types. For instance, the general constructions of the simple element, repetition by a known number, and recursion correspond in program statements to the assignment, *for* or *do* statement, and (recursive) procedures, respectively, and correspond in data types to the scalar type, the array, and the recursive data type, respectively. This analogy can be extended to the viewing of a program as a special type of data. Furthermore, there may be dependences and relations between certain data and/or programs. For example, a record representing a speech utterance and a record representing an industrial part may have certain common components and be different in certain components. Therefore, the notion of a (software) object has been developed in *object-oriented programming*. In this programming style every entity in a software system is an *object* which may contain data and procedures. An object is an active entity and there is no distinction between active and passive objects (or procedures and data). A software system is developed as a structured set of objects which form a hierarchy of classes. The problem is to define the relevant objects, their common properties, and their differences. An object can send a message to another object. This may respond by changing its internal state and/or by sending a message to a third object. The sender of the message will receive an answer, either from the addressee of the message or from some other object. Every object is an instance of a certain class of objects and has the properties and the behavior defined for this class. The class defines which messages an object may accept and how it will react to them. It also defines the structure, but not the actual size or content, of the data available to the object. If the definition of a class is changed, this is effective also for all instances of the class. A class is itself an object. New objects can be created as instances of classes. It is also possible to create new classes which may inherit a part or all of the properties of other classes and in addition may have new properties. This gives rise to a hierarchy of classes, subclasses, and so on. An example of the general structure of an object is given in Fig. 5.5.

```
class type 1.1: subtype of class 1
┌─────────────────────────────────────────────────────┐
│ declaration of public components                    │
├─────────────────────────────────────────────────────┤
│ synchronisation conditions                          │
├─────────────────────────────────────────────────────┤
│ declaration of private components                   │
│ initialization of public and private components     │
├─────────────────────────────────────────────────────┤
│ procedural part                                     │
└─────────────────────────────────────────────────────┘
```

Fig. 5.5. An outline of the structure of an object in the sense of object-oriented programming

Obviously, the concept of an object in the above sense is also very useful in a system for pattern analysis. For example, the structures defined in (3.1.1, 2, 4.5.1, 7.1.1) may be realized as objects, and the different types of segmentation objects may be viewed as a hierarchically ordered set of software objects. In addition it is also possible to attach the procedures necessary to compute the requested data to an object, for example, procedures for computing line elements, structural relations, or certainty factors. The structured or hierarchical organization of data will be considered further in Chap. 7 on knowledge representation and utilization. Some useful classes of objects in pattern analysis are the preprocessed pattern

```
class type 1 (any pattern analysis object)
    class type 1.1 (preprocessing method): subtype of type 1
        class type 1.1.1 (convolution): subtype of type 1.1
        class type 1.1.2 (erosion): subtype of type 1.1
        ...

    class type 1.2 (result of preprocessing): subtype of type 1

    class type 1.3 (segmentation method): subtype of type 1
        class type 1.3.1 (Canny operator): subtype of type 1.3
        class type 1.3.2 (phonetic segmentation): subtype of type 1.3
        ...

    class type 1.4 (initial segmentation I): subtype of type 1

    class type 1.5 (rule): subtype of type 1
        class type 1.5.1 (decision rule): subtype of class 1.5
        class type 1.5.2 (instantiation rule): subtype of class 1.5
        ...

    class type 1.6 (concept): subtype of type 1
        class type 1.6.1 (object model): subtype of type 1.6
        class type 1.6.2 (pragmatics of speech): subtype of type 1.6
        ...

    class type 1.7 (instance): subtype of type 1

    class type 1.8 (control algorithm) subtype of type 1
        class type 1.8.1 (top-down control): subtype of type 1.8
        class type 1.8.2 (bidirectional control): subtype of type 1.8
        ...

    class type 1.9 (learning algorithm): subtype of type 1
        class type 1.9.1 (parameter learning): subtype of type 1.9
        class type 1.9.2 (concept learning): subtype of type 1.9
        ...
```
Fig. 5.6. An outline of an object hierarchy for pattern analysis

in Chap. 2, the initial description and the method (procedure) in Chap. 3, the word and object model in Chap. 4, the control algorithm in Chap. 6, or the rule, concept, and instance in Chap. 7. An example of an object hierarchy is given in Fig. 5.6. This only gives the general idea, but it is not worked out in detail; for example, additional levels of sub-subtypes and so on could be introduced.

The object is a combination and generalization of the concepts of *program abstraction* and *data abstraction*. Program abstraction is achieved by procedures and coroutines. The procedure allows one to encapsulate a set of statements performing a well-defined action (e.g. convolution or FFT) and to reference them by the procedure name. The coroutine distinguishes between the declaration of a procedure and its executable incarnation. Each call of a coroutine causes the creation of a new executable incarnation and its actual execution using the same body of procedures but different data. Data abstraction is achieved by *abstract data types* consisting of the definition of the data and all of the operations defined on the data. A user can operate on the data only via those predefined operations. In addition an object in the above sense allows one to create hierarchies and exchange messages between objects.

5.1.4 Remarks

Programming details have been completely omitted because they depend on the language used. The structures of Sect. 5.1.1 are available in any high-level language, but the structures of Sect. 5.1.2 are only available in some of them and so far there are only a few examples of object-oriented languages such as SMALLTALK or objective-C. For example, FORTRAN has no explicit structures like graphs, allows no dynamic storage allocation, and no recursive data or procedure definitions, whereas PASCAL does. Advanced structures require explicit command of pointers to data and this is not supplied by some languages. Of course, in any language pointers to data may be simulated because they contain just the address of the respective data. It was already mentioned in Sect. 5.1.2 and illustrated in Fig. 5.4 that, for instance, a graph may be represented by an adjacency matrix, thus avoiding the use of pointers. Although it is usually possible to avoid advanced data structures, it is often very convenient to use them, e.g. in connection with dynamic storage allocation. For example, in PASCAL the tree in Fig. 5.3 could be defined by using the variant record data type and could be expanded or reduced by creating or deleting nodes.

5.2 Data Bases

5.2.1 A General Outline

Data structures may be viewed as storage space (in main storage and/or mass storage) which is allocated to one program. It may be useful or even necessary to also have centralized data which may be referenced by many users and/or programs. Such centralized data common to different users are called a *data base*. In pattern analysis the need for data bases arises for at least two reasons. First, it is not efficient if every user stores his own sample of patterns and, second, a results data base shared by the modules of a pattern analysis system was introduced in Sect. 1.4. Several requirements result from these two reasons.

It should be possible for different users to reference arbitrary portions of the data base, for example, two overlapping subsets of a sample of simple patterns or two (or more) windows of an image. References to data should not be constrained to physical adjacency of data in storage. Rather it should be possible to have access to arbitrary subsets of data which may be located at separated storage places. Of course, on mass storage devices physical adjacency will result in faster access to the data. Since many users may be involved, *data independence* is important. This means that an individual user should not need to care about how data are stored and accessed. If the storage structure and access mechanism change, this should not affect the user's programs. For instance, one user may wish to access sample values of an image row by row, another one column by column, and still another one in square blocks. Again, the actual storage structure may facilitate one access mode more than the other. This makes it useful to require *reconfiguration* of data in order to adapt the storage structure to varying needs. Because of the assumed data independence such a reconfiguration will affect only the performance of programs but will not require changes of programs. It should also be possible to add, delete, and modify (update) data in the data base. Because different users may be operating on the same data, it is necessary to provide a means for protecting portions of the data base from any changes. For instance, if in an analysis system several modules operate on a data base, it may not be tolerable that data, which are used by one module, can be changed by another module.

Since there are so many problems with data bases, it seems appropriate to point out that it is worthwhile solving them. Having centralized data it is possible to avoid redundancy of storage. This is particularly obvious with samples of simple or complex patterns. If such a sample is stored by every user, the same data will occur many times. Redundancy may also lead to inconsistency of data. This may arise if one occurrence of the data has been updated and the other has not (or not yet). Centralized data also allow the introduction and enforcement of standards which in their turn facilitate data interchange. It also becomes economical to develop a high-level language for data manipulation because it will be advantageous for many users, and it becomes economical to provide complex facilities for avoiding the loss of data. Finally, a results data base is an integral part of the analysis system in Fig. 1.4. It is the central device used by system modules to communicate with each other.

A general structure of a data base is shown in Fig. 5.7. It allows several users to reference different, but possibly overlapping sections of the data base. The individual user's view of the data base is described by his *external model*, which constrains his command of the data. The data model is a description of the entire information content of the data base. It is defined by a *schema* containing descriptions of the records in the data model; the model then consists of multiple occurrences of records. In order to achieve data independence no specifications of storage structure or access mechanisms are allowed in the schema. However, checks of authorization to use data and of correctness of data may be included. The interface between external models of the users and the data model is provided by appropriate mappings. The *internal model* is a description of the stored data base (but not in terms of physical blocks of data; this level was omitted in Fig. 5.7). If the structure of the stored data base is changed, the internal model changes, but the data model should not. Again, the interface between data model and internal model is provided by a mapping. This also changes if the internal model changes. Thus data independence can be achieved by preventing changes of storage structure which affect the data model. All access requests of the users are handled by the *data base management*. A user's request is accepted by the data base management which inspects various stages of the data base representation, performs authorization checks, and executes the necessary operations. Although individual users may issue requests to the data base in parallel, it is evident that the data base management will handle them sequentially.

If the "users" are indeed human operators of a large (interactive) pattern processing system, an organization similar to Fig. 5.7 will be very useful. Of

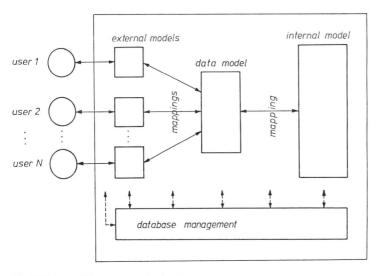

Fig. 5.7. A possible structure of a data base

course, in this case every user will need an appropriate amount of private storage to perform his operations on sections of data. If the "users" are just modules of an analysis system as indicated in Fig. 1.4, an organization which is more adapted to the task domain will probably be used. For instance, data items and storage structure in the internal model may be constrained such that the various mappings are nearly reduced to identity mappings in order to allow faster access. But it is also evident that the results data base in Fig. 1.4 has some internal structure including, in particular, the data base management.

5.2.2 Hierarchical Data Model

Three types of data models have been developed which are in common use. They are known as hierarchical, network, and relational models. Each one will be introduced by a simple example.

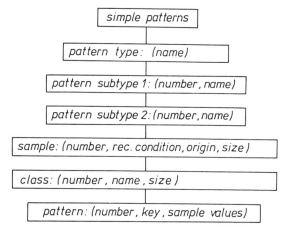

Fig. 5.8. A schema of a hierarchical data base

Suppose that a collection of different samples of simple patterns is to be stored. In this case a *hierarchical data model* seems to be a natural choice. A possible *schema* for definition of records is indicated in Fig. 5.8. It starts with the name of the data base which is "simple patterns". Next the root of the tree representing the hierarchy is defined to be "pattern type"; this is further labeled by a name for the pattern type. The next two statements indicate that two levels of subdivision of a pattern type are provided. Then individual samples are listed; they may be referenced by a unique number and, furthermore, the recording conditions, origin, and size of the sample are given. The sample is then subdivided into classes, and the last level are individual patterns. It is important to note that, for instance, the pattern type may be detailed by more than just one successor in the next level. Figure 5.9 gives an example. Here it is assumed that in addition to the subtype 1 "descriptive information" is also appended to the pattern type.

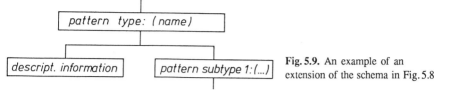

Fig. 5.9. An example of an extension of the schema in Fig. 5.8

This descriptive information may contain a set of relevant references concerning pattern recognition as applied to this pattern type. As mentioned previously, on any level there may be many occurrences of the data defined by the schema. This results in the *data model* illustrated in Fig. 5.10. It also becomes clear that the data may be – but, of course, need not be – stored and retrieved by physical adjacency if the tree is inspected from top to bottom and from left to right as indicated in Fig. 5.10. Another possibility would be to use pointers from one record to another. In this sense the model shows data content, but not physical storage.

It is apparent from the above illustrations that the hierarchical model is asymmetric because there are superiors and dependents in the model. This causes problems with data retrieval which depend on the model and not on the queries to the data base. The hierarchical model is appropriate if the data have an intrinsic hierarchical structure as is the case in Fig. 5.8. In other cases it may be questionable which data should be superior and which dependent.

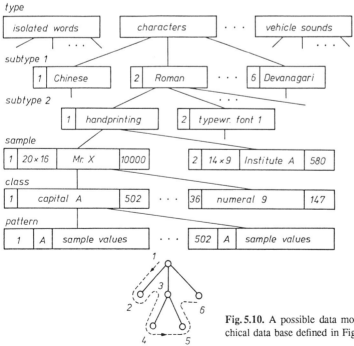

Fig. 5.10. A possible data model of the hierarchical data base defined in Fig. 5.8

5.2.3 Network Data Model

Viewing Fig. 5.10 as a structure containing records and pointers from one record to another it becomes clear that in the hierarchical model the pointers are restricted by the requirement that any dependent has precisely one superior. The *network data model* does not contain this restriction. As an example assume that a set of medical images is to be stored. The images may be of different types (for instance, chest X-rays or blood smears) and have different diagnoses. In this case one dependent, the image, has two superiors, the diagnosis and the type. A possible (DBTG) schema of a network data base is shown in Fig. 5.11. Although such schemas are to be stated by a suitable language, graphical representation is preferred here in order to avoid the introduction of such a language. An important aspect of the schema is that it consists of two DBTG sets DIAGIM and TYPEIM and indicates that, for instance, in set DIAGIM the diagnosis is the *owner* and image the *member* of the DBTG set (a DBTG set obviously is different from the set as understood in mathematics). Each DBTG set may have several occurrences, where one occurrence of a set has one occurrence of its owner and an arbitrary number of occurrences of its members. The data model, then, may look like Fig. 5.12. It shows four occurrences of the DBTG set DIAGIM, one of which has only the owner but no member, and three occurrences of TYPEIM. Owner and member records are linked by pointers and it is seen that this network data model is not a hierarchy, but a hierarchy is a special case of a network. Thus the network model is more general than the hierarchy and allows representation of many-to-many correspondences more directly. The comments and sample value part of the image record is not specified further. For example, the comments may include information about image size, scale, hospital, patient, and so on. In the

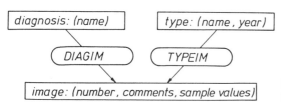

Fig. 5.11. Schema of a network data base

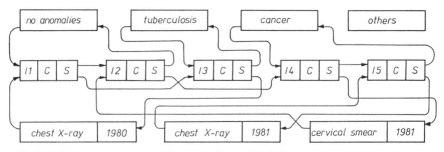

Fig. 5.12. Example of a data model corresponding to Fig. 5.11

DBTG approach it is not allowed for a record type to be both the owner and a member of a DBTG set. To achieve this a second record type *Link*, which need not contain any data, may be used. Then, for instance, two DBTG sets IMLINK and LINKIM can be defined where the former has images as owner and links as member and the latter vice versa.

5.2.4 Relational Data Model

The *relational data model* is based on the mathematical notion of a relation. If n sets $S_1, \ldots, S_n$ which need not to be distinct are given, a relation R is defined as a subset of the product set

$$S_1 \times S_2 \times \ldots \times S_n \supset R . \tag{5.2.1}$$

An element $r \in R$ is an ordered n-tuple

$$r = (s_1, s_2, \ldots, s_n), \quad s_i \in S_i, \quad i = 1, \ldots, n; \quad n \geq 1 . \tag{5.2.2}$$

The n-tuple may be viewed as a row of data, and the entire relation R is a table of data where the ith column of R contains only elements $s_i \in S_i$. The sets $S_i, i = 1, \ldots, n$ are the domains of R, that is they are the set of values out of which actual values appearing in a column of R are taken. Since R is a set, its members are unordered, but the rows of the table representing the set appear in a certain order. Although from a formal viewpoint the order of the rows is irrelevant, it is usually convenient to assume the rows to be ordered by the values of one (or some) domain(s). On the other hand, the order of the sets (5.2.1) is significant for R, but may be insignificant for the user of the table if he references columns of the table by a symbolic name and not by position in the n-tuple.

As an example take the simple object of Fig. 5.13 which is to be analyzed by an automatic system. Assume that the figure is segmented into segmentation objects in the hierarchical manner shown. If we assume a straightforward bottom-up approach, the first step might be to find small line segments (by one of the methods given in Chap. 3). The results of this and the following processing steps are stored in a results data base. Although a hierarchical organization would be possible in this example, a relational data base is chosen. It is organized into two tables, one containing segmentation objects or simple constituents (SC), the

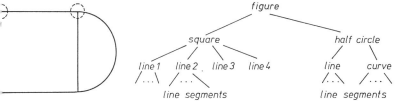

Fig. 5.13. A simple physical object and its segmentation into simpler constituents

SC

level i	number j	name	a_1	a_2	a_3	a_4	a_5	a_6	reliab.
1	1	sl	1	1	2	1	-	-	7
1	2	sl	2	0.9	3.1	1.1	-	-	6
1	3	sl	3.2	1.1	4	1	-	-	4
1	4	sl	4.1	1	5.3	1.1	-	-	3
⋮									
2	1	ll	1	1	5.3	1.1	-	-	8
⋮									

IR

number	part 1	part 2	relation	result	reliab.
1	1,1,sl	1,2,sl	left to	1,5,sl	7
2	1,5,sl	1,3,sl	left to	1,6,ll	7
3	1,6,ll	1,4,sl	left to	2,1,ll	8
⋮					
20	3,1,sq	3,2,hc	left adj.	figure	9

Fig. 5.14. Two tables containing results of analysis in relational form

other containing interrelationships (IR) between simple constitutents. Parts of both are shown in Fig. 5.14. The segmentation objects are identified by a number i indicating the level from Fig. 5.13, a number j giving the running number of a segmentation object on a particular level, and a name distinguishing constituents on one level. If all line segments are determined first, the table SC will at first contain only members from level 1 while table IR is empty. The domains of the three quantities are specified by

$$i \in S_1 = \{1,2,3\} ,$$

$$j \in S_2 = \{1,2,\ldots\} ,$$

$$\text{name} \in S_3 = \{\text{sl, sb, ss, ll, cl, sq, ci, hc, tr}\} , \qquad (5.2.3)$$

sl: small line (or line segment),
sb: small bar,
ss: small square,
ll: longer (straight) line,
cl: curved line,
sq: square,
ci: circle,
hc: half circle,
tr: triangle.

Even if presently there were no method available to extract, for instance, ss, it inclusion into domain S_3 is possible, but an ss would never occur as a result in table SC. The segmentation objects on all levels are further characterized by parameters $a_1,\ldots,a_6$ with the understanding that the meaning of the parameter depends on the simple constituents and that some parameters may be unused in some cases. Three examples of possible parameters are given. For the small line sl one might choose (a_1, a_2) to be the (x, y) coordinates of the lower left endpoint, (a_3, a_4) to be the coordinates of the upper right endpoint of sl. This completely determines a particular sl, but it may be convenient to include the angle between sl and the x axis and the length of sl as further parameters (which is not done here). The other parameters a_5, a_6 remain unused. For the longer line ll the same parameters are adequate. For the (full) circle (a_1, a_2) are the coordinates of the center, a_5 the radius, and a_3, a_4, a_6 are unused. The domains of the parameters in question are evident. The last column of the SC table is a measure of the reliability attributed to the segmentation object.

After extraction of line segments, interrelationships may be determined and entered in the IR table. This consists of a consecutive number, an identification of two simple constituents (part 1, part 2) whose interrelation is considered, a specification of the kind of relation, a result, and a reliability. This amounts to taking two constituents α_1, α_2 which obey a certain spatial relationship σ and attributing a new name (result) β to this triple

$$\beta = (\alpha_1, \alpha_2, \sigma), \qquad (5.2.4)$$

with the only modification that a number and a reliability are added. It is a particular realization of a structure I as introduced in (3.1.2). At first, the simple constituent sl with number 1 on level 1 is determined to be left of sl with number 2 on level 1; the new identification of the two is (1, 5, sl), that is, the two small lines are also considered to be a small line. In the second step this new small line is combined with (1, 3, sl), but the result is now considered to be a longer line ll on level 1. The distinction between sl and ll may be based on the length of the line. In number 3 of the IR table the first longer line on level 2, (2, 1, ll), is detected. Distinction between ll on level 1 and ll on level 2 may be based on the requirement that a ll on level 2 is limited by a junction of several lines (see dashed circles in Fig. 5.13). The newly found results are also entered in the SC table. Finally, at some step a complete "figure" is found and the process stops.

The rows of the SC table may be considered as *entities*, the rows of the IR table as *relationships* between entities. A row in the IR table has as a primary key the entities being related, which are replaced in the figure by an arbitrary enumeration. It is requested that an attribute of a primary key of an entity in SC is not undefined (or NIL); this is the *entity integrity rule*. It is also requested that if an attribute A of one relation (e.g. SC) is an element of the primary key of another relation (e.g. IR), then the set of values of A occurring in IR is a subset of the set of values of A occurring in SC; this is the *referential integrity rule*.

It is seen that all results of processing can be represented in tabular form. More generally, in the relational data model all information is represented in a uniform manner – the table – and associations between rows can only be made by use of data values in columns of the same domain. For instance, to find the simple constituents of (2, 1, ll) in the SC table one has to search (2, 1, ll) in the result column of the IR table and then obtain (1, 6, ll) and (1, 4, sl) from the corresponding row. There are no pointers from one record (row) to another as is the case with the hierarchical and network models. It was mentioned that tuples of a relation have no implicit order. On the other hand it is evident that in the above example a certain order of the rows, for instance, first according to level, second according to a_1, a_2 will reduce the amount of search necessary to find associations in data values.

A schema for the data model shown in the SC table has the form

relation: simple constituent (level i from S_1, number j from S_2,
   name from S_3, a_1 from S_4, ..., a_6 from S_9, reliab. from S_{10}), (5.2.5)
   domain: $S_1 = \{1, 2, 3\}$,
   domain: $S_2 = \{1, 2, \ldots\}$,
   ...
   domain: $S_{10} = \{1, \ldots, 10\}$. (5.2.6)

The definitions of (5.2.5,6) will be stated in some appropriate language, but consideration of such languages is not of interest here. The schema will also contain additional information, for instance, access control. It is important to note that according to the schema the values occurring in the same column of a table have to be drawn from the same domain. Therefore, it is impossible to change the domain depending on the value of some other column. This change might seem a reasonable requirement if, for instance, one would like to represent by a_6 the angle between an sl and the x axis, the arc length of cl, or the area of an sq, ci, hc, tr in (5.2.3). Since angle, arc length, and area are drawn from different domains, they must not occur in the same column. A solution might be to include additional parameters some of which remain unused or to split the SC table into several tables, one for small lines sl, one for curved lines cl, and so on, such that in every table all columns are used and their values are from the same domain.

The tables in Fig. 5.14 have the property that each one possesses one or more sets of columns the values of which uniquely identify the corresponding row; these sets are the *candidate keys* of the relation. For the SC table the two candidate keys are (level, number, name) and $(a_1, \ldots, a_6)$, for the IR table the three candidate keys are (number), (result), and (part1, part2). At least one such key always exists because a relation is a set and a set has no duplicate elements, so at least the complete n-tuple uniquely defines a row. If there is only one candidate key it is the *primary key*, otherwise the primary key may be chosen from among the candidate keys.

5.2.5 Normal Forms

So far no restrictions have been imposed on the relations. However, it us useful to require that the values have certain properties which may be obtained by normalization of the relation; the result is a *normal form*. Denoting the set of relations obtainable from (5.2.1) by {relations} and the set of relations in the ith normal form by $\{i\text{NF}\}$ one has

$$\{\text{relations}\} \supset \{1\text{NF}\} \supset \{2\text{NF}\} \supset \{3\text{NF}\} \supset \{4\text{NF}\}. \tag{5.2.7}$$

A relation is 1NF if the values in each column are atomic or nondecomposable. In this sense the SC table is 1NF, but the IR table is not. The latter contains triples of values in the part 1, part 2, and result column. To make the IR table 1NF one simply has to split the columns contradicting the 1NF requirement into several columns containing only atomic values. So consideration of 1NF relations is no real restriction.

Before considering 3NF and 4NF relations we introduce the concept of functional dependence which is related to the meaning or the semantics of data. A column y of relation R is said to be *functionally dependent* on a column x of R if and only if to each x value belongs precisely one y value. If such dependences are known and included, for instance, in the schema, they may be used to check the validity of data. Whenever a particular x value occurs it must be accompanied by one and the same y value, otherwise there is an error. Such a dependence may be represented graphically by an arrow pointing from x to y. If x is composite, that is, consists of values from several columns, y is said to be fully functionally dependent on x if and only if it is functionally dependent on x, but not on any subset of x. An x with the above properties is called a determinant. A weaker restriction is a multivalued dependence which exists if each x has a well-defined set of values from y associated with it. Realization of dependences of any kind is not possible by observation of values occurring in the relational tables because these values are time dependent (rows may be added, deleted, and updated); rather, the dependences express a constraint existing in real world phenomena which are modeled by the data. The functional dependences of the SC and IR tables are shown in Fig. 5.15. To simplify matters the notion "part 1" and so on was retained. It may be viewed as an example of a composite x having three columns to yield a 1NF relation by letting part 1 = (level 1, number 1, name) and so on. For the SC table the figure shows that (reliab.) and $(a_1, \ldots, a_6)$ are both functionally dependent on (level, number, name); in turn, (reliab.) and (level, number, name) are also dependent on $(a_1, \ldots, a_6)$. No column or set of columns is dependent on (reliab.). The dependences are full, that is no subset of $(a_1, \ldots, a_6)$ or (level, number, name) could be used. For the IR table similar interpretations hold. The dependence of (relation) on ((part 1), (part 2)) is also indicated full, that is no occurrences of, for instance, ((1, 1, sl), (1, 2, sl), left of) and ((1, 1, sl), (1, 2, sl), above of) are allowed. In this case a new relation "left above of" is to be introduced or otherwise a multivalued dependence would exist.

With the above remarks the 2NF, 3NF and 4NF property are defined as follows:

1. A relation is 2NF if it is 1NF and every non-key column is fully dependent on the primary key.
2. A relation is 3NF if it is 1NF and every determinant is a candidate key.
3. A relation R is 4NF if in all cases where a multivalued dependence of column y on column x exists it is true that all columns of R are also functionally dependent on x.

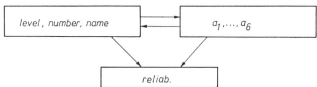

Fig. 5.15. Dependences of the columns in the SC and IR tables

The notion of 2NF only applies to relations with one candidate key, but since 2NF is of no special interest here it is not considered further. Relations in 3NF and 4NF are desirable because certain problems with storage operations are avoided. The above definitions and also a modified definition of 3NF and 4NF together with several examples and a discussion of desirable properties of 3NF and 4NF relations are found in the references. According to the above definitions both the SC and IR tables are in 3NF and also in 4NF.

5.2.6 Data Sublanguages

Data in a data base are accessed by users who employ a certain programming language. The subset of this language concerned with manipulation of the data base is the *data sublanguage*, which often consists of a set of subroutine calls. Rather than defining sublanguages in detail, an idea is given of what they should accomplish. Perhaps the most basic operations which a sublanguage should be able to perform are insert, delete, update, find. For a relational data base they should have the form

> Insert the row on scratch file SF into relational table RT.
> Delete the row with primary key value equal to PK from RT.
> Update in table RT and row PK the column C by replacing it by the new value NV.
> Find in RT the row with primary key PK and return it to SF. (5.2.8)

These statements are nonprocedural in the sense that they only exhibit *what* the user would like to be done, but they give no indication of *how* it is really done. This allows the user to concentrate on the information content of his data and frees him from representational details. Routines of this kind will have to be provided by the data base management system in order to allow efficient access

5.2 Data Bases

In general it is useful to allow operations which not only affect one row, but a set of rows, and which are restricted by more general conditions. For instance, the "Find" operation above will return a single row because the primary key is unique. However, the request

Find in SC rows with name = sl and return them to SF (5.2.9)

usually will be met by a set of rows. A more general condition is, for example,

Find in SC rows with name = sl and reliab. > 5 and return them to SF . (5.2.10)

For the purposes of the example in Fig. 5.14 it is useful to have a command of the form

Find in SC all pairs $R1, R2$ of two rows with name = sl and

$[a_3(R1) - a_1(R2)]^2 + [a_4(R1) - a_2(R2)]^2 < \Theta_2$, and return them to SF . (5.2.11)

Such a statement selects the set of line segment pairs having endpoints within a specified distance. The first two of the above statements are standard operations of set-oriented data sublanguages whereas the last is motivated by the problem considered here. Next we discuss some other standard operations and shall see that the last statement may be simulated by them.

Since relations are sets, the operations of union, intersection, difference, and Cartesian product should be provided by a set-oriented sublanguage. Of course, for union, intersection, and difference the operands must be compatible, that is the n-tuples (5.2.2) must have the same number of elements and corresponding elements must be drawn from the same domain. If two (compatible) relational tables RT1, RT2 are given, the *union* of RT1 and RT2 is a new table NT containing all (different) rows belonging to either RT1 or RT2 or both, the *intersection* is a new table containing all rows belonging to both RT1 and RT2, and the *difference* RT1 minus RT2 is a table with the rows belonging to RT1 but not to RT2. For the *Cartesian product* tables RT1 and RT2 need not be compatible. The result is a table with rows being the concatenation of a row from RT1 and a row from RT2, in that order. It thus generates the set of all pairs of rows from RT1 and RT2.

These standard set operations are enlarged by the relational operations Find (or Select), Project, Join, and Divide. The *Find* operation was used already in the above examples. The *Project* operation allows extraction of a specified subset of columns from a table. For instance, to find the results and their reliability in the IR table the statement

Project IR table over (result, reliab.) and return outcome to SF (5.2.12)

is used. The SF table then contains only two columns with result and reliab. The Join operation allows constructing a concatenation of two rows from two tables (which is similar to the Cartesian product), but concatenation is conditioned by the comparison of values from two columns, one from the first table and the other from the second (of course, the values must be from the same domain). Any one of the common comparison operators $=, \neq, <, \leq, >, \geq$ are allowed. Assume that we want to construct a table containing results on level 1, their constituents (part 1, part 2, relation), and their parameters. This may be done by a Join of the SC and IR tables. To refer to a table in 1NF we assume that the result column of the IR table is split into the three columns: level r, number r, name r. The desired result is obtained by

> Join SC and IR tables where level i of SC equals level r of IR and return the outcome to SF. (5.2.13)

All rows from SC and IR meeting the condition level i = level r are concatenated in SF, all others are discarded. The *Divide* operation between a binary relation RTB ($n = 2$) in (5.2.1) and unary relation RTU ($n = 1$) gives another unary relation RTQ, the quotient. Let the columns of RTB be X and Y from domain (X) and domain (Y), and let the column of RTU be Z from domain $(Z) =$ domain (Y). Then the domain of the quotient obtained from

> Divide RTB by RTU over Y and Z and return the outcome to RTQ (5.2.14)

is domain (X). A value $x \in X$ appears in RTQ if for all $z \in Z$ the pair x, z is in RTB. An example is not provided here. Now the aforementioned operation (5.2.11) may be stated in terms of the standard set operations by

> Find in SC rows with name = sl and return them to SF1,
> Take the Cartesian product of SF1 and SF1 and return it to SF2,
> Find in SF2 rows with $(a_3' - a_1'')^2 + (a_4' - a_2'')^2 < \Theta_2$
> and return them to SF. (5.2.15)

In the last statement it is assumed that parameters from the concatenated rows in SF2 are distinguished by a_i' and $a_i'', i = 1, \ldots, 4$. The last version, however has the disadvantage that the Cartesian product really generates the product set. This is unnecessary if a special routine is provided for the first version (5.2.11).

Finally, it is useful to have a statement which *locks* some parts of a data base, that is prevents certain users from accessing certain data. For instance while performing the last three statements above by a program named PA one may wish to allow access to small lines only by PA. This is done by

> Lock in table SC all rows with name = sl to be accessed by program PA only. (5.2.16)

After PA finished its operation the locked data is made available by

> Unlock in SC rows with name = sl. (5.2.17)

Special lock statements which allow read but not write operations may be included.

The discussion shows that standard data base techniques are very useful in pattern analysis, but also that special modifications and additions can be used. Only the relational data model was considered as an example. Similar statements may be developed for hierarchical and network models, but in general "a set-level hierarchical or network language is necessarily more complex than a set-level relational language" (quotation from [5.1]). Presently queries to a data base are handled by the central processing unit (CPU) which becomes the bottleneck in the system. Attempts have been made to remove details of operations on the data base from the CPU by so-called data base machines. The idea is similar, for example, to attributing a disk controller to several disks and thereby avoiding CPU control of the disks.

5.2.7 Semantic Data Model

In this section the development of important extensions to the above data models are mentioned briefly. The purpose is to provide data models capable of expressing more directly the user's view of the data. In order to accomplish this, any conception in the data base is considered as an object, objects are organized into classes which in turn are related via generalization and specialization hierarchies; an inheritance of attributes is possible. In addition, constraints, heuristics, temporal relationships, and further levels of hierarchy may be introduced. This amounts to representing *knowledge* about a task domain, and this topic is treated in more detail in Chap. 7; in particular, the *semantic nets* introduced in Sect. 7.4 may be viewed as an example of a *semantic data model*.

5.3 Pattern Data

5.3.1 Data Structures for Patterns

Some examples illustrating the use of the data structures introduced in Sect. 5.1 will be given here.

Starting with a recorded pattern (an image, image sequence, utterance, etc.) the basic structures of Fig. 5.1a, b, are widely used to arrange and store the relevant data. The sample values are preceded by descriptive information which contains all the relevant details about the pattern. Since preprocessing takes a pattern into another pattern, for example, an image into another image, the structures of Fig. 5.1a,b are also used for the preprocessed patterns, but now it is advantageous to add details about preprocessing steps to the descriptive information. Sample values from images are stored line by line, column by column, or in rectangular blocks covering the image. The resulting (logical) records are fairly long and are usually subdivided into several (physical) records for storage on magnetic tape, disk, or optical disk. In order to achieve fast access to rows, columns, or rectangular blocks of an image special storage structures have been suggested.

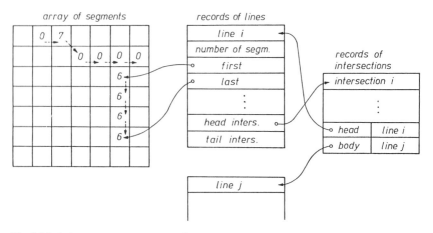

Fig. 5.16. A data structure to represent line segments, lines, and intersections

An important step of image segmentation was to obtain contour lines (Sect. 3.3, 4). Information about consecutive lines of polygonal approximation can be represented by linear lists shown in Fig. 5.2 or by rings. This can also be done if chain codes are used. A somewhat more involved structure is shown in Fig. 5.16. It shows two levels of processing. The first level contains line segments which may be obtained, for example, by masks or Hueckel operator (Sect. 3.3); the second level contains longer lines obtained from the segments. The details of the algorithms are not of interest here, but only the data structure which is shown in Fig. 5.16. It consists of an array of line segments, records for longer lines, and records for line intersections. Pointers are used to establish relations. The array of segments covers the image and each element of the array covers several sample values (say, 4×4) of the image. The element contains information about a line segment, if present, such as orientation, reliability, location in the image, and so on. A longer line is built by segments and stored in a record of lines. This contains among others a line identification, pointers to the first and last segment of the line, and an indication of intersections with other lines. From the pointer to the first line segment of line i in Fig. 5.16 the other segments in the array are obtained by tracing the chain-coded segments of the line. The chain code (see Fig. 2.4) of the direction to the next segment is given in the figure and is illustrated by a dashed arrow. An intersection between two lines may also be a junction of the beginning (head) or end (tail) of a line with the beginning, end, or body of another line; it is represented by records of intersections which contain information about the type of intersection and the lines involved. A data structure consisting of linked records may also be used, for instance, to represent geometric objects to be generated on a display. The structure consists of a semantic map, which is a list of pointers to the geometric objects, and a semantic net, which contains descriptions of the objects in terms of predicates and simpler constituents.

Besides the chain code mentioned above, the border (or contour) may be stored as an ordered sequence of coordinates. For instance, coordinates may be stored in the order of clockwise (or counterclockwise) tracing of the contour. An alternative is to store them in sets where each set contains only points having equal y coordinates. Within a set points are ordered according to increasing x coordinates. This representation allows, among other things, rapid evaluation of the minimum distance between a point and the boundary or of the boundary resulting from the union of two regions. The *generalized cones* are an approach to represent three-dimensional objects. They consist of an axis which is a curve in three-dimensional space and a set of contours obtained from cross sections of the object normal to the axis. The data structure is a list of coordinates of the axis with pointers to the coordinates of the cross sections.

Another important approach to segmentation was to obtain homogeneous regions of an image. It was discussed in Sect. 3.5 and illustrated in Fig. 3.22 that regions may be represented by a tree structure, in particular the *quad-tree*, and volumes by an *oct-tree*; viewing regions as leaves of a tree is a common technique. The tree is usually stored not in the form of the linked nodes shown in Fig. 5.3, but by coding the regions in a form adapted from Fig. 3.22. If we have an $M \times M$ image with $M = 2^p$, there may be at most p levels below the root of the tree. Then each square region is uniquely identified by a number of p digits, each digit ranging from 0 to 4. For instance, assuming $p = 3$ the tree in Fig. 3.22 is determined by the numbers 100, 210, 220, 230, 241, 242, 243, 244, 300, 400, which is the sequence given in Fig. 5.3. Another possibility is to store for each square the row and column number of its upper left corner and the length of its side. Assuming again $p = 3$ for simplicity, this would result for Fig. 3.22 in the code $\{(1,1,4),(1,5,2),(1,7,2),(3,7,2),(3,5,1),(3,6,1),(4,6,1),(4,5,1),(5,5,4),$ $(5,1,4)\}$. Still another scheme is the coding of each rectangular region by $\{$Intensity, $x_{min}, x_{max}, y_{min}, y_{max}\}$ with "Intensity" being the average intensity of the region.

The tree structure describes a hierarchical ordering of regions and the various levels of the tree represent pictorial details with various resolutions. Low-resolution images may be used at first to obtain a coarse idea of the image. This approach was first introduced for edge detection but it is equally suited for other segmentation operations. The result is a hierarchy of images of decreasing resolution or an *image pyramid* containing the images at different resolutions. An $M \times M$ image with $M = 2^p$ is stored in the 0th level of the hierarchy, an $(M/2 \times M/2)$ image in the first level and so on until a 2×2 "image" in the $(p-1)$th and a 1×1 "image" in the pth level is obtained (see also Sect. 2.3.7).

Besides the tree structures mentioned above, graphs are used to represent adjacency relations in segmented images. If the image is segmented by processing it line by line, the *line adjacency graph* (LAG) is used; and if a two-dimensional segmentation is performed, the result is a *region adjacency graph* (RAG) which is shown in Fig. 5.17. In line-by-line processing each line is segmented into several intervals, for instance, by a threshold operation or piecewise linear approximation (see Sects. 3.2 and 3.4). The intervals become nodes of the LAG. Two nodes are

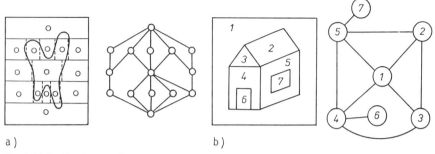

Fig. 5.17a,b. Graphs may be used to represent adjacency relations in images. (a) Line adjacency graph (LAG). (b) Region adjacency graph (RAG)

linked by an edge if they correspond to intervals in adjacent lines and if these intervals overlap in their x coordinates. Edges may be weighted by a measure of similarity between overlapping intervals in adjacent lines. If similarity is too low, an edge may be omitted. To separate a figure from the ground an edge is drawn only if two black intervals in adjacent lines overlap. In the RAG each region corresponds to a node of the graph and an edge between two nodes indicates that the two regions are adjacent, that is, they have a common border. These graphs may, for instance, be stored in the node adjacency list of Fig. 5.4d. It is mentioned in passing that the result of region extraction – a labeling of image sample values by their region number and a region adjacency matrix – can be used to find points on the border between two regions.

Control structures for pattern analysis may be represented by graphs in a natural way. The nodes correspond to states of analysis, the edges to events or actions causing a state transition. This topic will be discussed in more detail in Chap. 6. It should be noted, however, that the control graph is an abstract concept and is not stored as a data structure in physical storage. Rather it is a compressed view of the analysis steps. It is an example of a graph which is defined implicitly.

In Sect. 1.3 it was outlined as a basic assumption of pattern analysis that a pattern $f(x)$ has a certain structure. Therefore, representation of this structure is necessary for further analysis. From the various structural methods only the rules of a formal grammar as introduced in Sect. 4.3 and relations will be considered here. Other approaches are given in Chap. 7. The rules of the grammar may be used to represent structural properties; they may be stored in arrays, trees, or general graphs according to the type of rule. As an example consider the rules

$$t_0 \rightarrow t_1 t_2 t_3 t_4 ,$$

$$\begin{aligned} t_1 &\rightarrow n_1 & n_1 &\rightarrow Q_2 , \\ t_2 &\rightarrow n_2 & n_1 &\rightarrow 22 , \\ t_3 &\rightarrow Q_1 & n_2 &\rightarrow 33 , \end{aligned} \tag{5.3.1}$$

which were developed for fingerprint recognition. The rules are stored in nodes linked by pointers as shown in Fig. 5.18.

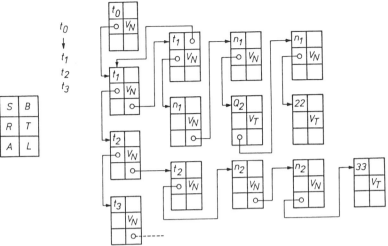

Fig. 5.18. Data structure to represent the productions (5.3.1). Each node contains the following fields: (S) this field contains the name of the symbol; (R) a pointer to the symbol just right of S; (A) a pointer to the next alternative of the production; (B) a pointer back to the node whose L field points to the present node containing B (not all of these pointers are drawn in the figure); (T) indicates whether the symbol in S is terminal (V_T) or nonterminal (V_N); (L) a pointer to a production whose left side is the symbol in S

A representation of rules in a table or graph is also possible. General graphs can be used to represent prototypes for matching patterns as discussed in Sect. 4.5. Relations in the sense of (5.2.1) are suited for representing the structural constraints of patterns and for finding consistent labelings for simple constituents. If $V_T = \{s_1, \ldots, s_n\}$ is a set of simple constituents and $L = \{l_\alpha, l_\beta, \ldots, l_\omega\}$ a set of possible labels or meanings, then, for example, the triple $\{(s_1, l_\alpha), (s_4, l_\beta), (s_2, l_\omega)\}$ indicates that a particular triple of constituents is constrained to be associated with a particular triple of labels. Let

$$R = \{\{(1, \alpha), (2, \alpha), (3, \beta)\}, \{(1, \beta), (2, \delta), (3, \alpha)\},$$

$$\{(2, \beta), (3, \alpha), (4, \alpha)\}, \{(1, \alpha), (3, \beta), (4, \gamma)\},$$

$$\{(1, \beta), (2, \delta), (4, \delta)\}, \{(1, \alpha), (2, \alpha), (4, \alpha)\},$$

$$\{(2, \beta), (3, \gamma), (4, \beta)\}, \{(2, \beta), (3, \gamma), (5, \alpha)\},$$

$$\{(1, \beta), (2, \delta), (4, \alpha)\}, \{(2, \beta), (3, \beta), (5, \alpha)\}\} \quad (5.3.2)$$

be a relation expressed for simplicity only by the indices of elements from V_T and L. Then this relation may be stored efficiently in the tree of Fig. 5.19. Since elements are ordered in the tree, a search of the relation is facilitated.

In order to obtain more general structures several recursive data structures have been defined. As an example we consider the iconic/symbolic data structure (ISDS). In this structure information about an image may be intensity values (numbers) or symbolic descriptions. Therefore, an ISDS may contain either pictorial information in an array of picture elements or relational information in a list of property values. The picture elements may be intensity values or pointers to another ISDS – thereby introducing a recursive structure. An example of an ISDS is given in Fig. 5.20.

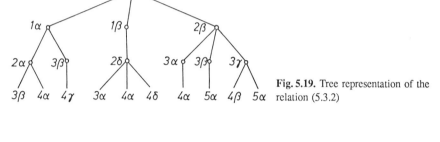

Fig. 5.19. Tree representation of the relation (5.3.2)

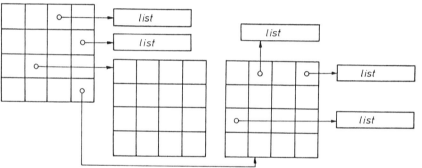

Fig. 5.20. An example of an iconic/symbolic data structure

5.3.2 Data Bases for Patterns

A data base for patterns in the sense of Sect. 5.2.1 is not just a collection of data concerning patterns, but also implies the availability of a data base management system which supplies some kind of high level routines for manipulation of data (a data sublanguage). An example of such data bases for image processing are systems to represent map data. Data may be organized in a *spatial data structure* as shown in Fig. 5.21 which basically is a relation. It consists of a header containing identifying information – such as the name of a city or county and the type of spatial data (see below) – and a table with two columns. The first column contains the names of relations specifying the spatial data, and the second contains pointers to these relations. Since the relations may themselves contain spatial data structures, a recursive definition results. The first row of the table is a binary relation whose two domains S_1 and S_2 are attributes (coordinates,

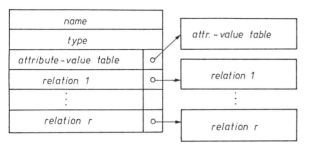

Fig. 5.21. A spatial data structure

population, length, ...) and their values, respectively. Some basic types of spatial data are the point (a pair of coordinates), the node (a point with an attribute-value table, for instance, a city or a road junction), the chain (an ordered set of points with a start and a stop node), the line (an ordered set of chains), the boundary (a line with common start and stop node), and the region (an area with a boundary). These concepts allow the representation of geographical items of a map in the spatial data structure. Within this structure an appropriate system can generate answers to a variety of queries such as "find all cities in a certain region".

Another example is an interactive system for extraction of objects from aerial images. Objects may be pointlike (junction, tower, ...), linelike (road, river, ...), or regionlike (forest, island, ...). The object description contains geometric and semantic properties and relations to other objects. It is assumed that images are taken in a temporal sequence, and data concerning objects are stored in a data base. This is organized into several files for different kinds of information. Each file is divided into pages such that each page holds information about all images belonging to one temporal unit. A header file contains global information about the images and pointers to files with the set of objects and with relations. From these files there are pointers to other files with further information. The system handles queries such as "find all roads crossing a specified road" or "find all objects which have a certain relation to other objects".

It is often useful to represent various types of images from different sources in a uniform manner, such as images from multispectral scanners or side-looking radar. In one example of such a case the data base is decoupled from particular implementations of pattern analysis and used for performance evaluation, error analysis, knowledge representation, and the learning of structural descriptions. Images are stored in two representations, a signal pyramid and a symbolic pyramid. The former is similar to the pyramid data structure described in Sect. 5.3.1, the latter is a symbolic description of the image at various levels of abstraction. An example is the description tree in Fig. 5.22 which is stored in an image description file (IDF). This contains general information about image segmentation and a block describing each entity at the various levels. The descriptive block contains, for instance, a symbolic name of the entity, coordinates of a surrounding rectangle, and the parent and descendants in the description tree. The signal and symbolic representations are related by mappings which yield associations between both. There is a scene file for each image in the data base which contains

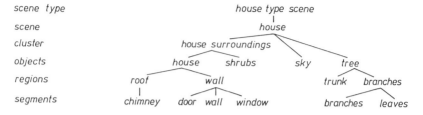

Fig. 5.22. The image description tree contains complete descriptions on varying levels of abstraction. Most details are provided at the lowest level

general information about the image, names of files containing the signal pyramid, and pointers to one (or several) IDF of the image. In addition, a relational data base is generated from the scene files and IDF in order to support queries to the data base such as "find all images from 11-channel scanners containing forests". Relations in the data base are, for instance, data file number, sensor type, scene type or IDF number, symbolic level, or symbolic name. Incidentally, the data base is an integral part of a larger system which also contains procedures for picture processing (methods in the sense of Sect. 1.4) and is implemented by a specialized hardware configuration. An extension is the inclusion of nondeterministic relations such as "region R has area 1 with probability p_1, area 2 with p_2, and area 3 with p_3".

An example of a complete result data base was designed and implemented in connection with the recognition of connected speech. It is a network-type data base organized in the three dimensions time of occurrence of a hypothesis, level of a hypothesis, and alternatives of hypotheses. At any time the data base contains the latest hypotheses about an utterance, but there is no information concerning the history of creation, modification, or deletion of hypotheses. This is stored, if necessary, in so-called local data bases. Hypotheses are uniformly represented by nodes and nodes are interrelated by links. An example of a section of the data base and of the information contained in a node is given in Fig. 5.23. Links between a hypothesis H1 on a higher level and hypotheses H2, H3 on a lower level may be of three types. An OR link means that H1 is supported by H2 or H3, an AND link means that H1 needs H2 and H3 for its support, and a SEQ link is an AND link with the additional requirement that H2 and H3 occur in a certain order, for example, a temporal order. The data base is monitored for changes to detect events which may require further processing, it has centralized read and write facilities, and it is equipped with locking and tagging facilities. Fairly arbitrary subsets of hypotheses in the data base can be restricted by locking to be accessed only by a particular module of the analysis system. Tagging does not restrict access, but it notifies a module that other modules also use and perhaps change the data. This data base provides a good example of a results data base shared by the modules of a pattern analysis system. Other approaches to the design of such result data bases are mentioned in Chap. 7.

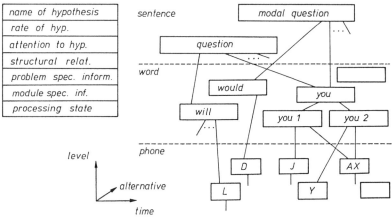

Fig. 5.23. A section of a results data base consisting of nodes (hypotheses) and links, [5.28]. The information contained in the nodes may be structured into the seven classes given in the figure; this format is the same on all levels. Examples of the information in the classes are *(from top to bottom)*: name and level of the hypothesis; estimated reliability; computational effort expected or spent to obtain the hypothesis; type of links to other hypotheses; temporal location of the hypothesis; an entry used by modules to store information about the hypothesis; and an entry containing an evaluation of the processing state, e.g., the hypothesis is "verified"

5.4 Bibliographical Remarks

Text books on data structures are [5.2–5], data structures for image processing in C are treated in [5.6], abstract data types are introduced in [5.7], and object-oriented programming is treated in [5.8]. General aspects of data models and data bases are covered in [5.1, 9–13]. Extensions of the relational model to incorporate non-normal forms are treated in [5.14], object-oriented data bases in [5.15, 16], and semantic data models and knowledge data bases in [5.17, 18].

Surveys of data structures for image processing are [5.19, 20]. The examples of data structures developed and used in pattern analysis are from [5.21–30]. The examples of data bases are from [5.31–35].

6. Control

Having discussed a variety of methods for preprocessing, extraction of simple constituents, and classification we proceeded with considering some details of representation and manipulation of data, in particular data which are results of the aforementioned steps of processing. Now we turn to a discussion of some general principles for the choice of processing methods and the order of their application to subsets of competing intermediate results; specification of these steps is referred to as *control*. A distinct control module is introduced into a pattern analysis system in order to allow a flexible system structure as indicated in Sect. 1.4. It became apparent that in analysis of (complex) patterns a module should be at hand which makes the best possible use of available processing methods for every pattern $f^r(x) \in \Omega$. One sequence of methods or processing steps, which is suited for a particular pattern $f^r(x)$, need not be optimal for another pattern $f^s(x)$. The control module should be able to find this optimal sequence, or at least a fairly good sequence of processing steps depending on the pattern offered to the system.

To put it another way: given the problem of finding a correct symbolic description of a set of sample values, the control module should guide the processing such that the solution of the problem is found; the guidance should take into account all relevant results of the processing and also all relevant knowledge about the structure of the patterns and the task domain. Such a system may exhibit behavior which to an observer shows some aspects of intelligence. In this chapter the following topics are discussed:
1. The Problem – clarify what is meant by control.
2. Some Common Structures and Their Representation – some ideas on control.
3. Design and Planning – restriction of the solution graph.
4. Judgement – estimate the quality of alternatives.
5. Searching – find an optimal path in the solution graph.

6.1 The Problem

Pattern analysis may be viewed as a *problem-solving activity*. The *initial state* of the problem is defined by the input pattern $f^r(x) \in \Omega$ of the system. By a sequence of actions (processing modules) which are ordered according to sequence (or step) number, the initial state undergoes a sequence of state transitions which lead to a sequence of new states, the *states* being defined by the content of the results data base. The system stops if either a *goal state* (a symbolic description B^r) is reached or no further action is possible. This process is depicted in Fig. 6.1 which is to be understood as a state transition schema similar to the hierarchical data base schema in Fig. 5.8. At any step there may be multiple occurrences of the nodes in the schema. Thus at any step the schema indicates that a set of states may exist, each state defined by a set of data. The processing modules available in the system yield a set of possible transformations $\{T\}$ of the data. The current set of data $\{\text{Data}\}$ in the results data base restricts $\{T\}$ to a set of applicable transformations $\{T/\{\text{Data}\}\}$. For instance, if a transformation $T_i \in \{T\}$ operates on a string of symbols with certain properties and $\{\text{Data}\}$ contains no such string, then T_i is not applicable to $\{\text{Data}\}$ and $T_i \notin \{T/\{\text{Data}\}\}$.

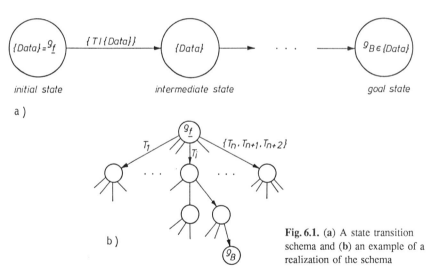

Fig. 6.1. (a) A state transition schema and (b) an example of a realization of the schema

From among the set $\{T/\{\text{Data}\}_i\}$ of applicable transformations in the ith step of processing a particular transformation has to be selected. For example, in speech recognition there may be intermediate data consisting of phonemes and some word hypotheses. A choice has to be made whether to generate additional word hypotheses or whether to check the syntactic and semantic consistency of already available word hypotheses. Furthermore, it may be useful or necessary to apply a transformation to a promising subset rather than to all available data. For example, in image analysis a contour detector may generate a large set of line elements and one may first try to obtain larger lines from reliably detected line elements only.

Application of all possible transformations $\{T/\{\text{Data}\}\}$ at any step yields the complete graph of the problem. An example of the graph is shown in Fig. 6.1. It is an implicitly defined graph in the sense of Sect. 5.1.2. The following points are evident: One may select a single transformation or several transformations in parallel (like T_n, T_{n+1}, T_{n+2}); the same data in the results data base may be obtained in several different ways; the system may stop with meaningless data because either no transformation is applicable or a time limit has elapsed; several similar but non-identical descriptions of an input pattern or answers to a user's question may be acceptable for the user. A useful system activity is some path in the graph leading from the input to an acceptable output.

In a well-designed pattern analysis system one would expect that for many (in the limit, for all) patterns f^r belonging to a particular task domain Ω there is at least one path in the solution graph which leads from the output f^r to the symbolic description B^r. It is the task of the control module to find this path. If the same sequence of actions is suited for all (or nearly all) patterns f^r, this sequence may be specified once and for all, and no control module is necessary. In this case the system has a fixed control structure, for example, the hierarchical structure in Fig. 6.4a. Although there are relevant applications where this approach has proven successful, this case is of less interest here. If the same sequence of actions does not yield satisfactory results, a distinct control module is definitely necessary. It has to find a path which is optimal in some sense or at least fairly good.

The following questions arise with the above stated control problem:

1. What is a suitable set $\{T\}$ of transformations?
2. What are suitable algorithms to find an optimal path to a solution?
3. Is it possible to find an optimal path with reasonable effort?
4. Given $\{T\}$, how should the set $\{\text{Data}\}$ be represented to obtain an efficient system?

The first question merely stresses the fact that a system initially must be given an appropriate set of transformations (or methods and knowledge with reference to Fig. 1.4). If the set is too small or too weak, there may be too many inputs or problems where no acceptable output or solution can be determined; if the set is too large, the search space implied by Fig. 6.1 will become excessively large. Usually, the selection of $\{T\}$ requires a thorough experimental investigation of various alternative algorithms with a representative sample of problems. Questions two and three indicate that two points have to be distinguished. An optimal path allows one to compute a solution to a given problem with minimum cost, for example, minimum time of computation, provided a solution does exist at all. How much effort it takes to find this optimal path, is another point as indicated by question three above. Since it may be excessively costly to find an optimal path in the graph of Fig. 6.1, one may restrict the search to a suboptimal path. Such a suboptimal path will usually require more time to carry out the transformations defined by it, but less time to search for it than the optimal path. Furthermore, by restricting the search to a suboptimal path it may happen that there are inputs (problems) where no acceptable output (solution) is found at all. Of course, in any practical situation one will have to infer the solution of the above questions from evaluation of a finite sample ω. Questions may be generalized by requiring a solution for not all, but for many patterns. The representation of the data in the set {Data} is a problem since, as evident in Fig. 5.4, one data structure, in this case a graph, may be represented in different ways. Answers to the above questions which are based on sound theory are desirable but not a prerequisite for designing a pattern analysis system. By initial experiments one tries to obtain a reasonable set $\{T\}$ and tries to find the remaining answers more or less heuristically. The performance of the system does, or does not, justify the approach of the designer.

The nodes of the graph in Fig. 6.1 are ordered by step number (not by time) and may give the impression of a hierarchical system in the sense of Sect. 6.2.2. However, this is not the case since a transformation may equally well generate a hypothesis about simple constituents which are to be expected in the pattern. After the generation of hypotheses and their inclusion in the set {Data}, transformations which test these hypotheses will become applicable. The generation of hypotheses requires knowledge about the structural properties of patterns and about the task domain; the incorporation of such knowledge is discussed in Chap. 7.

It is apparent from the above discussion that appropriate techniques for searching a path in the graph shown in Fig. 6.1 are of extreme importance if the control module is to be successful in directing the activities of the system in Fig. 1.4. There are four basic general ideas involved in control: *designing* and *planning*, that is, restriction of the search space; *judgement*, that is, evaluation of alternatives; and *searching*, that is, finding a path in a graph or tree. The goal of all these ideas is to avoid a blind search, which is unfeasible.

6.2 Some Control Structures and Their Representation

6.2.1 Interaction

It was recognized early that the execution of a fixed sequence of processing steps on a set of complex patterns often resulted in systems with poor performance. A good way out in such cases is to use an *interactive system* combining a human operator and a computer. The human's intelligence and experience (knowledge) are used to select a transformation T, a subset of the data, and possibly also to interpret some portions of the patterns; the computer's processing power is used to execute the selected transformation. The pair {human, computer} performs a sequence of actions on the pattern as indicated in Fig. 6.2. Thus the operator acts as a control module and selects one path through the graph of Fig. 6.1. By trying several alternative paths a fairly good sequence of operations may be

Fig. 6.2. State transitions in an interactive system. (D,I,S) relevant results are displayed to an operator, are interpreted by him, and he selects a processing method (transformation). (T) the operator-selected transformation is executed by a computer

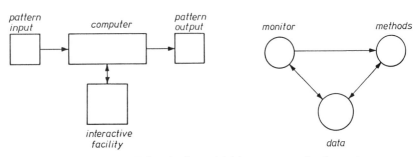

Fig. 6.3. Common hardware (*left*) and software (*right*) components of an interactive system

developed. Often an interactive system also provides a means for storing this sequence and executing it automatically. So, if a fixed sequence of operations is adequate in a task domain, an interactive system facilitates its discovery and allows its execution.

Some aspects common to many systems are shown in Fig. 6.3. The software exhibits some resemblance to Fig. 1.4 of Sect. 1.4. The knowledge and the control module are missing in the interactive system; instead a monitor is provided which handles the operator actions. Apparently, if the operator is removed from the system, he has to be replaced by appropriate software modules. The components in Fig. 6.3 include facilities for archiving of data, a convenient user interface, and may even include an expert system giving advice about the processing of certain types of patterns.

6.2.2 Hierarchical and Model-Directed Systems

Two important types of systems are the *hierarchical* and the *model-directed* or generative systems shown in Fig. 6.4. Both consist of n subsystems or processing modules. In the hierarchical system the output of the ith module is the input of the $(i+1)$th module. Processing starts at the lowest level, the sample values of the pattern, and proceeds bottom-up to the highest level. Therefore, this processing strategy is also called a *bottom-up* strategy. Systems of this type are common in pattern classification tasks, with typical modules being those of Fig. 1.3. The sequence of transformations is defined by the system structure.

The model-directed system starts with the generation of a hypothesis at some high-level module which is passed down for testing to a lower level module. Therefore, this strategy is also called a *top-down* strategy. Testing may involve generation of new hypotheses and/or operations on available data. The result of

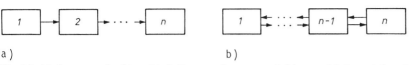

Fig. 6.4. (a) Structure of a hierarchical (*bottom-up*) system and (b) a model-directed (*top-down*) system, each with n processing modules

the test is passed back to the higher level module. Purely model-directed processing systems have been developed, for instance, in syntactical pattern recognition. Depending on the details of the strategy this type of system exhibits some flexibility in processing.

Combination of both approaches results in the *hypothesize-and-test structure*. One or several hypotheses about the interpretation of a pattern are generated bottom-up by a coarse analysis of the data, then the hypotheses are carefully verified by a model-guided phase.

6.2.3 Heterarchical and Data-Base-Oriented Systems

Two types of systems allowing fairly arbitrary control strategies are the *heterarchical* and the *data-base-oriented* systems of Fig. 6.5. The heterarchical system is a network where in the extreme case every module is connected to all other modules. The problem with this approach is the large number of connections. The number of connections between modules is reduced in the data-base-oriented system where each module is connected to one central exchange (the data base), and communication between modules is restricted to occur only via this central exchange. In fact, Fig. 6.5b is very similar to Fig. 1.4. In the latter the modules are grouped into methods, control, and knowledge, whereas in the former the modules have been left unspecified. The control module is viewed as an independent entity like all the other modules. It may be treated as a plug-in of the whole system.

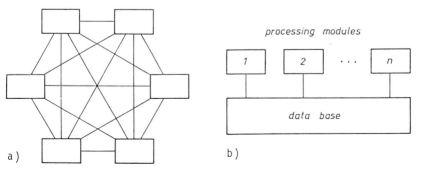

Fig. 6.5. (a) A heterarchical system, and (b) a data-base-oriented system

6.2.4 Network of States

Another class of systems is based on specifying all the possible states and state transitions in advance for all processing steps. States and transitions are stored in a *network of states*. Analysis of a particular pattern then amounts to finding a path through the set of states; this problem is treated in Sect. 6.5. In these terms there is nothing new with respect to Sect. 6.1. The important point is that in the systems of Fig. 6.2 the state set is not specified in advance; rather, there are algorithms, the processing modules, which allow the states to be computed. If states are specified in advance, a good deal of computation is saved. Which approach is preferable is a question of a time-space trade-off, since the specification of states requires that they be stored. The remaining computations concern the optimal path through the states (the judgement and search problem) and computations requiring the input pattern. An explicitly represented network of states is always finite; the implicit definition of a set of states by the specification of a set of transformations for their generation may well define a potentially infinite set of states. A graphical sketch of this type of system is given in Fig. 6.6 with the example taken from speech recognition. In this case one may think of the processing steps as time instances where at the νth instance a parametric representation c_ν of a frame of speech is compared to the set of possible prototypes associated with phones. State transitions are transitions from one phone to another; they are restricted by knowledge about words, sentences, and meaning of sentences. By restricting state transitions knowledge is directly applied to reduce the complexity of the network and to reduce the amount of searching. Figures 6.5 and 6.6 may be viewed as examples of implicit and explicit definitions of a search graph, respectively.

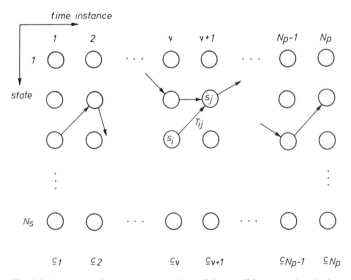

Fig. 6.6. A system using a precomputed set of the possible states. A path through the set of states is indicated by arrows. It is the result of analysis of a pattern

6.2.5 Representation by Abstract Programs

If the control module is conceived as an independent component of an analysis system, it is possible to treat it as a plug-in unit which may be specified, altered, and replaced. This idea may be based on the notion of an *abstract program* (or control graph or flowchart schema). The theory of these abstract programs will not be developed in detail here. An important property is that "an *abstract program* depicts the *control structure* of the program, leaving much of the detail to be specified in an *interpretation*." To put it differently, it helps to distinguish between the sequence of processing steps and the algorithms applied at each step. This allows one to change the control module, for instance, by reprogramming it or by treating the abstract program as the input of a supervisor program. In this latter case the core of the control module is the supervisor program which does not change. The supervisor executes processing steps in the sequence specified by the abstract program whose statements are read in as the data of the supervisor.

An abstract program P_a is made up of an *alphabet* A and *statements* S over A. The alphabet is a finite subset of *constants* and *individual variables*. The constants may be function constants or predicate constants. The individual variables may be input variables, program variables, and output variables. The values of those variables constitute the set {Data} in Sect. 6.1. Each of the function constants maps a domain D^n into D; each of the predicate constants is a predicate mapping of D^n into $\{T = \text{true}, F = \text{false}\}$. However, the mappings remain unspecified in the abstract program P_a. *Statements* of the abstract program P_a over A use *terms* $t(f, y)$ containing only input variables or input and program variables, respectively. The statements are START, ASSIGN, TEST, HALT, LOOP (undefined), and JUMP.

An abstract program is constructed from statements such that there is one START statement and every ASSIGN and TEST statement is on a path from the START to a HALT or LOOP statement. A particular abstract program may be represented by a sequence of statements or in graphical form by a *flowchart schema* (or pattern analysis graph). It is assumed that graphical representation is obvious and need not be considered here.

A *program* P is obtained from an abstract program P_a if an *interpretation* I of P_a is given. An interpretation consists of

1. a domain D which is a nonempty set of elements,
2. assignment of a total function mapping D^n into D to each function constant,
3. assignment of a total predicate mapping D^n into $\{T, F\}$ to each predicate constant.

A program is thus a pair

$$P = \{P_a, I\} . \tag{6.2.1}$$

If an initial value f^r is assigned to the input variables, the program P can be executed.

The above abstract programs P_a allow the specification of fairly arbitrary control structures. Nevertheless, their power may be enhanced by introducing *recursive abstract programs* P_{ra}, but we omit this point here. It can be shown that recursive abstract programs are more powerful than abstract programs, that is there are P_{ra} having no equivalent P_a.

Although this formalism is fairly general, there are three deficiencies. First, there are no explicit means for structuring and modularizing P_a or P_{ra}. For instance, no provision is made to use a $P(1)_{ra}$ as a statement of a $P(1)_a$ and vice versa. This is not *necessary* because both may be summarized in one larger $P(2)_{ra}$, but it is *useful* to have this option. Second, it is possible to represent a P_a in a straightforward manner by a graph, but this is not true for a P_{ra}. Therefore, a more general graph definition – the hierarchical graph – is introduced in the next section. The third deficiency concerns parallel actions and is deferred to Sect. 6.2.7.

6.2.6 Hierarchical Graphs

Informally, a *hierarchical graph*, or h graph, is a graph whose nodes may be graphs, with the latter graphs containing nodes which again may be graphs, and so on. They were introduced to represent the semantics of programs, and we will not define them more formally. Rather, the h graph is illustrated by a simple example in Fig. 6.7.

The h graph in Fig. 6.7 shows a part of a recursive picture parsing procedure where most of the details are not presented because only the concept of the h graph is of interest here. It gives a comprehensive presentation of the h graph which is supposed to be self-explanatory; in addition a compressed version is given with the nodes marked only symbolically. On the highest level there is only one node $v0$. However, $v0$ is not primitive because its content ($v0$) is itself

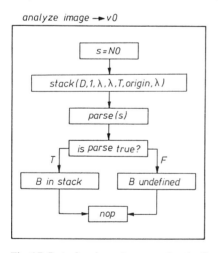
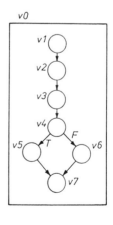

Fig. 6.7. Part of an h graph representing the flow of control (the sequence of processing steps) in an image parsing system

a graph with seven nodes $v1,\ldots,v7$. From the figure we obtain the information that there is a system which analyzes images. Furthermore, from the content of node $v0$ it is evident that a stack is initialized first, then in $v3$ a parse is done, and if the parse is successful, the description B is found in the stack; otherwise B is undefined. This gives a rough idea of what is going on. It is possible to go to any level of detail by examining nodes. It is possible, for example, that a node v contains other nodes among which is a node v' containing v again, so recursion is possible in an h graph.

6.2.7 Petri Nets

The examples of the last two sections were strictly sequential, that is one module was activated at any processing step. A limited number of possible parallel actions may also be represented, for instance, by h graphs if nodes are introduced containing a set of parallel processes. A different node is necessary for any possible set. If there are m modules (or processing methods), in the most general case any subset out of the 2^m possible subsets may be active at a given time. To represent this general case a graph with an exponentially growing number of nodes is necessary. This may be avoided by using Petri nets which are introduced by the example shown in Fig. 6.8.

A *Petri graph* has two types of nodes, which are called *places* V and *transitions* T; in addition it specifies two functions, the *input function* φ_{in} from V to T, and the *output function* φ_{out} from T to V. A *Petri net* in addition provides an initial labeling L of the places. In the graphical representation a circle is attributed to any $v \in V$, a bar to any $t \in T$, an arrow from v to t for any v, t with $\varphi_{\text{in}}(v, t) = 1$, an arrow from t to v for any t, v with $\varphi_{\text{out}}(t, v) = 1$, and a dot is placed in any circle representing a v with $L(v) = 1$. The application of these rules

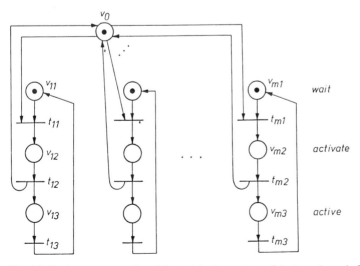

Fig. 6.8. Petri-net representation of the activity in a system of the type shown in Fig. 6.5b

results in a graphical representation of the type shown in Fig. 6.8. A transition is *active* if all the places connected to it by arrows are labeled (or contain a dot). An active transition may *fire*. Firing of a transition causes labeling of the places connected to the firing transition by outgoing arrows and removal of the labels connected to the transition by incoming arrows.

Figure 6.8 may be viewed as a representation of parallel activities of the kind possible in the system shown in Fig. 6.5b. Place v_0 is the control module and places $v_{ij}, i = 1, \ldots, m$ are m processing modules in various states of activity. Initially all modules are waiting in place v_{i1} for activation by v_0 and all transitions t_{i1} are active. At some time v_0 will select a module, say the first one, to be activated. This causes t_{11} to fire, removes the dot from v_0 and v_{11}, and puts a dot into v_{12}. The only active transition now is t_{12}. If t_{12} fires, the dot is removed from v_{12} and dots are put into v_0 and v_{13}. Now module 1 is active and does some kind of processing, module v_0 is ready to select some other module, and transitions $t_{i1}, i = 2, \ldots, m$ and t_{13} are active. This process apparently allows any subset of the m processing modules to be active at any time. Also any subset of the active modules may finish its activity at any time. But the control module can activate only one module at any time.

One point should be emphasized. The Petri net is useful for representing the states of activity of the system in Fig. 6.5b and it is true that this representation would be intricate, to say the least, with a graph or h graph. But, remembering the implicit and explicit specification of a graph in Sect. 5.1.2, it is not mandatory to represent these states; rather it is sufficient to specify an algorithm which generates the states of system activity. Incidentally, the state of system activity given by the activity status of the processing modules should not be confused with the state of problem solution given by the content of the data base. Specification of such an algorithm is quite possible by the methods of Sects. 6.2.5, 6. This gives us reason to consider h graphs again in another example.

Let us consider the system in Fig. 6.5b again and assume that we have available m processing modules and p processors for executing the processing algorithms of the modules. It should be possible that any combination of up to p processes can be executed on any number of the p (possibly all) processors. The case where a module is active on more than one processor is admissible. This may occur if, for instance, an edge-detecting module is operating on two separate sections of an image. Figure 6.9 shows an example of a possible algorithm initiating this kind of activity. It does not show the possible states of system activity as in Fig. 6.8, but it gives a condensed idea of how they are generated. Of course, nodes $v5, v6$, or $v9$ would require further specification. Thus the h graph or flowchart is adequate to represent the complicated control actions of a control module. A Petri-net representation of the states of system activity is also possible, but it is fairly complicated because all combinations of processes and processors are allowed.

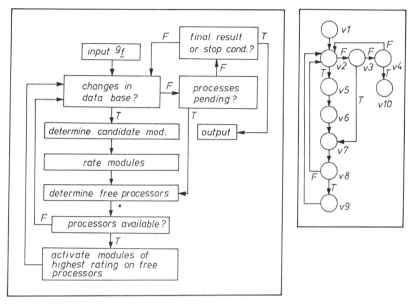

Fig. 6.9. Hierarchical graph representation of an algorithm which activates processing modules on available processors

6.3 Designing and Planning

It is unfeasible and probably also unnecessary for a system to treat pattern analysis completely as a problem of finding a path in a graph. Rather one tries in a prior *design phase* of the system to reduce the graph as much as possible. This can be done by selecting a limited set $\{T\}$ of transformations, by reducing $\{Data\}$ to constraine subset selection, and, of course, by trying to find parts in the graph where the same sequence of processing steps is executed for (almost) all input patterns. The latter results in an analysis system which executes some of the processing steps in a fixed order and only selects among alternative transformations and/or subsets of data where this is really unavoidable. A powerful interactive image processing system is very useful in this design phase.

In the *planning phase* one first tries to obtain a rough idea of the dominant segments of a pattern, image or speech, and to use this to select the processing methods and data segments which should be used next for a detailed analysis. This also results in a reduction of the search space by reducing the number of nodes and edges in the graph of Fig. 6.1. Often in image and speech understanding the planning phase is executed as a fixed sequence of processing steps yielding initial guesses or hypotheses about segments, objects, or words. The planning phase proceeds from a quick, incomplete, rough first look to a deliberate, thorough, and detailed scrutiny. With this in mind it is convenient in image processing to use a pyramid data structure or a resolution hierarchy.

To summarize, the design phase is used to constrain the graph in Fig. 6.1 *for all* input patterns of a task domain Ω; the planning phase is used to obtain additional constraints for *one* particular input pattern.

6.4 Judgements

6.4.1 General Aspects

In order to avoid a blind search in the search graph indicated in Fig. 6.1 it is necessary to *judge* the available processing alternatives. As discussed previously such alternatives may arise from

1. different *processing modules* which can compute the same type of result (e.g. the computation of line segments by different algorithms or by the same algorithm but using different parameters),
2. competing or *uncertain results* for the interpretation of data (e.g. several class labels for some segment of speech or a "homogeneous" region obtained from region growing and a contour line found inside this region by an independent line finder),
3. *imprecise results* obtained from preprocessing and segmentation (e.g. the line ending is "very near" to the junction; or the depth of the object corner is between 1.00 and 1.05 m).

The consideration and quantitative treatment of points 1–3 is summarized as *judgement*. The quantitative evaluation of results is often termed *scoring*, where a score is usually limited to measuring the reliability of a result.

In order to cope with these problems there are in general three sources of information available which may be used by an automatic system:

1. Problem specific knowledge which is represented in the system to model the task domain Ω – a priori judgement.
2. General knowledge of the system designer concerning, for example, the properties of processing algorithms – a priori judgement.
3. Intermediate results computed by the system during the analysis of an input pattern f^r – a posteriori judgement.

The process of computing intermediate results often implies (at least heuristically) an approach to the estimation of the *reliability* (or certainty) of a result. For example, the reliability of a line may be based on the line contrast; the reliability of a phone may be derived from the test variables of the classifier. In addition, a priori knowledge may be used to estimate the *precision* of a result, for example, to rate the precision of depth estimates from a stereo algorithm or from a laser range finder. Task specific knowledge may be used to rate the *importance* of a result, for example, to estimate that a certain word has a high pragmatic importance in the understanding of an utterance or that a long line like structure is important in the interpretation of satellite images. Reliability, precision, and importance combined make up the *priority* of a node in the search graph in Fig. 6.1. This means that from among several nodes the one having the highest priority should be considered next.

6.4.2 Importance

Importance is related to the contribution of a result or transformation to the final success of the analysis. Since we assume a knowledge based approach to pattern analysis a *model of the task domain* is available. Possible representations of such models are discussed in Chap. 7, for example, production rules or semantic nets. From the model one can derive measures of "closeness" (or "distance") of a result or an action to the sample values f^r or to elements of an initial description I^r; one can also derive measures of closeness to a desired goal of processing. In a data-driven processing strategy the actions that reach a goal state most quickly are the most important; in a model-driven strategy the actions which lead quickly to hypotheses which can be verified by observed data are the most important. In a flexible system phases of data-driven and model-driven control may alternate with each other. The selection of these phases may be based on the importance of the expected results.

In the phase of recognizing objects in a model-driven strategy one has to decide which segmentation objects to compute or to inspect. The most important segmentation objects (or simple constituents, properties, features) are those which best discriminate among a set of objects. A "good" segmentation object is one which is present in about half of the objects and absent in the other half. This segmentation object may be found in an observed pattern or not, but in any case the set of potential objects will be reduced by about one half. Obviously, an important segmentation object may be detectable only with low reliability and its attribute values may have low precision.

More sophisticated measures of importance can in principle be derived from estimating the information content of a result or transformation.

6.4.3 Reliability

One of the standard approaches to measuring the reliability of a result is probability theory and the estimation of statistical parameters. Let S be a set of propositions (or events, results, hypotheses). If $s, t \in S$ are two pairwise disjoint events, then with $s \cap t = \phi$, the relations

$$P(s \vee t) = P(s) + P(t) ,$$

$$P(s) + P(\neg s) = 1 \qquad (6.4.1)$$

hold where "$\neg s$" means the negation of s or "NOT s". It is beyond the scope of this book to review probability and estimation theory, but some points are discussed in the following. The advantages of probability theory are that it provides a well-developed theory of probability measures, it gives precise operations for the estimation of probabilities and statistical parameters, and, with certain assumptions, it gives confidence intervals for the estimates. In addition, classifiers operating according to Theorem 4.2.3 in Sect. 4.2.1 give directly the a posteriori probability of a decision. One of the disadvantages is that in some cases it does not seem reasonable or possible to assign a probability to a proposition. For example, assume the proposition is "I am fairly convinced that the visible object is a car". How should one assign or estimate the probability that this proposition is correct? Another disadvantage arises from (6.4.1). Assume a set $T = \{A, B, C, D\}$ of pairwise disjoint events, and assume that there is no evidence to show which event is correct. In this case one usually assumes equally probable events, that is $P(A) = P(B) = P(C) = P(D) = 0.25$. From (6.4.1) it is now "known" that $P(A \vee B \vee C) = 0.75 > P(D) = 0.25$, that is, we "know" that the event $A \vee B \vee C$ is more probable that the event D, although initially no knowledge of this type was given. This problem caused the introduction of the so-called fuzzy measures discussed below.

The conditional probability $P(H|E)$ that a hypothesis H is correct if certain evidence E is given can be computed in analogy to (4.2.19). If multiple independent sources of evidence $E_1, \ldots, E_n$ are available, the probability that H is correct is

$$P(H|E_1, \ldots, E_n) = \frac{P(E_1|H)}{P(E_1)} \cdots \frac{P(E_n|H)}{P(E_n)} P(H) . \qquad (6.4.2)$$

This is an example of the combination of evidence using probability measures. In general one would have to compute the conditional probability from

$$P(H|E_1, \ldots, E_N) = \frac{P(E_1, \ldots, E_n|H)P(H)}{P(E_1, \ldots, E_N)} , \qquad (6.4.3)$$

which requires the estimation of an n-dimensional conditional probability $P(E|H)$.

The above mentioned problems with probabilities can be avoided by fuzzy measures. A *fuzzy measure* V has to meet the restrictions

1. $V(O) = 0$,
2. $V(I) = 1$,
3. IF $s \in S$ entails $t \in S$, THEN $V(t) \geq V(s)$, $\qquad (6.4.4)$

where O and I are the impossible and certain event, respectively. This definition has to be specialized in order to be practically useful. A specialized fuzzy measure is the *credibility* CR which can be derived from a *basic probability assignment* (BPA) M. This has to meet the requirement

$$\sum_{s \in S} M(s) = 1 . \qquad (6.4.5)$$

The credibility of a proposition t is defined by

$$CR(t) = \sum_{s:\, s \text{ entails } t} M(s) . \qquad (6.4.6)$$

This definition avoids the problem in the above example which used the set T of disjoint events. Let the set of possible events be the power set of T, that is, $S = 2^T$. The relation "s entails t" then corresponds to "s is a subset of t". If one does not have or does not assume any knowledge about the probability of events, one assigns $M(A, B, C, D) = 1$, and $M(\ldots) = 0$ in any other case. This avoids a preference for any particular proposition. In general (6.4.5, 6) exclude (6.4.1). In addition to the credibility CR the *plausibility* PL is defined by

$$\text{PL}(t) = \sum_{s: s \text{ does not entail } \neg t} M(s) = \sum_{s \not\subseteq \neg t} M(s) . \tag{6.4.7}$$

Without proof we quote the relations

$$\begin{aligned} \text{PL}(s) &= 1 - \text{CR}(\neg s) , \\ \text{CR}(s) + \text{CR}(\neg s) &\leq 1 , \\ \text{PL}(s) + \text{PL}(\neg s) &\geq 1 , \\ \text{CR}(s) &\leq \text{PL}(s) . \end{aligned} \tag{6.4.8}$$

Intuitive insight into these relations is obtained from considering the above example using the set S. It may occur that there are two (or more) sources of evidence concerning the same set S of events. If BPAs M_1 and M_2 are assigned to the two sources, it is necessary to compute the combined BPA M resulting from the combination of both sources. The combined BPA is

$$M(s) = \frac{\sum_{t \cap u = s} M_1(t) M_2(u)}{1 - \sum_{t \cap u \neq \phi} M_1(t) M_2(u)} , \quad s, t, u \in S . \tag{6.4.9}$$

From the resulting M the values PL and CR of an event s supported by both sources of evidence can be computed.

Propositions s with $M(s) > 0$ are called *focal propositions*. The probability measure then is a special BPA, called the *dissonant BPA*. For any focal s and any t, where s does not entail t, it must be true in a dissonant BPA that s and t are disjoint. In this case only A, B, C, and/or D can be focal propositions, the relation

$$\text{PL}(s) = \text{CR}(s) = P(s) \tag{6.4.10}$$

holds, and the first relation in (6.4.8) reduces to the second relation in (6.4.1). The *consonant BPA* is another special BPA where focal propositions form a *hierarchy*. For example, focal propositions in the above example can be $\{A\}, \{A, B\}, \{A, B, C\}$, and $\{A, B, C, D\}$. In this case

$$\begin{aligned} \text{CR}(s \wedge t) &= \min \{\text{CR}(s), \text{CR}(t)\} = \text{NE}(s) , \\ \text{PL}(s \vee t) &= \max \{\text{PL}(s), \text{PL}(t)\} = \text{PO}(s) . \end{aligned} \tag{6.4.11}$$

Where $\text{NE}(s)$ and $\text{PO}(s)$ are called the *necessity* and *possibility* of s, respectively. For all s we have

$$\min\{\mathrm{NE}(s), \mathrm{NE}(\neg s)\} = 0\,,$$
$$\max\{\mathrm{PO}(s), \mathrm{PO}(\neg s)\} = 1\,,$$
$$\mathrm{PO}(s) < 1 \to \mathrm{NE}(s) = 0\,, \qquad (6.4.12)$$
$$\mathrm{NE}(s) > 0 \to \mathrm{PO}(s) = 1\,,$$
$$\mathrm{NE}(s) = 1 - \mathrm{PO}(\neg s)\,.$$

From credibility and plausibility the *belief interval* is defined by

$$\mathrm{BI}(s) = (\mathrm{CR}(s), \mathrm{PL}(s))\,, \qquad (6.4.13)$$

which is not to be confused with the confidence interval of statistics. From possibility and necessity one may derive a *certainty factor*

$$\mathrm{CF}(s) = [\mathrm{PO}(s) + \mathrm{NE}(s)]/2\,, \quad 0 \leq \mathrm{CF} \leq 1\,. \qquad (6.4.14)$$

The definition of a basic probability assignment includes the probability measure and the possibility and necessity as special cases. This generalization does not guarantee a better approach to the judgement of the reliability of a result since the problem is to model the properties of a task domain adequately and not to provide a theoretically sophisticated definition. However, it can be expected that a more general approach offers more potential for handling realistic problems.

6.4.4 Precision

The standard approach to judging imprecise propositions is the use of *fuzzy sets*. An element a can belong "more or less" to a fuzzy set A, that is, the characteristic function of the fuzzy set A may have values in the interval $[0, 1]$. For example, a proposition of the type "the student is young" or in general "X is A" constrains the set of possible values of the variable X. Let the unconstrained set of values of X be U, the set of values constrained by "X is A" be A. The set U is the *universe of discourse*. The *characteristic function* of the set A is denoted by μ_A. A finite fuzzy subset of U is defined as the set

$$A = \{u_i, \mu_A(u_i)\,, \quad i = 1, \ldots, n\}\,, \quad u_i \in U\,, \quad \mu_A(u_i) \in [0, 1]\,. \qquad (6.4.15)$$

If $\mu_A \in \{0, 1\}$, a *crisp subset* of U or a subset in the ordinary sense results. Some value $\mu_A(a)$ is viewed as the *possibility* that the proposition "X has value a" is correct if one assumes "X is A". The *possibility distribution* $\pi_x(a)$ gives the possibility that the proposition "X has value a" is correct assuming that "X is A" is correct. Equivalently, one may formulate that an X having value a has the *grade of membership* $\mu_A(a)$ to the set A. Therefore, the relation

$$\pi_x(a) = \mu_A(a) \qquad (6.4.16)$$

holds. The negation "X is not A" has the possibility distribution

$$\pi_x(a) = 1 - \mu_A(a) = \mu_{\bar{A}}(a)\,. \qquad (6.4.17)$$

6.4 Judgements

The combination of propositions by AND or OR, that is "X is A AND Y is B" or "X is A OR Y is B" has the possibility distributions

$$\pi_{x \wedge y}(a,b) = \mu_{A.B}(a,b) = \min \{\mu_A(a), \mu_B(b)\},$$
$$\pi_{x \vee y}(a,b) = \mu_{A+B}(a,b) = \max \{\mu_A(a), \mu_B(b)\}. \quad (6.4.18)$$

Figure 6.10 shows an example of a possibility distribution. It gives, for example, $\mu_{\text{young}}(22) = 0.66$, which may be interpreted that a student of age 22 years has the grade of membership $\mu = 0.66$ to the set of young students or that the proposition "the student is 22" has the possibility $\pi = 0.66$ of being correct provided the proposition "the student is young" is correct.

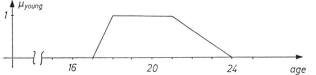

Fig. 6.10. An example of a possibility distribution

If a possibility distribution π or a characteristic function μ is given, it is possible to assign possibility PO and necessity NE to certain propositions. Let the proposition "X is A" (e.g. "the student is young") be given. Then from (6.4.11, 12) the possibility and necessity of a proposition "X is A_1" (e.g. "the student is aged 20–22") is

$$\text{PO}(A_1) = \max_{a \in A_1} \mu_A(a),$$
$$\text{NE}(A_1) = \min_{a \notin A_1} \left[1 - \mu_A(a)\right]. \quad (6.4.19)$$

The notion of fuzzy subsets also allows one to define *linguistic variables*, that is variables whose values are words or words chains expressed in a natural language. The set of words or word chains being values of the linguistic variable is called the *term set* T. For example the linguistic variable "age of student" might have the term set

T(age of student)

= {young, average, old, above average, not young}. (6.4.20)

Each value in T has to be defined by a fuzzy subset A, and an example of a definition of the value "young" is given in Fig. 6.10.

6.4.5 Priority

The above measures of reliability and precision may be used as a measure of priority, for example, in the selection of subsets of alternative results. However, a reliable result may be fairly insignificant for further processing because it contributes little to the success of further analysis, and vice versa. Several measures of priority MP have been developed. The principle of one approach is to compute several alternative results over a segment of the input pattern (speech or image) and their scores of reliability or precision. From these one can estimate the maximal score MSI achievable over the whole interval. Then for each partial result covering a certain subinterval two terms are available: the first is the score SRS of a partial result covering the subinterval, the second is an estimate of the additional score achievable over the remaining interval. An optimistic estimate is the maximum of the additional score which is the difference of the MSI and the maximal score MRI of a result covering the subinterval. The resulting measure of priority is

$$MP = SRS + (MSI - MRI) . \qquad (6.4.21)$$

6.5 Search Strategies

6.5.1 An Algorithm for a State-Space Search

In Sect. 6.1 it was stated that the task of the control module is to find a path leading to a solution in the graph of Fig. 6.1. Two basic approaches to problem solution may be distinguished. The first is *state-space search*, which works in a forward (or bottom-up or data-driven) direction and applies transformations to the initial state (in our case the input pattern f^r) until the goal state (in our case a symbolic description B^r) is obtained. The second is *problem reduction*, which works backwards (or top-down or goal-driven) and reduces the original problem to one or more subproblems, sub-subproblems and so on, until a set of primitive problems is obtained. The solution of primitive problems is assumed to be straightforward or well known and yields successively the solution of the higher level problems. No distinction between these approaches has been made so far because the transformed initial states as well as the derived subproblems may be considered as elements of {Data}, the content of the results data base. In any case, a state in the problem solution graph is defined by {Data} and may also contain an arbitrary mixture of transformed initial states and subproblems. The question is which techniques are available for the selection of paths in general and in the specific context of pattern analysis. It turns out that general search techniques are different for both approaches and, therefore, they will be distinguished in the following.

In a *state-space search* it may be feasible for simple problems to systematically apply all possible transformations until a stop occurs. A stop may occur

6.5 Search Strategies

if a solution is found, if no more new states can be generated, or if a prespecified expenditure is exceeded. Such a blind search, which may be done as a breadth-first or depth-first search, is not feasible in pattern analysis because of the combinatorial explosion of the solution graph. Rather, a technique is required which explores promising paths. A short outline of one such technique is given in the following. Referring to Fig. 6.1 again, assume that the search reached some state v_i. The application of all possible transformations to v_i yields all possible successor nodes (states) of v_i; this process is called the *expansion* of the node v_i.

It is required that for any node v_i a *cost function* $\varphi(v_i)$ gives the cost of an optimal (minimal cost) path leading from the initial node v_0 through node v_i to the goal node v_g. If there is a set of goal and/or start nodes, then $\varphi(v_i)$ refers to those yielding minimal cost. Of course, one may also use the value, score, or gain instead of cost, and search for the path of maximum value. Denoting the cost of an optimal path from the initial node v_0 to v_i by $\psi(v_i)$ and of an optimal path from node v_i to the goal v_g by $\chi(v_i)$, the cost of the optimal solution path constrained to go through v_i is assumed to be

$$\varphi(v_i) = \psi(v_i) + \chi(v_i) \,. \tag{6.5.1}$$

It should be noted that costs or scores need not to be additive in general, but we will only consider this case in the following since it allows simple and efficient search algorithms. The above functions will usually be unknown and available estimates have to be used instead of the costs. If $\hat{\varphi}, \hat{\psi}, \hat{\chi}$ are estimates of φ, ψ, χ, respectively, (6.5.1) is replaced by

$$\hat{\varphi}(v_i) = \hat{\psi}(v_i) + \hat{\chi}(v_i) \,. \tag{6.5.2}$$

An obvious estimate $\hat{\psi}$ of ψ is the cost of the path found from v_0 to v_i which is obtained by adding the edge costs. The availability of an estimate $\hat{\chi}$ is assumed for the moment; some remarks follow below. The algorithm given below requires a *consistent* or *monotone* computation of $\hat{\chi}$, that is

$$\hat{\chi}(v_j) - \hat{\chi}(v_k) \leq r(v_j, v_k) \,, \tag{6.5.3}$$

where $r(v_j, v_k)$ is the actual cost of a minimal cost path from node v_j to its successor v_k (provided, of course, that a path exists). This means that a cost estimate has to be "optimistic" in the sense that the estimate must not be larger than the actual costs.

An algorithm for state-space search is as follows:
The task is to find an optimal path in a graph, subject to the cost assignments mentioned above.
Put the start node v_0 on a list OPEN and evaluate $\hat{\varphi}(v_0)$ by (6.5.2).
While OPEN is not empty do:
    Remove from OPEN the node v_i with minimal $\hat{\varphi}(v_i)$ and put it on CLOSED.
    If v_i is a goal then do:
        Stop with SUCCESS (the optimal path is found and may be obtained

from tracing back the pointers).
Else do:
Expand v_i and evaluate $\hat{\varphi}$ for successors of v_i by (6.5.2).
Put those successors not yet on OPEN or CLOSED on OPEN and set pointers from them back to v_i.
Attribute the smaller of the $\hat{\varphi}$ values just and previously computed to the successors already on OPEN and redirect pointers to v_i from nodes with lowered $\hat{\varphi}$.
End if.
End while.
Stop with FAILURE (no path to a goal was found).
End search.

An example is deferred to Fig. 6.14 of Sect. 6.5.3. The properties of this algorithm, which is called A^* *algorithm*, are summarized in the following theorem.

Theorem 6.5.1.

1. The above algorithm always *terminates* for finite graphs; it *terminates* for infinite graphs if a solution path exists.
2. If $\hat{\psi}$ is chosen as mentioned above and if $\hat{\chi}$ is a lower bound on χ, the above algorithm is *admissible*, that is, it always finds an optimal path to a goal state if a path exists.
3. If the above algorithm has available a lower bound on χ which is everywhere strictly larger than the bound used by some other algorithm, then the above algorithm never expands more nodes than the other algorithm.
4. If A^* expands a node $v_i, \hat{\varphi}(v_i) \geq \varphi(v_0)$, that is costs are monotone.
5. If the monotone assumption (6.5.3) holds, A^* has already found an optimal path to a node v_i when it selects v_i for expansion and $\hat{\psi}(v_i) = \psi(v_i)$. Furthermore, the $\hat{\varphi}$ values of a sequence of nodes expanded by A^* are nondecreasing.

Proof. See, for instance, [6.1], pp. 57–65 or [6.17]. □

Incidentally, the consistency assumption thus assures that a node on CLOSED will never have to be put on OPEN again. Without (6.5.3) this might occur and would have to be taken into account in the algorithm. The nodes and pointers generated by the above algorithm form a search tree even if a general graph is searched. This is indicated in Fig. 6.11. The algorithm is optimal in the sense that it expands a small number of nodes. However, the number of nodes expanded is only one point affecting the value of the algorithm. Another point is the expenditure for computing estimates of $\hat{\varphi}$. An algorithm with $\hat{\chi} \equiv 0$ is admissible since this is a lower bound on χ. The resulting algorithm is called a *uniform-cost algorithm* because, with the estimate mentioned below (6.5.2), it expands nodes at uniform cost contours in the search tree. If the edge cost is set to unity, the cost of going from v_0 to v_i is equal to the depth of v_i in the search tree and, therefore, a *breadth-first search* is performed. It is mentioned in passing that general graph

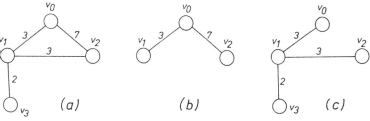

Fig. 6.11. (a) A graph with start node v_0 and associated costs at the edges [6.1]. (b) The tree generated after expansion of v_0. (c) The tree generated after expansion of v_1 and redirection of pointers

search algorithms can be formulated avoiding the assumption (6.5.1–3). They are not treated here because they are usually too complex to be used in control algorithms for pattern analysis.

An important point is the selection of appropriate estimates $\hat{\varphi}, \hat{\psi}$, and $\hat{\chi}$ in (6.5.2). An estimate of $\hat{\psi}$ was proposed already in connection with (6.5.2). An estimate $\hat{\chi} \equiv 0$, although admissible, will result in an inefficient search. A better lower bound on $\hat{\chi}$ may in some cases (for instance, in certain games) be obtained from heuristic information. Sources of such heuristic information in pattern analysis have been mentioned in Sect. 6.4.1. It is possible to use an estimate $\hat{\chi}$ which is not a lower bound on χ. This may reduce the search effort at the expense of losing admissibility, that is, a nonoptimal path to a solution may be found. If only any path to a goal is to be found, it is possible to use $\hat{\psi} \equiv 0$ as an estimate of ψ. In this case only the remaining costs are considered in the search, admissibility is lost, and it may even occur that no path at all is found if $\hat{\chi}$ is not a good estimate of χ. It is mentioned again that two questions of efficiency of a search algorithm have to be distinguished (see also questions 2 and 3 of Sect. 6.1):

1. What effort does it take to carry out the transformations (computations) required by the (optimal) solution path?
2. What effort does it take to find an (optimal) solution path in the graph of Fig. 6.1?
   This effort in turn is influenced by two factors:
   2.1. The number of nodes expanded when searching a path.
   2.2. The effort to compute an estimate $\hat{\varphi}$ of φ.

If a fixed sequence of transformations is appropriate for all patterns of a task domain Ω, then question 1 is the most important point, and the control module is unnecessary because a fixed control structure suffices. Otherwise question 2 becomes essential and it may altogether be cheapest to search for a nonoptimal path if this reduces the overall search effort.

Since heuristic considerations are necessary anyway in finding an estimate of φ, it is reasonable to apply them elsewhere as well. A problem is that the search tree may become very large; in this case one can use *pruning techniques* to reduce the complexity of the search tree and to retain only the most useful part of it. This is discussed in Sect. 6.5.3. Heuristic information should also be used to further restrict the set $\{T/\{\text{Data}\}\}$ of possible transformations at each step introduced in Sect. 6.1. The extreme case would be to restrict $\{T/\{\text{Data}\}\}$ such that only one T_i remains at each step, which may depend on the pattern.

6.5.2 An Algorithm for Searching AND/OR Trees

Next, a basic search technique for the *problem-reduction approach* is discussed. If a problem P is reduced to subproblems, the situation is as depicted for a simple example in Fig. 6.12. It shows that P is solved if either P_{11} and P_{12} and P_{13}, or P_{21} and P_{22}, or P_{31} and P_{32} can be solved. The AND condition is marked graphically by an arc. An equivalent reduction is shown in Fig. 6.12b, where P is solved if either P_1, or P_2, or P_3 can be solved. Each of the $P_i, i = 1, 2, 3$ is solved if all the successor problems of P_i can be solved. This shows that reduction of a problem can always be represented by a special tree, called an AND/OR *tree*, which on each level has only nodes of either type AND or type OR. An AND node has only OR nodes as its successors, and an OR node has only AND nodes as its successors. It may happen that problem reduction yields nodes with more than one parent node. In this case a general AND/OR *graph*, not a tree, results, but discussion is restricted here to the simpler AND/OR trees. The start node representing the original problem may be an AND or an OR node depending on its successors. The process of replacing problems by subproblems is repeated until a *primitive* (solved) *problem* is reached which need not be further reduced, or no more successors of a node can be generated. A node containing a primitive problem is called a *terminal node*; all other nodes are nonterminal nodes. The aim is to show that the initial problem is either solvable or unsolvable. This is done by using the following recursive definitions:

1. Determine nodes with *solvable problems* (solvable nodes):
   1.1. The terminal nodes are solvable.
   1.2. A problem represented by a nonterminal AND node is solvable if anyone of the problems in the successor OR nodes can be solved.
   1.3. A problem represented by a nonterminal OR node is solvable if the problems in all the successor AND nodes can be solved.
2. Determine nodes with *unsolvable problems* (unsolvable nodes):
   2.1. Nonterminal nodes having no successors are unsolvable.

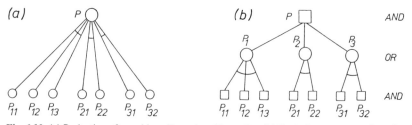

Fig. 6.12. (a) Reduction of a problem P to subproblems P_{ij}. (b) Equivalent representation of problem reduction by an AND/OR tree

2.2. A problem represented by a nonterminal AND node is unsolvable if all of its successor OR nodes are unsolvable.
2.3. A problem represented by a nonterminal OR node is unsolvable if anyone of its successor AND nodes is unsolvable.

In state-space search a path leading to a solution is desired; here a *solution tree* is looked for which, by applying the above definitions of solvable nodes, proves that the start node (the root of the AND/OR tree) is a solvable node. No solution exists if by application of the definitions for unsolvable nodes the start node can be shown to be unsolvable. Three examples are given in Fig. 6.13.

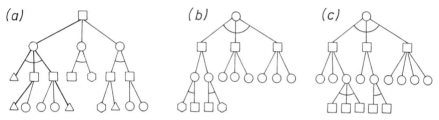

Fig. 6.13a–c. Examples of AND/OR trees. (a) The start node is solvable, a solution tree is indicated by heavy lines. (b) The start node is unsolvable. (c) It is still open whether the start node is solvable or not. (□) AND node, o OR node, (△) terminal node, (○) nonterminal node without successors

To find a solution tree in an efficient manner, costs have to be assigned to a subtree rooted at a given node v_j. As in (6.5.3) let $r(v_i, v_{ij})$ be the cost of an edge between node v_i and its jth successor node $v_{ij}, j = 1, \ldots, n_i$, or equivalently $r(v_i, v_{ij})$ is the cost of transforming v_i into v_{ij}. An estimate $\hat{\varphi}(v_i)$ of the cost $\varphi(v_i)$ of any subtree rooted at v_i may be obtained from the following recursive definitions:

3. Estimate costs $\hat{\varphi}(v_i)$ of a tree rooted at v_i:
   3.1. If v_i is a terminal node, $\hat{\varphi}(v_i) = \varphi(v_i) = 0$.
   3.2. If v_i is a nonterminal node with no successors, $\hat{\varphi}(v_i)$ is undefined.
   3.3. If v_i is a nonterminal AND node with OR successors $v_{ij}, j = 1, \ldots, n_i$,

$$\hat{\varphi}(v_i) = \min_j \{r(v_i, v_{ij}) + \hat{\varphi}(v_{ij})\} . \quad (6.5.4)$$

   3.4. If v_i is a nonterminal OR node with AND successors $v_{ij}, j = 1, \ldots, n_i$,

$$\hat{\varphi}(v_i) = \max_j \{r(v_i, v_{ij}) + \hat{\varphi}(v_{ij})\} \quad \text{for max costs}, \quad (6.5.5)$$

$$\hat{\varphi}(v_i) = \sum_{j=1}^{n_i} [r(v_i, v_{ij}) + \hat{\varphi}(v_{ij})] \quad \text{for sum costs}. \quad (6.5.6)$$

In (6.5.5, 6) max costs and sum costs are distinguished. The former give the maximal costs of one path from the root to a terminal node, the latter give the sum of all edge costs in the subtree. The costs $\hat{\varphi}(v_{ij})$ are needed in (6.5.4–

6). If the successors of v_{ij} are already evaluated, $\hat{\varphi}(v_{ij})$ follows from repeated application of definition 3. If the successors are not yet available, $\hat{\varphi}(v_{ij})$ is some heuristic estimate of $\varphi(v_{ij})$. If no information is available, an optimistic estimate is to use minimal cost (or maximal value) of $\varphi(v_{ij})$. During the search process two types of nodes are distinguished. The first are nodes known to be terminal, nonterminal, or with no successors yet generated; these nodes are called *tip nodes*. The second are nodes known to have either AND or OR successors; these nodes are called *nontip nodes*.

Assume that during the search the tree in Fig. 6.13 was generated, but a solution tree has not yet been extracted. The search tree contains several subtrees rooted at the start node, each of which might turn out to be part of a solution tree. These subtrees are called *potential solution trees*. The potential solution tree rooted at the start node v_0 and estimated to be part of an optimal (minimum cost) solution tree is denoted by t_0. It is extracted from the search tree by use of the following definition:

4. Determine the potential solution tree t_0.
    4.1. The start node v_0 is in t_0.
    4.2. If v_i is a node in t_0 and has OR successors $v_{ij}, j = 1, \ldots, n_i$ in the search tree, then the successor with minimal $\{r(v_i, v_{ij}) + \hat{\varphi}(v_{ij})\}$ is in t_0.
    4.3. If v_i is a node in t_0 and has AND successors in the search tree, then all the successors are in t_0.

An ordered search algorithm can now be given. The task is to search an AND/OR tree and determine whether the start node v_0 (the initial problem) is either a solvable node or an unsolvable node.

Put v_0 on a liste OPEN and compute $\hat{\varphi}(v_0)$ by definition 3.
While v_0 is not shown to be solvable and not shown to be unsolvable do:
    Compute t_0 by definition 4.
    Select from t_0 a tip node v_i that is on OPEN and put it on CLOSED.
    If v_i is a terminal node, then do:
        Determine solvable nodes by definition 1.
        If v_0 is shown to be solvable, then note SUCCESS, else
        remove from OPEN nodes with solved ancestors.
    Else do:
        Expand v_i.
        If v_i has successors, then do:
            Put successors on OPEN and provide pointers back to v_i.
            Compute $\hat{\varphi}$ of successors and recompute $\hat{\varphi}$ of v_i
            and ancestors by definition 3.
        Else do:
            Determine unsolvable nodes by definition 2. If v_0
            is shown to be unsolvable, then note FAILURE, else remove
            from OPEN nodes with unsolvable ancestors.
        End if.
    End if.

End while.
End search.

An example is given in Fig. 6.15 of the next section. The properties of this algorithm are summarized in the following theorem.

Theorem 6.5.2. If $\hat{\varphi}$ is a lower bound on φ for open nodes, then the above algorithm is admissible, that is it finds an optimal solution tree provided a solution tree exists.

Proof. See, for instance, [6.1], pp. 130–136. □

This theorem is similar to Theorem 6.5.1, but nothing is said about the number of nodes expanded. It seems reasonable to select that tip node from t_0 for expansion which is most likely to refute the assumption that the present t_0 is in fact part of an optimal solution tree. Such a tip node might be the one with the highest estimated costs. There are results from constraint satisfaction tree search which support this idea. The above algorithm may be modified by heuristics similar to those discussed in Sect. 6.5.1.

6.5.3 Remarks on Pruning

The algorithms of Sects. 6.5.1, 2 search until success or failure causes a termination. In many cases it is not feasible to insist on finding the optimal path because the search effort would exceed any reasonable bound. The game tree of chess, for instance, was estimated to have 10^{120} nodes and the search tree for a moderate size image containing 7×100 sample values and 50 simple constituents was estimated to have 10^{12750} nodes, which might be reduced to 10^{4500} nodes by application of knowledge. However, it will usually be feasible to search for the next best transformation to be applied under the constraint that the search effort is bounded. Such a bound may be computer time, depth of search, or number of nodes expanded.

An example of a search tree which might occur in a state-space search is given in Fig. 6.14. The numbers outside the nodes are $\hat{\varphi}$ values computed from (6.5.2) when the node was generated. The numbers inside the nodes give the order of their generation by the algorithm of Sect. 6.5.1. The order of nodes resulting from expansion of the same node may be assigned arbitrarily. Having explored the tree four levels deep one may now decide to choose the first transformation on the basis of the available information and to prune the rest of the tree. This is done by removing nodes $\{2, 4, 5, 6\}$ from the lists OPEN or CLOSED. The path found must now go through nodes 1 and 3, provided such a solution path exists. The search then continues by expansion of node 15 as indicated by the dashed lines; the best transformation applied to node 3 is selected again on the basis of a search to at most four levels below node 3 and so on. Several modifications of this *pruning technique* are possible. The tree may be generated to the full breadth and a depth of four levels; this full tree, including nodes without numbers, also is indicated

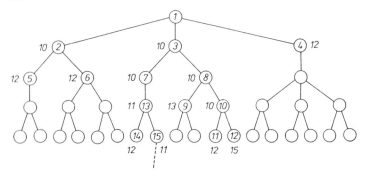

Fig. 6.14. A search tree generated by the algorithm of Sect. 6.5.1

in Fig. 6.14, but it is not generated by the algorithm. The depth of evaluation of the search tree before pruning may be altered. Instead of selecting the best transformation one may be cautious and select the m best transformations. For instance, in Fig. 6.14 one might also retain the path to node 2 and only remove node 4.

Similar pruning is possible with AND/OR trees. Figure 6.15 shows an example of a tree generated by the algorithm of Sect. 6.5.2. Numbers inside the nodes give the order of their generation by the algorithm. Numbers on the edges give the edge costs $r(v_i, v_{ij})$ in (6.5.4–6) and numbers outside the nodes give the costs $\hat{\varphi}(v_{ij})$ in these equations. Expansion of the start node 1 yields nodes 2–10. Initial cost estimates $\hat{\varphi}$ of solution trees rooted at nodes 5–10 are assigned; these are the leftmost numbers outside the nodes. The $\hat{\varphi}$ values are used to assign costs to nodes 2–4 by (6.5.5); again, these are the leftmost numbers outside the nodes. Now the potential solution tree contains nodes 1, 3, 7, 8. Expansion of node 8 adds nodes 11–16. With the $\hat{\varphi}$ values of nodes 13–16 the $\hat{\varphi}$ value of node 8 is modified to 8. The notation 5/8 outside the node 8 thus means that an initial cost of 5 was assigned which after expansion of node 8 was altered to 8. Now the potential solution tree is changed and contains nodes 1, 2, 5, 6. Expansion of

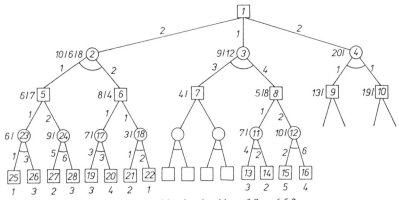

Fig. 6.15. An AND/OR tree generated by the algorithm of Sect. 6.5.2

nodes 6 and 5 adds nodes 17–28 and modifies the initial costs of 10 at node 2 to 6 and finally to 8. Next node 20 might be chosen for expansion, but to reduce the size of the AND/OR tree one may also decide to prune it first. If only the best potential solution tree discovered so far is to be retained, nodes 3 and 4 and all their descendants are removed. Further search is now restricted to the subtree containing nodes 1, 2, 5, 6 and their descendants. Similar to a state-space search, one may retain the m best potential solution trees and change the depth of search before pruning. Incidentally, after the generation and evaluation of, for instance, node 15, one need not generate node 16. Already after evaluation of node 15 it is known that inclusion of node 12 instead of 11 would yield higher costs at node 8. Therefore, it suffices to use node 11 and omit node 12.

It was always assumed that the AND/OR trees arising from problem reduction require the solution of all problems represented by AND nodes. This means that no AND nodes, which are descendants of a nonpruned OR node, can be pruned. The situation is somewhat different when searching AND/OR game trees. Without going into details it is mentioned that in this case the consideration applied to nodes 15 and 16 in the last paragraph can be extended to the *alpha-beta pruning technique*. This arises from the fact that in playing two-person games with perfect information (such as chess or checkers) one wants to "prove" that player A can win (or draw) against player B from position X if it is his turn to move. Therefore, there must be at least one move from X which finally results in a win for A; so the descendants of X are OR nodes. If A has moved from X to a new position, say Y, it is B's turn to move. Now it must be shown that A can win (B must loose) from all moves B can choose in position Y; so the descendants of Y are AND nodes. In this way a game is represented by an AND/OR tree. Since both of the players have to choose a single move if it is their turn, a path in this tree is followed during a game, and no subtree is selected. It is expected that A will always choose the move maximizing his score or minimizing his costs, whereas B will always choose the move minimizing this score or maximizing these costs (which are attributed to A).

Of course, any of the pruning techniques may result in a nonoptimal path or the missing of any path. This depends crucially on the quality of the available cost estimates. To a certain extent one may trade off quality of cost estimates and depth of search. In many cases one will have to rely on a fairly shallow search because it is the only thing which can be afforded.

Some rules have been developed in the context of production systems to reduce the number of transformations (productions). These may be used to reduce the number of successors generated for a node. The first idea is to impose an *a priori* ordering on the transformations and use only the m with the highest order out of the applicable transformations. Another idea is to prefer the more special cases, that is a transformation which is a specialization of another one or which is working on more specialized data. A third idea is to prefer transformations working on more recently generated data. Finally, transformations are not allowed to be active twice in succession. However, it is still an open question which task

6.5.4 Dynamic Programming

Another basic search algorithm is *dynamic programming* (DP) which may be considered as a breadth-first search procedure, considered here in the context of Fig. 6.1. It is assumed that there are N nodes in the graph of Fig. 6.1 and they are listed in a column $0, 1, \ldots, \nu, \ldots, N$ in Fig. 6.16. For any two nodes v_i, v_j of the graph in Fig. 6.16 the cost of an edge is denoted by $r(i, j)$; if there is no edge between two nodes, the cost is set to a very large value ($+\infty$). The cost of a path between two nodes is assumed in this case to be the sum of the edge costs along the path. Again, more general cost functions are possible. The control problem now is to find a minimal cost path from the start node v_0 to a goal node v_g, and this is the same as above. A DP solution of this problem is based on the *principle of optimality* which states that if the optimal path from v_0 to v_g passes through node v_j, then the path from v_0 to v_j is itself optimal (note the analogous statement for the A^* algorithm in the case of the monotone restriction).

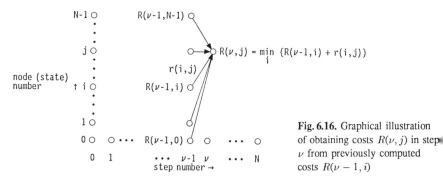

Fig. 6.16. Graphical illustration of obtaining costs $R(\nu, j)$ in step ν from previously computed costs $R(\nu - 1, i)$

The *forward DP algorithm* computes costs recursively in steps $\nu = 0, 1, \ldots, n \leq N$ for all N nodes (or states) $v_j, j = 0, \ldots, N - 1$. Let $R(\nu, j)$ be the minimal cost of going from v_0 to v_j with ν steps. If we are initially in node v_0, $R(0,0) = 0$ and $R(0, j) = \infty, j = 1, \ldots, N - 1$. In the next step, $\nu = 1$, the minimal cost $R(1, j)$ of going to some node v_j from v_0 in one step may be obtained by exploring for every node v_j all possible paths to v_j (there are at most N) and selecting that path leading to v_j with minimal cost, that is

$$R(1, j) = \min_{i \in \{0, \ldots, N-1\}} \{R(0, i) + r(i, j)\}. \tag{6.5.7}$$

A pointer is set back from j to that $i = i^*$ yielding the minimum, and only this optimal partial path is remembered. Therefore, the above argument may be used at any step ν and one obtains the minimal cost $R(\nu, j)$ of going to v_j in ν steps from

$$R(\nu, j) = \min_i \{R(\nu - 1, i) + r(i, j)\}. \tag{6.5.8}$$

This is depicted in Fig. 6.16. The DP forward search is summarized as follows:

1. Specify initial costs $R(0, j)$, $j = 0, 1, \ldots, N - 1$, edge costs $r(i, j), i, j = 0, 1, \ldots, N - 1$, assume the start node to be v_0 and the goal node v_g.
2. For step number $\nu = 1, \ldots, N$ do:
   2.1. For node number $j = 0, 1, \ldots, N - 1$ do:
      2.1.1. Compute $R(\nu, j) = \min_i \{R(\nu - 1, i) + r(i, j)\}$
      $= R(\nu - 1, i^*) + r(i^*, j)$
      2.1.2. Set pointer back from j to i^*.
   End do
   End do
3. Obtain an optimal path by tracing the pointers back from v_g to v_0. If there are N nodes, the optimal path will (hopefully) have less than N nodes. This may be detected from a constant sequence $R(n, v_g), R(n + 1, v_g), \ldots R(N, v_g)$. In this case the optimal path has only n steps. It should be noted that costs in general are dependent on the input pattern.

Comparing the A^* and DP algorithms the following points should be noted:

1. The DP principle of optimality is valid only for restricted cost functions, namely those which are separable and monotone. This is true, for instance, if node costs are the sum of edge costs as assumed above. An equivalent assumption was made for A^*. If this is not true, the more general graph search algorithms can be used.
2. The DP algorithm is extremely simple, provided computation of $r(i, j)$ is simple. This does not mean, however, that DP should be preferred to A^* whenever DP is applicable. In fact, A^* only explores the most promising node at each step. In addition it only requires an implicit definition of the search graph, whereas DP requires an explicit definition. Therefore, A^* may be faster than DP. On the other hand, the A^* algorithm is more complicated and requires more overhead than DP. The actual trade-off depends on the particular problem.

6.5.5 Heuristic Strategies in Image Analysis

Usually it is not feasible to treat image analysis as a problem of searching in a state-space, of problem-reduction, or a mixture of both, where one starts with the input image and ends with the description, and probably it would also not be useful in general. However, treating certain parts of the analysis by search methods is possible, for instance, the parsing of a symbol string, the selection of subsets of intermediate results, or the use of a semantic net. Other parts may be processed by a fixed sequence of processing steps. The problem-reduction approach can be used in image analysis by the application of knowledge which allows one to generate hypotheses about possible objects in the image.

It was mentioned in Sect. 6.5.1 that search algorithms may be modified by heuristics to reduce the search space. A powerful control strategy to achieve this is the use of planning as mentioned in Sect. 6.3. It was introduced into image analysis in the context of model-guided edge detection. Basically, a rough idea of what the image contains is obtained first – often by taking a coarse resolution – then a plan is made of further analysis and this plan is used to guide the analysis, often by switching to a finer resolution. The plan may reduce the number of possible alternatives (the number of OR nodes in a problem reduction), for instance, by giving evidence about the location of an object or providing a limited number of labels for a region. It may also allow an informed guess about useful transformations, for example, by indicating that a contour line most probably occurs at a certain location. Then one should apply a specialized module for contour extraction and avoid an "expansion of a node" which would generate all successors of a node. Using this plan a mixture of a bottom-up and top-down – or a data-driven and model-driven – control structure can be achieved. The plan is generated by bottom-up processing, and then top-down processing takes over and makes use of this plan.

It is most important in search algorithms to have reasonable estimates of the costs introduced in (6.5.1, 4). Such estimates may be based on expected processing time for a transformation, on a rating of the reliability of results obtained by a transformation, or on experience gathered from analyzing a sample ω of images f^r belonging to the task domain Ω under consideration. An example will be discussed in connection with semantic nets in the next chapter.

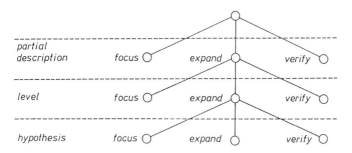

Fig. 6.17. Hierarchical decomposition of the control module for expanding the search space

In one approach the control module itself is subdivided further into a hierarchy of modules each deciding upon a particular action. In Sect. 7.1 it is shown that knowledge may be organized into different levels, and in Sect. 5.3.2 an example of a results data base was discussed where results were arranged in the three dimensions level, alternative, and time. Now the control module is subdivided into two sorts of aspects, termed vertical and horizontal decomposition. The vertical decomposition yields the three components, working on a partial description of the image (that is a subset of the results data base making up a still incomplete description), working within that description on a certain level, and working within that level on a particular hypothesis. The horizontal decomposition yields the components focus, expand, and verify; this means, for instance, that the system focuses on a particular level, expands that level (which in turn requires focusing on a hypothesis at that level, expanding the hypothesis, and verifying the new hypotheses), and finally, it verifies the new results obtained from the expansion. Figure 6.17 shows the control decomposition for expanding the search space. It results in a restriction of control options, but makes the choice of a control strategy more transparent.

A model based system for inspection of industrial parts illustrates some general points made above. The system has available a small set of transformations or processing algorithms selected during the design phase; these allow the system to find the part's contour, to detect holes, lines, textures, and small holes. The value or reliability of each algorithm is subjectively estimated and is constant, independent of the input. For example, the algorithm for detecting the contour of an object is given maximal reliability because the contour is a large feature and supposed to be reliably detectable. A list of part models contains the task specific knowledge. The difference between an inspected part and the part models is accumulated in a difference table which is a cumulative score of the hypotheses that the inspected part is an instance of a certain part model. The control algorithm of the system is as follows:

1. Perform preprocessing of the image.
2. Obtain the contour and its attributes and update differences. This may be viewed as a planning stage to obtain a rough idea about the part. Initially the difference table has a zero for each part model. Differences are updated by adding the reliability of the contour finding algorithm to the difference of all those models whose contour is not similar to the inspected one.
3. Select as candidate models those with smallest differences.
4. If one model is identified, then determine descriptive information (which is necessary only for the assembly of parts, but not for recognition) and STOP.
5. Select the most promising next simple constituent for recognition. This is done by using four criteria: select constituents from those models most similar to the inspected part; select the constituents most important for classification; select constituents detectable with the highest reliability; select constituents requiring least computer time. If more than one constituent remains after this selection, an arbitrary one is chosen.
6. Activate an appropriate transformation to detect that constituent.
7. If the constituent is not found, then update differences by adding the reliability to all models containing this constituent, else store it in a description table and update differences by adding the reliability to all models not containing this constituent.
8. Start a new cycle at step 3.

This control algorithm is in the form of an abstract program. Although no admissible strategy is employed, the "best first" philosophy of a search algorithm is clearly adopted in steps 3 and 5. The potentially enormous graph in Fig. 6.1 is drastically reduced by selecting a small number of useful transformations in the design phase of the system, by constraining the number of possible models in the planning phase (contour extraction), by allowing only limited selection among alternatives in steps 3 and 5, and by using only two two-way branches in steps 4 and 7.

A production system (see Sect. 7.3) may be chosen as a general system architecture and has been employed for the analysis of aerial photographs. Processing always starts with smoothing and segmentation of an image into elementary regions by region growing. The basic properties of elementary regions are stored in a central data base (the so-called blackboard). The next step is the extraction of characteristic regions, for example, a textured area, a line-like region, a green area, or a rectangular area. To reduce the processing time a pyramid data structure (a resolution hierarchy) is used. Characteristic region extraction algorithms may be applied in parallel to the image and results are stored in the data base. Finally, specialized object detectors which are organized as production rules are used to detect objects such as houses, roads, or forests. As outlined in the next chapter, a production system consists of a set of data (a data base), a set of production rules with a condition and an action, and a monitor which selects the next rule to be executed. The approach used here is to apply all rules whose condition part is met by the data in parallel and to store all results supplied by the actions of those rules in the data base. This may cause conflicts of the type that the same region is attributed to different objects and these conflicts are resolved in a subsequent step where the most reliable alternative is selected. For example, the reliability of a road is assumed to increase with its length. Furthermore, object detectors may indicate errors in the initial segmentation, and in this case the system tries to obtain a refined segmentation not contradicting object detectors.

6.5.6 Heuristic Strategies in Speech Understanding

There are examples of systems where the value or score of a hypothesis – this corresponds to the value (not the cost) of a node in the above terminology – is derived consistently from the conditional probability that the hypothesis is correct if certain evidence was used to obtain it. Among the different possible scoring methods the so-called shortfall density score has turned out to be very useful. It is, in fact, an example of a priority measure as described in Sect. 6.4.5. The input utterance is segmented and scanned by a lexical retrieval component to obtain a set of best-matching words and their scores. The score φ_w of a word w is allocated to scores φ_s of the segments spanned by the word proportional to the length of the segment. Since any segment may be covered by many words, there will be one maximal score φ_{sm} arising from some word in any segment. Let φ_{wm} be the sum of the maximal segment scores φ_{sm} over those segments spanned by the word w, let l_w be the length of those segments, and let φ_m be the sum of the maximal segment scores over all segments spanned by the whole utterance. Then the *shortfall density score* φ_d is defined by

$$\varphi_\mathrm{d} = \frac{\varphi_\mathrm{w} + (\varphi_\mathrm{m} - \varphi_\mathrm{wm})}{l_\mathrm{w}}. \tag{6.5.9}$$

This is a special case of (6.4.21). Note that $\varphi_\mathrm{sm} - \varphi_\mathrm{s} \geq 0$ by definition and, therefore, $\varphi_\mathrm{d} l_\mathrm{w}$ cannot increase if the number of segments considered increases, that is if computation of $\varphi_\mathrm{d} l_\mathrm{w}$ is extended to more than one word. The value of φ_d is used to determine the order of processing, that is the word or the sequence of words with maximal φ_d is processed first. It can be shown that this scoring method, and in fact some others, allows one to obtain an algorithm for finding a set of words spanning the whole utterance, which is admissible in the sense that the first such set found has the highest possible score. The basic control strategy for finding such a set of words is to attribute so-called seed events to an initial best scoring set of words found by lexical retrieval (this is similar to the planning approach in image analysis). The seed events are extended by adding words to the left or right of them. The syntactic, semantic, and pragmatic aspects of a newly generated sequence of words are evaluated by a linguistic component.

A heuristic search method, called *beam search*, has turned out to be very efficient in many cases. It provides a special pruning technique. Referring back to Fig. 6.6 let a_i be the prototype of a phone associated with state s_i and c_ν be the parametric representation of the utterance at time ν. Furthermore, let T_{ij} be the likelihood that state s_j follows immediately after s_i and let $P_{i,\nu}$ be the likelihood of state s_i at time ν. Experience has confirmed that only the values 0 or 1 for T_{ij} are sufficient. Now let

$$I_j = \{i | T_{ij} = 1\}, \tag{6.5.10}$$

that is, I_j gives the set of states s_i at time ν which may be followed by the state s_j at time $\nu + 1$. Finally, let $d(a_j, c_{\nu+1})$ be a measure of similarity between a_j and $c_{\nu+1}$. Then the likelihood or value of going to s_j is defined by

$$P_{j,\nu+1} = d\left(a_j, c_{\nu+1}\right) \max_{i \in I_j} \{P_{i,\nu}\}, \quad j = 1, \ldots, N_s. \tag{6.5.11}$$

If at time ν there are N_ν possible paths through Fig. 6.6, then at time $\nu + 1$ there would be $N_\nu B_\nu$ paths in a search tree, where B_ν gives the average number of elements in I_j at time ν. However, not all of them are retained, only the path from the best parent to successor s_j at time $\nu + 1$ as indicated by the max operation in (6.5.11). Furthermore, all successors and corresponding paths are pruned which have a value $P_{j,\nu+1}$ that differs from the maximum value by more than a threshold Θ. Since one parent may have more than one successor, but not vice versa from (6.5.11), after stepping through all time instances the final best path is obtained by tracing back the pointers from the end of the search to the beginning. The T_{ij} and I_j are constructed from the knowledge that only a certain set of sentences is allowed. These sentences constrain possible state transitions. Although this search may miss the optimal path, in practice it has turned out to yield good results.

Without going into detail the following summarizes an "average" strategy of several systems for continuous speech understanding:

1. Obtain a parametric representation of speech.
2. Segment speech into subword units.
3. Generate word hypotheses and select the m best scoring ones as initial hypotheses.
4. Check the syntactic, semantic, and pragmatic consistency of high scoring hypotheses.
5. Try to extend consistent hypotheses by either predicting possible continuations or by appending already hypothesized words.
6. Verify predictions using acoustic data.
7. Repeat steps 4 to 6 until either a high scoring sequence of words spanning the utterance is found or a time limit is exceeded.

Steps 1 and 2 correspond to a fixed sequence of processing steps which were found useful in many experiments, and step 3 may be viewed as an initial planning stage which reduces the number of alternatives and causes focusing of the search. Steps 4–6 involve search processes and selection among available transformations. The above control structure has the form of an abstract program from which a program is obtained by providing an interpretation. This would have to specify, for example, how to obtain a parametric representation, how to extend constituents, or how to compute scores of hypotheses.

6.6 Bibliographical Remarks

General treatments of the control problem are given in [6.2, 3]. The data-base-oriented system type is discussed in [6.4], the state network in [6.5].

Control as an abstract program was treated in [6.6], and a theoretical investigation of abstract programs is given in [6.7]. Hierarchical graphs are defined, for example, in [6.8], Petri nets in [6.9]. The example in Fig. 6.7 is according to [6.10]. Additional approaches not mentioned here are the actors [6.11] and programmed productions [6.12].

An overview of judging uncertain and imprecise results is given in [6.13], and fuzzy techniques are treated, for example, in [6.14]. These two references provide concise overviews and contain additional references. An approach to the automatic learning of fuzzy relations is given in [6.15, 16].

General graph search techniques are treated in detail in [6.1, 17], and dynamic programming in general is covered in [6.18].

The examples of heuristic techniques in image analysis are from [6.19–21]. An explicit use of the A^* algorithm is made in [6.22], and the state-space search is used extensively in [6.23]. The examples on speech understanding are from [6.5, 24, 25]. Several alternative strategies are investigated in [6.26]. A discussion of control problems for speech understanding using semantic nets is given in [6.27].

7. Knowledge

The discussion in the last chapter showed that in analysis of complex patterns a solution may require searching a graph or tree in a search space which is usually very large. The task is nearly hopeless unless the search space can be reduced. A powerful tool for constraining the number of alternatives and reducing the search space is the incorporation of knowledge.

Knowledge has become a current term in pattern analysis and artificial intelligence, but since it is also an everyday word it seems desirable to clarify the concept. Standard dictionaries provide several meanings, the one coming closest to what is meant here is *performance knowledge* which the individual requires to succeed in his physical environment; it may also mean information or understanding acquired through experience. These notions are readily adaptable to pattern analysis. Other notions are omitted here.

As a first-order approximation to a definition of knowledge in the context of pattern analysis we state that *knowledge* represents constraints and associations between objects and events occurring in the real world, reminiscent of Postulate 5 in Sect. 1.3 where the existence of structure in a complex pattern was required. An extremely important point is that this knowledge may be incomplete, imprecise, and uncertain. Since image and speech data are noisy and segmentation results are erroneous, the resulting inferences are themselves incomplete and erroneous.

In this chapter the following topics are discussed:

1. Views of Knowledge – some further clarification of the concept and a first idea of its representation.
2. Logic – a short view of its use for the representation and use of knowledge.
3. Production Sytems – conceiving knowledge as a set of (antecedent, consequent) pairs.
4. Semantic Nets – suitably defined graphs for the representation and use of knowledge.
5. Relational Structures – a short note on the adaptation of the relational data model to pattern analysis.
6. Grammars – finite schemes to define potentially infinite sets of objects.
7. Acquisition of Knowledge (Learning) – approaches to making a system gather the knowledge it needs.
8. Explanation – a short note on making a sophisticated system transparent to the user.

7.1 Views of Knowledge

7.1.1 Levels and Hierarchies

It was mentioned above that knowledge represents constraints in the real world. This is depicted in Fig. 7.1. Knowledge in pattern analysis, in particular in image or speech analysis, can naturally be segmented into a number of *levels* as discussed also in Sect. 1.2. A basic assumption for pattern analysis is to decompose a complex pattern into segmentation objects (or simpler constituents) and their relations. This process of segmentation in some cases may be reasonably applied also to the segmentation objects, thereby decomposing them into even simpler objects. A first idea of possible levels in image and speech analysis is given in Fig. 7.2.

```
physical environment (world 1) - time and space constraints
     ↓
structure (energy / matter)
     ↓
explicit representation of structural properties
     ↓
┌───────────┐
│ KNOWLEDGE │
└───────────┘
     ↑
explicit representation of rules, theorems, algorithms
     ↑
rules, theorems, algorithms
     ↑
symbolic world (world 2) - constraints and rules of mathematics,
     logic, experience, and common sense knowledge
```

Fig. 7.1. Knowledge in pattern analysis and understanding

From the schematic tables of Fig. 7.2 two problems become apparent:

1. What is an efficient way to represent, for instance, the set of verbs and nouns on the "word" level, and how to represent the relations among them and also among the next higher and lower levels? This is the problem of *representation of knowledge (declarative knowledge)*.
2. What are efficient transformations to move from one level to another, either up or down in the layers? This is the problem of transformation or *utilization of knowledge (procedural knowledge)*.

Of course, all these problems are highly interdependent. If, for instance, levels 3 and 4 in Fig. 7.2b are omitted, this implies that there must be algorithms to synthesize words directly from segments; a transformation relating levels 4 and 5 then becomes useless. There are no general solutions of the stated problems, but there is a variety of different approaches which have proved useful in certain tasks.

Experience from sufficiently involved task domains indicates that it is useful to provide a more refined structuring of knowledge as indicated in Fig. 7.3. It shows two conceptions, the "house" and the "phrase". Both may be specialized, in this case to "apartment house" and "interrogative phrase", respectively. When

7.1 Views of Knowledge 273

LEVEL		OBJECTS
0	type of image	landscape, room, city, ...
1	object	house, car, person, tree, ...
2	part	roof, window, wheel, ...
3	element	line, blue area, text. area, ...
4	primitive elements	line segment, texture primitive, ...
5	sample value	{black, white}, color, ...

(a)

LEVEL		OBJECTS
0	type of sentence	noun phrase, question, ...
1	sent. frag.	noun part, prepos part, ...
2	word	noun, verb, article, ...
3	syllable	start syll., center syll., ...
4	syll. part	start, vocal part, stop, ...
5	segment	voiced, fricated, ...
6	paramet. repres.	14 pole LPC, pitch frequ., formants, ...
7	sample value	$\{f_\nu | \nu = 1,2,...\,;\; f \in \{1,...,2048\}\}$

(b)

Fig. 7.2a,b. Examples of knowledge levels in pattern analysis and understanding. Relations between objects on the same or different levels are omitted

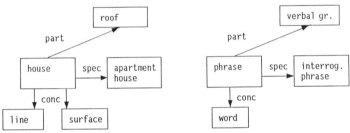

Fig. 7.3. Two examples of a three-dimensional hierarchy for representation of knowledge. *(part)* part relation, *(conc)* concrete relation, *(spec)* specialization relation

trying to find a house in an image or an interrogative phrase in an utterance one may first try to find conceptions which are more concrete in the sense that they are closer to the sample values. Those more concrete conceptions in the figure are the "line and surface" and the "word". Finally, the "house" and "phrase" may be segmented into parts, in this case parts like "roof, window" and "verbal group, noun group", respectively. This establishes a three-dimensional hierarchy of "parts", "specializations", and "concretizations" for the organization or structuring of knowledge.

To this structuring of knowledge another hierarchy can be added as shown in Fig. 7.4 for the "object" as an example. The most general conception is the "object" which is a *model scheme* and is defined by a list of boundary segments, a list of regions, and/or a list of vertices. A less general conception is a "particular object", for instance, a "car" or a "house". The "house" is defined by certain regions such as roof, front and so on, and represents a general *model* of something which may occur in the real world (see also Sect. 4.5.1). If this model is to be useful, it should not precisely describe one single house, but define general properties of a house. When analyzing an image containing a house, one *instance* of a particular object, in this case one instance of a house, is obtained and stored in the results data base; a specific description comprising, for example, the height and length, the number of windows, the color and so on is given.

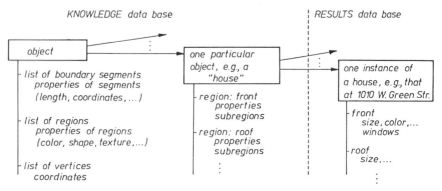

Fig. 7.4. A knowledge hierarchy of a model-scheme, a model, and an instance

There is an apparent similarity between a particular object and a description of a record contained in the schema of a data model introduced in Sect. 5.2.1. An instance of a particular object is then an occurrence of the record. The object may be viewed as a schema-schema. It is a general description of common properties of several schemas, and one schema is a general description of common properties of several instances of a schema (or of several records). The schema-schemas and the schemas are stored in a knowledge data base, the instances of a schema (or records or hypotheses) are stored in the results data base. This distinction is justified by the fact that results change if a new pattern is analyzed, but the knowledge does not; it changes only slowly if we include learning or automatic knowledge acquisition (Sect. 7.7).

In accordance with (3.1.2, 4.5.1) we can summarize the basic structure for representing some real world *conception*. The formal representation is a *concept*

C: name of conception D: type T;
(attribute A: type of attribute TA, value is a real number R
or an element from a finite set of names of attribute
values AV)*,
[(part P: structure C)*, (concrete K: structure C)*,
(specialization V: structure C)*, (schema M: structure C),
(instance I: structure C)*],
(structural relation between parts and concretes S (P, K):
real number measuring the fulfilment of the relation)*,
goodness G: vector of real numbers R),
C: (D: T; (A: (TA, R∪AV))*,
[(P: C)*, (K: C)*, (V: C)*, (M: C), (I: C)*],
(S (P, K): R)*, G: R). (7.1.1)

This gives a refinement of (3.1.2, 4.5.1) by introducing the distinction between parts, concretes, specializations, schemas, and instances. This distinction has not been necessary up to now because the results of segmentation in (3.1.2) and the object models in (4.5.1) were of a simpler structure than the general task specific knowledge structures considered here. It should be noted that there are *minimal concepts* having no parts and no concretes, and that the schema relation is useful for automatic knowledge acquisition and may be missing if this is not used in some task domain. The definitions (3.1.2, 4.5.1, 7.1.1) show that a uniform treatment of segmentation results, object modeling, and knowledge representation is possible. Some details of a possible implementation are given in Sect. 7.4.

7.1.2 Submodules

In general it will be useful and necessary to structure the knowledge module into a set of fairly independent submodules each performing a limited task. An example is given in Fig. 7.5 where the modules for knowledge and methods from Fig. 1.4 are further subdivided. The representations of Fig. 7.2, 5 support each other. In the former the emphasis is on "what is available", that is, on the representational part or *declarative knowledge*, and in the latter the emphasis is on "what is achievable", that is, on the transformational part or *procedural knowledge*. As mentioned above, one part is useless without the other, and both are considered as parts of the knowledge module.

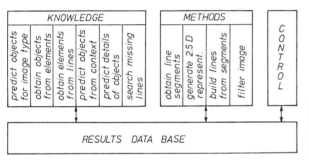

Fig. 7.5. Transformations between knowledge levels are organized into submodules

Consideration of the knowledge levels and modules discussed above makes the distinction between methods and knowledge in Sect. 1.4 clearer, but it is also evident that it is impossible to establish a unique border. Usually the higher levels in the (vertical) hierarchy are associated with high-level processing, knowledge, symbol manipulation, or nonnumeric computations, whereas the lower levels are associated with low-level processing, methods, signal processing, or numeric computations. For the medium levels, however, it is a matter of definition whether to consider them as belonging to methods or to knowledge. Furthermore, there are examples of transformations relating low levels and using a good deal of high-level knowledge. Such a direct incorporation of knowledge into low-level processing may occur, for instance, when extracting lines from an image. Low-level methods using no task specific knowledge were treated in Chaps. 2, 3. For low and medium level processing the numerical classification procedures discussed in Chap. 4 may also be of importance, for instance, when classifying speech segments containing a particular phoneme.

Certainly, a lot of specialized knowledge is also required on the designer's side to develop the low-level stages of processing. In this sense it would be justified to talk uniformly about levels or modules for any processing. The reason for maintaining the distinction between methods and knowledge in this text is that methods are less dependent on the task domain and (so far) no low-level knowledge is represented *explicitly* in the system. It seems that approaches to pattern analysis have followed – perhaps of necessity but not deliberately – the lines adopted for seeing (and hearing) in biological systems, where some "hard-wired" low-level processing is done already in the neuronal nets of the retina; unfortunately, the high-level processes (algorithms) leading to the conscious impression of "seeing" (or "hearing") are still unknown.

7.1.3 Aspects

When looking at an object, event, or pattern one may focus on different aspects of this object or event. To give a simple example, consider schematic diagrams. One aspect are the *pictorial properties* of such a diagram. They may be represented by an image (a low-level concept according to Fig. 7.2) or a symbolic description (a high-level concept). A transformation between these two representations may require operations using some intermediate levels, but these are only auxiliary levels and not of interest in themselves. A second aspect are the *electrical properties* of a schematic diagram. These may be represented by an image or, alternatively, by appropriate equations or a descriptive text. It is also possible to derive the higher level concepts from the image, probably by using some intermediate levels, but certainly by using levels quite different from those needed for the pictorial aspects. A third aspect might be the *physical properties* of a schematic diagram, for instance, its realization as a printed circuit board, an integrated circuit chip, or a television receiver. Again it is possible to design these concepts using the schematic diagram and other suitable resources. The pictorial properties of a schematic diagram – and in the broad sense the "perceptual" properties of a complex pattern – are of direct interest in pattern analysis, the other properties shift to problems of artificial intelligence in general. These become important when viewing a system for pattern analysis as part of a larger system performing a certain task, e.g. assembly of devices.

7.1.4 Frames

A general approach to the representation of knowledge are the so-called *frames*. Some of the approaches discussed in the next sections may be viewed as special cases of this one. A frame presents an elaboration of the idea that knowledge is structured and contains components of varying certainty and detail. An example of a frame structure is given in Fig. 7.6 and it can be seen that it bears some resemblance to the hierarchical graphs of Sect. 6.2.6. In general, a frame is a data structure representing a stereotyped situation, like being in a certain living room or seeing a house from a certain viewpoint. It is selected when encountering a

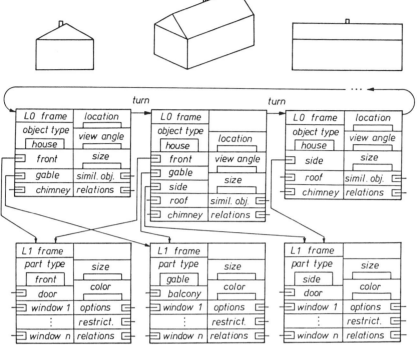

Fig. 7.6. An example of frames representing a house from different viewpoints

certain situation. A frame is a network of nodes and relations, where the top levels – the L0 frame in Fig. 7.6 – contain statements that are usually true and the lower levels have terminals or slots to be filled with specific data. For instance, when seeing a one-family house from a certain viewpoint one will expect to notice a gable, a front, and a chimney. Some parameters such as size and absolute or relative location may be given with typical ranges and specified for a particular situation. Important relations of objects are specified, such as "the front is below the gable". Pointers are provided to lower level frames, for instance, to the L1 frames "front" or "side". Restrictions, such as the windows being rectangular, can be specified. The number of windows seen in the front of a house will be less certain and has to be provided for a particular situation. In addition the frame may contain information about its usage, about similar objects, about objects or events being related in space or time, and perhaps about actions to be taken if the selected frame turns out to be in severe disagreement with the actual situation. Incidentally, programming languages were designed to implement the frame concept.

Frames may be related to each other by transformations and then summarized in a frame system. In Fig. 7.6 the three L0 frames are related by the fact that they represent the same three-dimensional object seen from different viewpoints. Therefore, some L1 frames are common to different L0 frames. Frame systems are linked by an information-retrieval network which provides information about what to do next if a selected frame does not match the situation. Although the example comes from image analysis, the frame concept is also applicable in recognizing connected speech or understanding natural language, and so on.

A comparison of Figs. 5.12 and 7.6 reveals that the frames have some similarity to the network model for data bases. In fact, both have similar requests. A large amount of information ("knowledge" in one case, "data" in the other) is to be represented such that interrelations become apparent, access is supported, changes are possible, and consistency is maintained. Thus representation of knowledge may also be viewed as a problem of designing a suitable data structure or data base. So far, a data base always was supposed to contain results of the analysis process. This is adequate because these are rapidly changing. The content of a knowledge data base usually will be fixed or at most slowly changing, thereby making the problem of updating the data base less important. Sometimes the knowledge data base is referred to as long-term memory and the results data base as short-term memory, but we prefer to avoid these terms because of their biological and physiological bias. Finally, it should be remembered that different representations of knowledge are possible depending on the purpose of its usage. Figure 7.7 shows three representations of a simple situation; the first representation is suited for visual inspection, the second for verbal communication, and the third for computer manipulation. In this sense, frames are a representation of knowledge which – possibly with simplifications, modifications, or enlargements – is suited for computer manipulations, but they were also supposed to be candidate models for human perception.

7.2 Predicate Logic

7.2.1 Some Basic Definitions

Formal logic, in particular *first order predicate calculus* (FOPC), is an important basic tool which may be used for the representation and use of knowledge. It provides a formal language which allows one to represent symbolic statements, to manipulate symbolic structures, and to make inferences using symbolic structures. It is not intended here to give an in depth treatment of logic, but only a short introduction to its application in the area of pattern analysis.

Representation of knowledge is by sentences or *formulas* which are defined by a *syntax* in FOPC. The allowed sentences are called *well-formed formulas* (WFFs). A formula is built from an alphabet of *symbols*. There are *predicate, constant, variable,* and *function symbols*. A *term* is a constant symbol, a variable symbol, or a function symbol having terms as arguments. This recursive definition of a term allows nesting of function symbols, e.g. $f(g(x,y),a)$. An *atomic formula* consists of a predicate symbol having terms as arguments. An atomic formula is a WFF (but not the only one).

7.2 Predicate Logic

There is a large square s
above a small triangle t
There is a small circle c
inside the square s. The
triangle t and the circle c
have nearly the same area.

{{square: s, circle: c,
triangle: t},
{inner: c, outer: s},
{above: s, below: t},
{same area: c, same
area: t}}

Fig. 7.7. Different representations of the same situation

More complex formulas may be constructed by *connectives*. Of special importance are the connectives AND (logical and, $\wedge$), OR (logical or, $\vee$), NOT (logical negation, $\neg$), IMPLIES (logical implication, $\rightarrow$), and EQUIVALENCE (logical equivalence, $\leftrightarrow$). Atomic formulas and their negation are called *literals*.

Similar to those of Sect. 6.2.5 the above definitions are of an abstract type, that is without relation to a certain task domain. The relation between a task domain and the symbols of FOPC is provided by an *interpretation* in the sense of Sect. 6.2.5. It relates an n-ary predicate symbol to an n-ary relation, a constant symbol to an object of the task domain, a variable symbol to an object variable of the task domain, an m-ary function symbol to an m-ary function of the task domain. For example, the predicate symbol $AB(u, v)$ is interpreted as "objects denoted by variable u are *above* objects denoted by variable v", the constant symbol MAR3 is interpreted as "the building located at Martensstrasse 3", the variable symbols u and v are interpreted as "antenna and house", respectively, the function symbol $ac(r)$ is interpreted as the "area of circular objects having radius r". Two WFFs having the same truth value regardless of their interpretation are called *equivalent*.

f_1	f_2	$\neg f_1$	$f_1 \vee f_2$	$f_1 \wedge f_2$	$f_1 \rightarrow f_2$	$f_1 \leftrightarrow f_2$	$f_1 > f_2$
0	0	1	0	0	1	1	?
0	1	1	1	0	1	0	?
1	0	0	1	0	0	0	0
1	1	0	1	1	1	1	1

Fig. 7.8. Truth tables of the standard logical connectives $\{\neg, \vee, \wedge, \rightarrow, \leftrightarrow\}$ and of the non-standard connective $>$.

Every formula with associated interpretation can be viewed as a statement about the task domain; this statement may have the *truth value* TRUE (T or 1) or FALSE (F or 0). The truth value of a formula containing variables depends on the objects bound to the variables. Truth values can be determined from the *truth tables* of the logical connectives shown in Fig. 7.8. An example of the *implication* relation $f1 \rightarrow f2$ is

IF (it is raining) THEN (streets are wet); rain $\rightarrow$ wet streets . (7.2.1)

The truth table states that the implication $f3 = f1 \rightarrow f2$ is true, unless the premise $f1$ is true and the conclusion $f2$ is false. The formula $f4 = \neg f1 \vee f2$

has the same truth value as $f3 = f1 \to f2$. Therefore, implication can be replaced by NOT and OR connectives using the relation

$$f_1 \to f_2 = \neg f_1 \vee f_2 . \tag{7.2.2}$$

There are more intricate connectives than just AND, OR, NOT and so on. An example are the so-called *counterfactuals*. The situation (7.2.1) is different in a counterfactual or in *counterfactual implication* $f1 > f2$, for example,

IF (it were not raining), THEN (the cars would not have crashed) . (7.2.3)

A counterfactual is a statement where the premise is either known or expected to be false. The truth table of counterfactual implication is also given in Fig. 7.8. Obviously, the truth of a counterfactual depends not only on the truth of the components. We have $(f1 > f2) \to (f1 \to f2)$ but *not* the reverse. Counterfactuals are *non-monotonic*.

Let b be a formula containing the variable x. If b is TRUE regardless of the assignment of objects to x, we denote this by $\wedge_x(b(x))$, where "$\wedge_x$" is the *universal quantifier*. If there is at least one assignment of an object to x such that b is TRUE, we write $\vee_x(b(x))$, where "$\vee_x$" is the *existential quantifier*. An unquantified variable is called a *free variable*, a quantified variable is called a *bound variable*. A formula without free variables is called a *sentence*. A WFF is a *clause* or is in *clause form* if it has the form $L_1 \vee L_2 \vee \ldots \vee L_n, n \geq 1$, with L_i a literal. A set W' of WFFs is said to be in *clause form* if every WFF in W' is a clause. Every WFF can be converted to clause form by the following steps:

1. Eliminate equivalences.
   Replace the formula $a \leftrightarrow b$ by the two formulas $a \to b, b \to a$.
2. Eliminate implication symbols.
   Use $x \to y = \neg x \vee y$.
3. Reduce the scope of a negation operation such that it applies to at most one atomic formula. The following relations are useful in this step:

   $\neg(a \vee b) = \neg a \wedge \neg b$,

   $\neg(a \wedge b) = \neg a \vee \neg b$, (de Morgan's law) (7.2.4)

   $\neg \vee_x (P(x)) = \wedge_x(\neg P(x))$,

   $\neg \wedge_x (P(x)) = \vee_x(\neg P(x))$, (7.2.5)

   $\neg(\neg x) = \neg\neg x = x$, (7.2.6)

   $a \vee (b \wedge c) = (a \vee b) \wedge (a \vee c)$ (distributive law). (7.2.7)

4. Rename variables such that every quantifier has a uniquely named variable bound to it, e.g. rewrite $\wedge_x[S(x) \wedge \wedge_x(T(x))]$ by $\wedge_x[S(x) \wedge \wedge_y(T(y))]$.
5. Eliminate existential quantifiers by *Skolemization*. For example, the formula $\wedge_x[\vee_y P(x,y)]$ is replaced by $\wedge_x[P(x,g(x))]$ where $g(x)$ is a new function symbol, the *Skolem function*, expressing the possible dependence of y on

x. In general every variable y in the scope of an existential quantifier is replaced by a new Skolem function $f(u,v,\ldots)$ having as arguments all the universally quantified variables which are bound by a universal quantifier whose scope includes the scope of the existential quantifier to be replaced by f. Existentially quantified variables not in the scope of a universal quantifier are replaced by a new constant symbol.

6. Move all universal quantifiers to the front of the formula; the quantifiers at the front of the formula are the *prefix*, the rest is the *matrix*.
7. Convert the matrix to conjunctive normal form using (7.2.4–7).
8. Eliminate universal quantifiers by adopting the convention that all variables in the matrix are universally quantified.
9. Elminiate the logical AND by replacing the WFF $x \wedge y$ by the two WFFs x, y.
10. Rename variables such that every clause has different variables. This step is not necessary for conversion to clause form but essential for the inference method to be described later.

A clause without variables is called a *ground clause*.

The predicate calculus defined here is called *first-order predicate calculus* because it does not allow quantification of predicate or function symbols.

7.2.2 Unification

When trying to prove a theorem it is often the case that two terms have to be matched. In general, let two terms consisting of constants, variables, and function symbols be given. A *substitution* is a set of ordered pairs

$$s = \{t_1/v_1, \ldots, t_i/v_i, \ldots, t_n/v_n\} ,\qquad(7.2.8)$$

where the t_i are terms and the v_i are variables; application of a substitution s to a WFF f, written fs, means that the term t_i is substituted for every occurrence of the variable v_i, or that every occurrence of v_i is replaced by t_i. If for a set $W = \{f_1, f_2, \ldots, f_m\}$ of WFF there is a substitution s such that

$$f_1 s = f_2 s = \ldots = f_m s = f ,\qquad(7.2.9)$$

the set W is said to be unifiable and s is called a *unifier* of W; the process is called *unification*. For example, let f, g be function symbols, A, B constant symbols, and x, y, z variable symbols and consider the two terms

$$t = f(x, g(A, z)) \quad \text{and} \quad u = f(g(y, z), x) .\qquad(7.2.10)$$

Then a unifier is

$$s_1 = (g(A, B)/x\, ; \; A/y\, ; \; B/z) .\qquad(7.2.11)$$

It says that $g(A, B)$ is substituted for x, or that x is replaced by $g(A, B)$ and so on, resulting in

$$ts_1 = f(g(A,B), g(A,B)) = us_1 \ . \tag{7.2.12}$$

In general there may be several possible unifications; in the above example $s_2 = (g(A,z)/x; A/y)$ is another unification. The unifier s_2 is more general than s_1 in the sense that s_1 may be obtained from s_2 by *composition* with $s_3 = (B/z)$ yielding

$$s_1 = (g(A,B)/x \ ; \ A/y \ ; \ B/z) = s_2 * s_3 = (g(A,z)/x \ ; \ A/y)*(B/z) \ . \tag{7.2.13}$$

It can be shown that for unifiable terms there is always a *most general unifier s* which cannot be obtained by composition of some unifier with another substitution.

7.2.3 Inferences

Among the most important properties of FOPC is that there are *inference rules* which can be applied to WFFs to produce new WFFs, or to produce new results from given data or to prove new theorems from given axioms and theorems. Some basic inference rules are *modus ponens, modus tollens, universal specialization,* and *reductio ad absurdum*, which are stated in Fig. 7.9. They may be regarded as intuitively obvious. A WFF derived from other WFFs is a *theorem*. The sequence of inference rules used to derive a theorem is a *proof*. Problems in image and speech understanding may be viewed as the problem of finding proofs in the above sense.

```
modus ponens
Rule:       IF s is TRUE, THEN t is TRUE.        s → t
Datum:      s is TRUE.                           s
Inference:  t is TRUE.                           t

modus tollens
Rule:       IF s is TRUE, THEN t is TRUE.        s → t
Datum:      t is FALSE.                          ¬t
Inference:  s is FALSE.                          ¬s

universal specialization:
Datum:      For all assignments of x, P(x) is TRUE.   (∀x)P(x)
Inference:  P(A) is TRUE for any constant A.          P(A)

reductio ad absurdum:
Rule:       IF s is TRUE, THEN t is TRUE.        s → t
Rule:       IF s is TRUE, THEN t is FALSE.       s → ¬t
Inference:  s is FALSE.                          ¬s
```

Fig. 7.9. Some basic inference rules in first order predicate logic

A more general inference rule is *resolution* which works on clauses. In fact, the basic rules of Fig. 7.9 are special cases of resolution. The resolution rule for ground clauses states that from two given clauses a *resolvent* can be obtained by

$$\text{given ground clauses: } P \vee \neg Q \ , \quad Q \vee R \ , \quad \text{resolvent: } P \vee R \ , \tag{7.2.14}$$

that is, the complementary pairs Q and $\neg Q$ are eliminated from the two clauses. For example, if the given clauses are P and $\neg P \vee Q$, the resolvent is Q, and this by Fig. 7.8 and (7.2.2) is *modus ponens*. An extension of resolution from ground clauses to clauses is possible using unification. Assume two clauses

$$a = (a_1 \vee a_2 \vee \ldots \vee a_m) \quad \text{and} \quad b = (b_1 \vee b_2 \vee \ldots \vee b_n) . \quad (7.2.15)$$

Let a_1 contain variables as arguments and let b_1 differ from a_1 only by a negation and possibly by some arguments. Assume that a_1 and b_1 may be unified by a unifier s, that is

$$a_1 s = \neg b_1 s . \quad (7.2.16)$$

Then from the *parent clauses* a and b above the *resolvent*

$$r = (a_2 \vee \ldots \vee a_m)s \vee (b_2 \vee \ldots \vee b_n)s \quad (7.2.17)$$

is obtained. This results in the *generalized resolution rule*

IF (the parent clauses a and b as in (7.2.15) are given AND $a_1 s = \neg b_1 s$) ,

THEN (the resolvent r in (7.2.17) can be obtained). (7.2.18)

The above rule can be applied repetitively to a set of clauses until either the empty clause NIL is derived or no more terms of the form (7.2.15, 16) can be found.

The resolution rule can be applied to proving theorems. Let a set W' of WFFs be given, and a theorem represented by the WFF G. The goal is to derive G from W', or to prove G using W'. An informal justification of the resolution method for theorem proving is as follows. If a WFF G can be derived from W', then every interpretation satisfying W' also satisfies G and none satisfies the negation $\neg G$. In addition, no interpretation can satisfy $W = W' \cup \neg G$. Therefore, if G satisfies W', it cannot satisfy W for any interpretation. It can be shown that if resolution is applied repeatedly to a set of unsatisfyable clauses, the empty clause NIL will eventually result. So if a theorem G can logically be derived from a set W' of WFFs, resolution will produce the empty clause from W, and if resolution produces the empty clause from W, G follows from W'. This approach is *complete* in the sense that every formula G which follows from W' in fact can be derived from W' by resolution.

The completeness guarantees that a proof of a WFF G will be found in a finite number of steps if G follows from W'. However, if G does not follow from W', two cases are possible. In the first case the process stops after a finite number of steps without producing NIL; in the second case the process generates new clauses without stopping. This is not a peculiarity of resolution, rather it is a property of any complete inference system in FOPC. The statement that a proof will be obtained in a finite number of steps if a proof exists does not exclude that this finite number may be very large and it does not state *how* this proof is to be computed. The efficiency of a system thus depends on the details of the processing strategy.

Finally we consider a simple example from image understanding. We distinguish three types of statements: 1) general knowledge which is assumed to be always true; 2) intermediate results computed from a certain image which are true only for this image; 3) questions concerning the content of the image. The questions are viewed as theorems which are to be proven (or disproven) using the general knowledge and the intermediate results. The three types of statements are first given in natural language.

1. General knowledge:
    1.1. A house and a river are adjacent if their horizontal distance is less than 1000 m and if their vertical distance is less than 5 m.
    1.2. A house adjacent to a river is endangered by high water.
2. Intermediate results:
    2.1. The current image contains a house denoted by $H1$.
    2.2. The current image contains a house denoted by $H2$.
    2.3. The image contains a river denoted by $R1$.
    2.4. The horizontal distance between $H1$ and $R1$ is 100 m.
    2.5. The vertical distance between $H1$ and $R1$ is 4.9 m.
    2.6. The horizontal distance between $H2$ and $R1$ is 980 m.
3. Questions:
    3.1. Is the house $H2$ endangered by high water?
    3.2. Is there a house in the current image which is endangered by high water?
    3.3. Which house is endangered by high water?

It should be obvious that the above questions can be viewed as theorems G_i which are to be proven by the set of WFFs given in 1.1. to 2.6. Questions 3.1. and 3.2. are yes/no questions. If the answer to 3.2. is "yes", one does not know which house is endangered. In the question 3.3. the name of the endangered house is asked for and this requires a *constructive proof* of the corresponding theorem.

The next step is to convert the statement to WFFs. Let $HO(x)$ and $RI(y)$ be predicates having the value TRUE if the objects represented by x and y are a house and a river, respectively. Let $HD(x, y)$, be a predicate having the value TRUE if the objects denoted by x and y have a horizontal distance of less than 1000 m; introduce a similar predicate $VD(x, y)$ testing for the vertical distance. The predicates HO, RI, HD, VD are assumed to be given and are not defined further. Furthermore, introduce two predicates $AD(x, y)$ and $EN(x)$ having the value TRUE if objects x and y are adjacent and if x is endangered by high water, respectively. The WFFs of the above statements are:

1.1. $(\wedge_x, \wedge_y)[(HO(x) \wedge RI(y) \wedge HD(x, y) \wedge VD(x, y)) \rightarrow AD(x, y)]$
1.2. $(\wedge_x, \wedge_y)[(HO(x) \wedge RI(y) \wedge AD(x, y)) \rightarrow EN(x)]$
2.1.–2.6. $HO(H1), HO(H2), RI(R1), HD(H1, R1), VD(H1, R1), HD(H2, R1)$
3.1. $EN(H2)$, negate: $\neg EN(H2)$
3.2. $\vee_x[HO(x) \wedge EN(x)]$, negate: $\wedge_x[\neg(HO(x) \wedge EN(x))]$
3.3. $\vee_x(HO(x) \wedge EN(x))$ <constr.>, negate: $\wedge_x[\neg(HO(x) \wedge EN(x))]$; add $AN(x)$ to account for constructive proof: $\wedge_x[\neg(HO(x) \wedge EN(x))] \vee AN(x)$.

The last WFF contains the term AN(x) for ANswer (x) and is introduced to save the variable x. If resolution using 3.2 produces the empty clause, resolution using 3.3 will produce AN(.) where the argument of AN is some constant denoting an endangered house. In this way a constructive proof of a theorem may be obtained.

Since resolution works on clauses, the above WFFs have to be converted to clause form by the steps given in Sect. 7.2.1. The result is

1.1. $\neg$HO(x) $\vee$ $\neg$RI(y) $\vee$ $\neg$HD(x,y) $\vee$ $\neg$VD(x,y) $\vee$ AD(x,y)
1.2. $\neg$HO(u) $\vee$ $\neg$RI(v) $\vee$ $\neg$AD(u,v) $\vee$ EN(u)
2.1.–2.6. is already in clause form
3.1. $\neg$EN($H2$)
3.2. $\neg$HO(w) $\vee$ $\neg$EN(w)
3.3. $\neg$HO(w) $\vee$ $\neg$EN(x) $\vee$ AN(w)

The set W' is now $W' = \{1.1, 1.2, 2.1, 2.2, 2.3, 2.4, 2.5, 2.6\}$, the three theorems are $G_1 = \{3.1\}$, $G_2 = \{3.2\}$, $G_3 = \{3.3\}$. Resolution on the three sets $W_i = W' \cup G_i$ will produce the answers to the three questions, in this case "no", "yes", and "$H1$", respectively. Finding appropriate substitutions s and application of (7.2.18) are straightforward in this example and, therefore, are omitted.

7.2.4 Remarks

The advantages of using FOPC for knowledge-based processing in image or speech understanding are that a rigorous and well-established body of theory exists and also programming languages have been implemented which allow one to write WFFs and provide a resolution algorithm. Some deficiencies of FOPC may be seen from the example of the last section. For example, the predicate HD(x,y) is a crisp predicate having only truth values 0 or 1; the consequence is that a house at a distance of 999 m fulfills the predicate, a house at a distance of 1001 m does not. In the given task domain it would be more reasonable to allow, for example, a gradual transition, say between a distance of 500 to 1500 m. This cannot be represented in FOPC. The intermediate result HO($H1$) may not be absolutely certain due to noisy image data and unreliable segmentation. So the answer to 3.2. might be "probably" or "maybe" instead of "yes" or "no"; such an answer cannot be obtained by FOPC. The answer to the question "is $H2$ endangered" is "no" since VD($H2,R1$) is not in the intermediate results. However, the reason for this may be either that this result cannot be computed because it is wrong (in which case the answer "no" is always TRUE), or that this result has not yet been computed but may occur later in the data base of intermediate results. In this latter case the answer "no" will have to be corrected later. Since FOPC is *monotonic*, a theorem which has been disproven once cannot be proven later, or vice versa. These few remarks demonstrate that there are serious deficiencies of FOPC in its application to pattern analysis.

7.3 Production Systems

7.3.1 General Properties

Production systems are another approach to representation and utilization of knowledge. They are a program structure performing data- or event-driven operations, as distinguished from conventional programs operating on expected data. In a production system the steps of data examination and data modification are clearly separated. A *production system* consists of three basic components.

1. A set of *production rules*, or *rules* for short, in the form of antecedent-consequent pairs. They define possible operations on data and may be viewed as the content of a constant or slowly changing knowledge data base (long-term memory).
2. A set of *data* which are inspected and modified by the production rules. This set is part of the results data base (short-term memory).
3. A *strategy* for selecting production rules to be executed. This is in some sense a control module, but it may be local to a knowledge module or submodule and need not be global to the whole pattern analysis system.

The power of the above approach rests in its generality. The antecedent and consequent pairs may range from simple tests and assignments up to arbitrarily complex programs for each component of the pair. Data may be of any kind, be it integer variables, general graphs, or rules. Similarly, the rule-selection strategy may be implicitly defined by inspecting rules in fixed order or also by an arbitrary set of metarules. These options, on the other hand, leave a lot of design choices to be filled in.

The format of a production rule is

$$\text{IF (premise or condition), THEN (action or conclusion),} \qquad (7.3.1)$$

or equivalently

$$\text{antecedent} \rightarrow \text{consequent,} \qquad (7.3.2)$$

which is in the format of the logical implication in (7.2.2). The IF term, *antecedent*, or left-hand side (LHS) of the rule specifies a condition which has to be met by the data if the rule is to be applicable. In case the condition is met the THEN term, *consequent*, or right-hand side (RHS) of the rule specifies the action to be taken. Execution of the action will result in a change of the data contained in the data base. Two simple examples are the rules

$$\begin{array}{ll} \text{IF} & \text{(there are two long, closely adjacent, nearly parallel lines,} \\ & \text{AND these are crossed by two other long, closely adjacent,} \\ & \text{nearly parallel lines),} \\ \text{THEN} & \text{(there is evidence that there may be a crossing of two highways,} \\ & \text{OR a highway crossing a river, OR a railroad crossing a river),} \quad (7.3.3) \\ \text{IF} & \text{(BCD), THEN (ABCDE).} \qquad (7.3.4) \end{array}$$

7.3 Production Systems 287

The production rule (7.3.3) requires that detailed algorithms are supplied to check the condition and to add the conclusion to the data base. The rule (7.3.4) only requires a simple checking for the occurrence of a substring in a character string and its replacement by another substring. The rule (7.3.4) is identical to a rewrite rule of a formal grammar as introduced in Sect. 4.3.

If a production system contains a large data base and a large set of productions, there will, in general, be several IF terms whose conditions are met by the data. So at any time there will be a set of productions which might be executed and a choice has to be made which one to execute. Furthermore, there may in general be several portions of data which all meet the IF term of one and the same production. So a choice on which data to execute a production may also be necessary. A straightforward solution is to order productions in a fixed sequence and also to impose a fixed order on the data. The IF term of the first production is compared to the data, and if it is not met, the next production is taken, and so on. The first production whose IF term is met by certain data is executed and other data which also meet this IF term are disregarded. After execution of a production the process starts again with the first one. The system halts if no productions in the set of rules are found to be applicable. An unsatisfactory point with this solution is that system behavior depends on the order of the rules. Alternatively, all productions whose IF terms are met may be executed in parallel. This may result in competing and inconsistent alternative results which have to be resolved later.

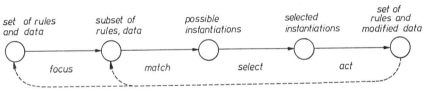

Fig. 7.10. An execution cycle of a production system consisting of a focus, match, select, and act phase. The focus phase may or may not be invoked after every cycle

Another approach is to perform the execution cycle shown in Fig. 7.10. As an example assume that the data and rules contain, among others, those shown in (7.3.5, 6). In the focus phase of the execution cycle an active subset of data and rules is determined. This may be facilitated by a hierarchically or otherwise structured arrangement of data and rules, and may be accomplished by a control action which is part of a hierarchical control module of the type shown in Fig. 6.17; this control module may be implemented by *metarules*. Selection of an active subset reduces the amount of search necessary in the match phase. For example, let the active subset be

$$\text{DATA} : (AV), \ (BV), \ (DE), \ (AZ), \ (DV), \tag{7.3.5}$$

$$R1 \quad : (A\,\hat{Y}),\ (B\,\hat{X}) \to\ (A\,W),\ (B\,W),$$
$$R2 \quad : (A\,C),\ (D\,\hat{X}) \to\ (A\,C),\ (D\,U), \qquad (7.3.6)$$
$$R3 \quad : (D\,E) \to\ (E\,D).$$

The data elements such as $(A\,V)$ or $(D\,V)$ in (7.3.5,6) are assumed to be unordered, and $\hat{X}, \hat{Y}$ in (7.3.6) denote a variable. A rule and the actual data supplied to a rule constitue an *instantiation* of this rule. The match phase searches for applicable rules and their instantiations I which in this case are

$$I_1(R1) \quad : (A\,V),\ (B\,V) \to \ldots,$$
$$I_2(R1) \quad : (A\,Z),\ (B\,V) \to \ldots, \qquad (7.3.7)$$
$$I_1(R3) \quad : (D\,E) \to \ldots.$$

So there are two instantiations of $R1$, one of $R3$, and $R2$ is not applicable. Let $I_2(R1)$ and $I_1(R3)$ be chosen in the select phase for execution, which may be done sequentially or in parallel. In the act phase the selected rules are executed, and the DATA are modified to

DATA $\quad : (A\,V),(B\,V),(D\,E),(A\,Z),(D\,V),(A\,W),(B\,W),(E\,D).$
$$(7.3.8)$$

The result is independent of the order of the rules, and depends only on the strategy for selecting rules for execution. Some general ideas for the selection of rules will be discussed below.

The operations can work "forward" or establish a *forward chaining* in the sense that data are inspected in order to derive operations on data or conclusions based on data. An inverse or *backward chaining* is also possible. In this case an initial hypothesis is given and one tries to reduce it to facts contained in the data base by applying rules in reverse order. The forward and backward chaining operations correspond to the bottom-up and top-down strategies of processing, respectively.

It is also possible to integrate rule selection into the production rules by adopting the approach of programmed grammars. In this case rules have the form

IF (premise or condition),
THEN (action or conclusion), $\qquad (7.3.9)$
TRUE (next rule or rules),
FALSE (next rule or rules).

If the condition in the IF term is met, the next rule follows from the TRUE term; otherwise it is taken from the FALSE term.

The standard production rule format (7.3.1) may be generalized by the introduction of a *censor*, which can block the rule in certain (rare) situations; the censor thus has the role of an exception. The rule then has the form

IF (condition), THEN (conclusion), UNLESS (censor). $\qquad (7.3.10)$

The censor condition is logically equivalent to

IF (condition) AND NOT (censor), THEN (conclusion) . (7.3.11)

The UNLESS part, however, expresses that generally the conclusion can be made if the condition is met. If no information is available about the censor, it is reasonable to use only the IF (...) THEN (...) part. For example,

IF (the animal is a bird), THEN (the animal can fly) ,

UNLESS (the animal is a penguin) . (7.3.12)

By attaching probabilities to the conclusion and the censor, these expectations can be made explicit and quantitative.

Two important properties of production systems are that they have the computational power of universal Turing machines and that production rules are independent, thereby allowing incremental addition of new rules, provided the monitor does not assume a certain order of the rules.

7.3.2 Inferences

The example at the end of Sect. 7.2.3 can also be stated using a rule based approach. The implication $A \to B$ is replaced by the rule IF (A), THEN (B). Using "modus ponens" as an inference rule one can infer B if A is an element of the data base. In the case of the above example:

1. rules:
   1.1. IF [HO(x) AND RI(y) AND HD(x,y) AND VD(x,y)], THEN (AD(x,y))
   1.2. IF [HO(x) AND RI(y) AND AD(x,y)], THEN (EN(x))
2. data: HO($H1$), HO($H2$), RI($R1$), HD($H1,R1$), VD($H1,R1$), HD($H2,R1$)
3. question (goal): EN(Z)

Adopting the convention that a variable is matched by any data element of the same type one can match HO($H1$)/HO(x), RI($R1$)/RI(y) and so on. Substituting appropriate results for variables yields AD($H1,R1$) and EN($H1$) as above. In forward chaining the steps of substitution (or matching) and application of modus ponens are repeated until either the goal is established or no new results can be computed. In backward chaining one starts with the goal and looks for a rule having the goal on its right side. Then one tries to establish the elements of the condition of this rule either from the data or from other rules. In this sense FOPC and the production system show few differences.

However, since the production system does not claim to realize FOPC, it is quite possible to modify the rules. For example, instead of using predicates HO(x) and so on being either TRUE or FALSE one can use [attribute, value, certainty] triples in the condition part of a rule. This allows one to represent a statement of the type "object $H1$ is a house with certainty (or grade of membership) 0.85" by [HO, $H1$, 0.85], or in general [house, x, CF(house(x))]. It is now necessary to combine the certainties of the components of a condition to obtain the certainty of the conclusion. Different approaches to measuring certainty were discussed in Sect. 6.4. A straightforward case results if certainties are modeled by fuzzy sets according to (6.4.15–18). If a rule of the type

IF $(A_1 \land A_2 \lor \neg A_3)$ THEN (B) (7.3.13)

is given and the certainties of A_1, A_2, and A_3 are given by $CF(A_1)$, $CF(A_2)$, and $CF(A_3)$, respectively, the certainty of B is by (6.4.17, 18)

$$CF(B) = \max \{\min \{CF(A_1), CF(A_2)\}, 1 - CF(A_3)\}. \qquad (7.3.14)$$

An example of a rule of this type is given in (7.3.28). This demonstrates that production systems can represent and manipulate statements which cannot be expressed in FOPC. The equations (7.3.13, 14) may be viewed as an inference rule of the type $[A, CF(A)] \to [B, CF(A)]$ or a straightforward generalization of the implication $A \to B$. Apparently, $CF(A)$ may be modeled by probability, bpa, or fuzzy sets.

The notion of fuzzy sets provides a further generalization of modus ponens. A short introduction to fuzzy sets was given in Sect. 6.4.4. Consider a rule

IF   (the length of the line if fairly large,
AND the average gradient is not low,
AND the reliability of the line is above average,
AND the line is surrounded by the same region,
AND the line endings are open),
THEN (extend the line slightly at both ends). (7.3.15)

The condition contains linguistic variables (see Sect. 6.4.4) like "length", "average gradient". Some are defined by fuzzy subsets (e.g. the "length" is "fairly large"), some by crisp subsets (e.g. the "line ending" is "open"). The logical implication $A \to B$ in FOPC can be generalized to a *fuzzy implication* or a *fuzzy conditional statement* of the type

IF (the fuzzy subset A is given),
THEN (the fuzzy subset B is implied) . (7.3.16)

For brevity this is denoted by $A \to B$ in this section. The universe of discourse is U, V, that is $U \supset A, V \supset B$. A *fuzzy relation* R between A and B is a fuzzy subset of the product set $U \times V$. The fuzzy implication (7.3.16) then is represented by the fuzzy relation R whose membership function is defined by

$$\mu_R(u,v) = \min \{\mu_A(u), \mu_B(v)\}, \quad u \in U, \quad v \in V. \qquad (7.3.17)$$

Let A' be another fuzzy subset of U which may be an observation or some available evidence. Given R, the evidence about V is described by the fuzzy subset B' of V obtained from the *compositional rule of inference*

$$B' = A' \circ R$$

$$\mu'_B(v) = \max_{u \in U} \min \{\mu_{A'}(u), \mu_R(u,v)\}$$

$$= \min \left\{ \max_{u \in U} [\min (\mu_{A'}(u), \mu_A(u))], \mu_B(v) \right\}. \tag{7.3.18}$$

In the special case of $R = A \to B$, and $A' = A$, $B' = A' \circ R = A \circ (A \to B) = B$. In this sense (7.3.18) is a generalization of modus ponens. Whereas in modus ponens it is necessary to confirm A in order to infer B, the compositional rule of inference does not require A' to be identical with A.

7.3.3 An Example of a Rule Format

The unrestricted rule format in (7.3.1) usually is restricted in a particular implementation of a rule based system. We first consider an example of a rule format designed for general production systems. A rule consists of the symbol P, the name of the production, the condition, IF-term or LHS, the symbol $\to$, the action, THEN-term or RHS and thus has the form

$$(P \text{ NAME (condition)} \to \text{(action))}. \tag{7.3.19}$$

The "condition" consists of a sequence of elements (or "patterns") having

1. a particular type, e.g. the type EXPRESSION,
2. a sequence of attribute–value pairs, where values may be numbers, constant symbols, variable symbols, or predicates.

The "action" consists of an unconditional sequence of actions having

1. a particular type, e.g. MAKE a new data element, MODIFY an existing element, or REMOVE one or more elements,
2. a sequence of attribute–value pairs.

An element of the "condition" in the rule memory is viewed as a partial description of a data element in the data base. A "condition" is satisfied if

1. every element of the "condition" which is not negated machtes a data element,
2. no element of the "condition" which is negated matches a data element.

An example of a rule is the following:

$(P$ NAME1 (GOAL Atype simplify Aobject $\hat{X}$)

(EXPRESSION Aname $\hat{X}$ Aarg1 0 Aop $*$)

$\to$ (MODIFY 2 Aop NIL Aarg2 NIL). $\tag{7.3.20}$

The above rule has the name NAME1; the "condition" has two elements of type GOAL and EXPRESSION, respectively; the "action" has one element of type MODIFY. An attribute is preceded here by a capital letter A (which is not a standard notation) followed by the name of the attribute; $\hat{X}$ denotes a variable. The above "condition" element of type EXPRESSION would match the data element (EXPRESSION Aname tom Aarg0 5 Aarg1 0 Aop $*$ Aarg2 1), but it would not match the data element (EXPRESSION Aname tom Aarg1 0 Aop +). The variable $\hat{X}$

is matched by any value; however, if there are several occurrences of $\hat{X}$ in the condition, the matching data element has to have the same value of the variable in all places.

The *monitor* or *interpreter* of a production system has to perform the following basic steps:

1. Match, that is compare all conditions of all rules to all data elements in order to find applicable rules.
2. Resolve conflicts, that is select among the applicable rules one whose action is to be executed. If there is no applicable rule, then END.
3. Act, that is perform the actions given by the "action" of the selected rule.
4. Continue at step 1.

A general problem is the efficiency of step 1, and the strategy in step 2. An efficient matching algorithm is the *RETE* algorithm. The purpose of the algorithm is to compare the "conditions" of the rules to the data elements. Its output is a set of pairs of the form

(rule name, list of data elements matched by the condition of the rule) .

(7.3.21)

Such a pair is called an *instantiation* of the rule. The set of pairs is called the *conflict set*. The RETE algorithm achieves efficiency by taking two actions:

1. Avoid iterating over data elements in working memory. The idea is to check for changes in working memory and store the resulting changes in the conflict set.
2. Avoid iterating over elements of "conditions" in the rule memory. The idea is to arrange the conditions of the rules in a tree-structured sorting network such that the same condition occurring multiply in different rules is checked only once. The details of the algorithm are omitted here.

Some general rules for *conflict resolution* are the following:

1. Impose a certain order on the rules (priority) and select the first applicable rule (or the rule having highest priority).
2. Do not use the same rule on the same data more than once.
3. Prefer the rule working on the most recent data; if there is more than one, prefer the rule working on the next most recent data, and so on. Recency can be measured in units of monitor cycles or in units of other actions.
4. Prefer the rule which is more specialized, that is it has more conditions including negations.
5. Prefer the more general rule.
6. Prefer the rule working on more specialized data, that is using as data a proper superset of the data used by other rules.
7. Select a rule which has not been applied so far.
8. Impose a certain order (priority) on the data and select the rule using the most important data.

9. Select one instantiation at random if no other choice can be made.
10. Execute applicable rules in parallel. This may result in competing and inconsistent data, a problem which has to be treated separately.

7.3.4 Examples from Pattern Analysis

Next we consider an example of a rule format designed for speech understanding. The data elements in the results data base are unordered and consist of a list structure. Two common forms of data elements are the *hypothesis* and the *link* between hypotheses (see also Fig. 5.23). For example, consider the hypothesis and the link element

$\langle$HYP WRD "TRAIN" BTIME/(20 5) ETIME/(70 3) VLD/60$\rangle$ (7.3.22)

$\langle$LNK WRD UHYP/H20 LHYPS/(H10 H11) UIMP/90 LIMP/30$\rangle$. (7.3.23)

The element in (7.3.22) represents the hypothesis (HYP) that the word (WRD) train appears in an utterance, that it begins (BTIME) at 20 ± 5 and ends (ETIME) at 70 ± 3 time units in the utterance, and that the reliability (VLD) of this hypothesis is estimated to be 60 units. Incidentally, the symbol WRD refers to the word lexicon, and the "TRAIN" is represented internally by an integer giving the position of this word in the lexicon. There are some additional fields in a hypothesis not shown above such as a reference to another data element or lists of references to upper and lower links. A link relates hypotheses on different levels, for instance, by giving an upper hypothesis (UHYP) which is supported by lower hypotheses (LHYPS). The element in (7.3.23) is a link (LNK) relating the word hypotheses H10, H11 to upper level hypothesis H20. The strength of the upper implication (UIMP) of the support is estimated to be 90, the lower implication (LIMP) 30 units.

Rules are based on a set of action elements which allow one to create new data elements and to delete or modify existing ones. An example is the action

$\langle$NEW $\hat{L} = \langle$LNK SYL UHYP/$\hat{W}$ LHYPS/($\hat{S}1\,\hat{S}2$) IMP/90$\rangle\rangle$. (7.3.24)

This example also illustrates the use of variables, which are denoted by a " $\hat{\ }$ ". With (7.3.24) a new (NEW) link is provided and bound to the variable $\hat{L}$ for later reference. The link is from two syllables bound to the variables $\hat{S}1$ and $\hat{S}2$ to an upper hypothesis bound to $\hat{W}$ and has an implication of 90 units. The actions also produce an indication of the change which is used in the following cycle of the production system. All conditions have, first, a test for the nature of a change. An example of a production rule named R is

R :(NGAP $\hat{S}$ = $\langle$ETIME/([> $\hat{B}$T + 15]*)$\rangle$) $\rightarrow$ NGPS ; (7.3.25)
NGAP :($\langle$NEW O$\hat{S}\rangle$ ÔS = $\langle$HYP OSEG SEGS/($\hat{S}$)$\rangle$ $\hat{S}$ = $\langle$HYP SEG " $-$ "
       BTIME/($\hat{B}T^*$)ETIME/($\hat{E}T^*$)$\rangle$) ; (7.3.26)

```
NGPS :(→ ⟨NEW M̂ = ⟨HYP SYL "GAP" BTIME/(B̂T 3) ETIME/(ÊT 3)⟩⟩
         ⟨NEW ⟨LNK OSEG UHYP/M̂ LHYPS/(ÔS) UIMP/100⟩⟩
         ⟨NEW Ŵ = ⟨HYP WRD "GAP" BTIME/(B̂T 3) ETIME/(ÊT 3)
            IVLD/100 SYLNUM/1⟩⟩
         ⟨NEW ⟨LNK SYL UHYP/Ŵ LHYPS/(M̂) UIMP/100⟩⟩) .            (7.3.27)
```

This example shows that production rules, conditions, or parts of conditions in a rule, and actions or parts of actions in a rule may be named. For instance, if a condition is used in several production rules, it may be referenced in the rules by its name. The rule R tests whether a new option-segment (OSEG) hypothesis occurred and contains the segment "-". Begin and end, bound to variables $\hat{B}T$ and $\hat{E}T$, may have an arbitrary tolerance; this is indicated by the symbol $*$ which is matched by any element. If in addition to condition NGAP the endtime is at least 15 time units after the begin time (ETIME/([> $\hat{B}T$ + 15]$*$)), the IF term in rule R is met and the action NGPS is executed. This creates four new data elements. The first is, for instance, the hypothesis bound to $\hat{M}$ that a gap exists on the level of syllables lasting from time $\hat{B}T\pm 3$ to $\hat{E}T\pm 3$, and the last is a link from $\hat{M}$ to $\hat{W}$ with implication of 100 units. Thus rule R propagates a gap on the segment level up to the syllable and word level. These few examples show a realization of production rules in a complex task, but no complete specification was intended. The control strategy of the system is very simple: all productions, whose conditions are met, are executed. But a production will only be executed if the relevant change has just occurred. This prevents multiple executions of the same production on the same data.

Another example is a rule employed in a system for the automatic interpretation of scintigraphic images of the human heart. On a high level of abstraction diagnostic terms are inferred by rules representing medical knowledge. For example, the "akinetic behavior of a segment of the left ventricle" is defined by the rule

```
IF        (LV_WEAK AND [IA_AKIN OR PL_AKIN OR B_AKIN OR S_AKIN]:
           THEN (AKIN),
CF(AKIN) = CF(LV_WEAK) AND [CF(IA_AKIN) OR CF(PL_AKIN)
           OR CF(B_AKIN) OR CF(S_AKIN)],
CF(AKIN) = min {CF(LV_WEAK), max {CF(IA_AKIN), CF(PL_AKIN)
           CF(B_AKIN), CF(S_AKIN)}} .                              (7.3.28)
```

This is a straightforward translation to a rule of the statement "if the motion of the left ventricle is weak and one of its four segments (i.e. the inferioapical, postero-lateral, basal, or septal segment) shows akinetic behavior, then the motional behavior of the left ventricle is akinetic". The different components of the condition, e.g. "the basal segment has akinetic behavior", are represented by concepts, e.g. the concept B_AKIN. The certainty of an instance of a concept is denoted by CF (.), e.g. CF(B_AKIN). Composition of certainties is by (7.3.14).

Finally we give an example of a rule in a medical expert system

```
IF        (the stain of the organism is gram positive,
           AND the morphology of the organism is coccus,
```

AND the growth confirmation of the organism is chains),
THEN (there is a suggestive evidence of 0.7 that the identity of the
organism is streptococcus). (7.3.29)

The examples (7.3.22–27) and (7.3.28) are from systems for pattern analysis and understanding in the sense that the original input to the system is a sequence of sample values, that is an image or speech signal or, in general, a sensor signal. Any evidence requested, for example, in the condition of (7.3.28) is automatically computed from the input signal. The system corresponding to the example (7.3.29) requires that an operator answers certain questions of the system; so this system cannot perform pattern analysis starting from an image or speech signal.

7.4 Semantic Nets

7.4.1 Introductory Remarks

A semantic net is a data structure which is a specialized graph and defines some concept. As an example take the following verbal definitions of "house", "bridge", and "city":

A house (*is a* building [*having* {one or more stories and a roof}]).
A bridge (*is a* building [*crossing* {a river or a street}]). (7.4.1)
A city (*is an* area [which *always has* {many buildings and streets}]).

These may be used as verbal definitions in speech processing, but the approach may as well be used to define geometrical properties in image analysis. Figure 7.11 gives a semantic net representation of (7.4.1). The nodes marked α, β and γ in the figure belong to the terms in parenthesis, square brackets, and curly brackets in (7.4.1), respectively. In the figure a 1 edge stands for the "is a" relation (A house "is a" building ...), a 2 edge denotes a modification of a noun (... a building having ...) or to put it another way it denotes a part of the concept, a p edge denotes a predicate (... having ...), a 3 edge specifies an argument of a predicate (... having one or more ...), a d edge indicates a determiner (... a building ...), an r edge a relation (... one or more stories ...), and a c edge a canonical form (have as infinitive of having). To avoid a mess of lines, some nodes – like those for "or", and "a" – are duplicated although this does not really occur. It is seen that some definitions refer to other concepts, like "river" or "street", which themselves might require a definition, and that some definitions make use of the same concepts. This allows one to find similarities between concepts and also to infer the meaning of new concepts. For instance, if the meanings of "garden" and "vegetable" are defined in the net, the meaning of "vegetable garden" may be inferred by a search procedure. Some of the concepts used in the net will have to be considered as *primitive*, that is they need no further definition in the semantic net. If the net is designed properly, it is thus

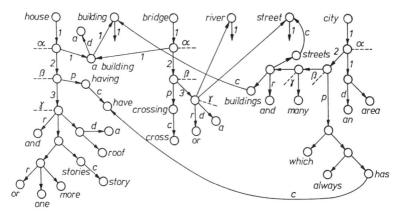

Fig. 7.11. A semantic net representing the concepts in (7.4.1)

possible to obtain a description of a high-level concept (or a complex pattern) by using only primitive concepts (e.g. segmentation objects).

The semantic net approach offers the potential of representing different *levels of abstraction* or different *conceptual systems* within the same framework, of integrating quite different procedural knowledge, and of handling imprecise and uncertain knowledge and results. A semantic net basically is a graph having only a restricted number of node and link types. Meanwhile there are quite a few definitions of the syntax of semantic nets, and the above figure was only an introductory example. Two types of nodes are common to most of them, the concept node and the instance node, and three types of links, the specialization link (or the "is a" relation), the part link (or the "has a" relation), and the instance link. In a general definition one has to define precisely the allowed components of a net (that is the *syntax* of the net), the meaning of the components (that is the *semantics* of the net), and the allowed use of the components (that is the *pragmatics* of the net). The number of components should be as small as possible in order to facilitate efficient control algorithms, but the components should be sufficiently general in order to allow the definition of rich task dependent knowledge structures. An example of a fairly general definition is given in the next section.

7.4.2 Components of a Semantic Net

An overview of the main components of the semantic net to be defined in the following is given in Fig. 7.12. The net has three types of nodes, the *concept*, the *modified concept*, and the *instance*. The concept node allows one to represent a model-scheme as well as a model in the sense of the horizontal hierarchy in Fig. 7.4. Since the model scheme is used for knowledge acquisition its discussion is deferred to Sect. 7.7. The modified concept contains a concept which is constrained due to instances of other concepts. The instance contains a result of processing. The *concept node* represents a class of objects, events, or situations.

It is an *intensional description* or the *intension* of the class. An *instance node* represents one particular member of this class; it is an *extensional description* or an *extension* of the class. Since we work in the context of image or speech understanding, instances always originate from some sensor data and are computed from these data. Different nodes of the same type or of different types are related by five types of *links*. In addition, various components of the net are defined by a set of substructures.

```
node
    NODE_TYPES: concept
                modified concept
                instance
                / each node is defined by slots /
link
    LINK_TYPES: specialization, specialization of
                part, part of
                concrete, concrete of
                model, model of
                instance, instance of
                / for every link the inverse link is defined by "..of"/
substructure
    SUBSTRUCTURE_TYPE: attribute description
                       link description
                       relation description
                       adjacency
                       modality description
                       value description
                       range
                       function description
                       identification
                       / each substructure is defined by items /
```
Fig. 7.12. The main components of a semantic net

A concept, for example, the concept "car", may have *parts* like "wheel", "luggage boot", "window", or "hood" related to the concept by the *part link*. In addition a concept may have specializations related to it by the *specialization link*. For example, a specialization of a car is the "sports car". Usually a specialization *inherits* all attributes, analysis parameters, relations, analysis relations, concretes, and parts of the more general concept except when this is explicitly excluded. This is marked by the item "modifies" in the corresponding substructure. It is possible to relate a concept to another one which is not a part of it in a physical sense, but which is related to it on another level of abstraction. For example, a car may be moving and thus be related to a concept like "motion on a street", or a wheel may be related to a concept "rotation". In order to infer a wheel in an image it is necessary first to find an "elliptic shape". This type of relation is expressed by the *concrete link*. It is called concrete because a concretization of a concept is closer to the sensor data and in this sense more concrete or less abstract. The specialization, part, and concrete are in direct correspondence to Fig. 7.3. The complete data structure is shown in Fig. 7.13.

When computing an instance of a concept, e.g. the "car", we have to have instances of all of its parts and concretes, in the above simple example instances

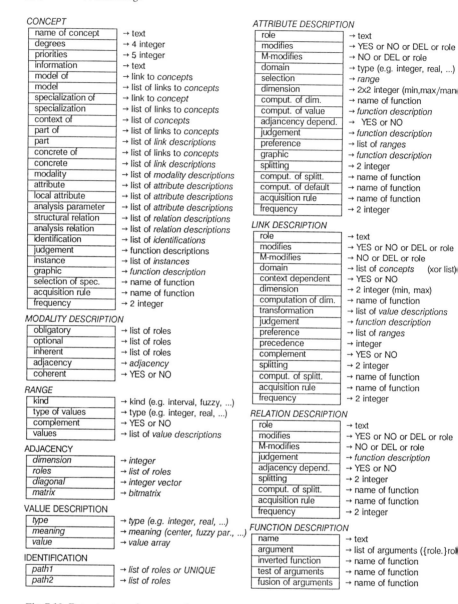

Fig. 7.13. Data structures in a semantic net

of "wheel", "window", "luggage boot", and "hood". To compute an instance of "wheel" we have to have an instance of "elliptic shape". However, in the context of sensor data interpretation it is not possible to identify an "elliptic shape" as a "wheel" without having the context of the superior concept "car" because an elliptic shape may have quite different meanings depending on the context. Therefore, it is necessary to distinguish two types of parts, the *context-dependent parts* like the "wheel" and the *context-independent parts* like a "hood". These two types of parts are treated differently in the instantiation of concepts as shown below. The notion of context-dependent parts amounts to the introduction of *context sensitivity* in knowledge representation. If, for example, two concepts X and Y have a concept K as a context-dependent part, the slot "context of" of K has the entries X and Y; the item "context depending" of the substructure "link description" has entry "YES" for both X and Y. This is also shown in Fig. 7.15 below.

It may be too demanding to require instances of all parts of a concept for its instantiation since due to occlusion or viewing angle only some parts may be visible. Furthermore, the same concept, for example, the "chair" may have different types or number of parts and concretes; a "chair" may have three or four legs, and it may have a cushioned seat or not. Therefore, parts and concretes can be grouped into *sets of modality* where each set is sufficient to compute an instance. A set of modality has obligatory parts which must be present and optional parts which may be present. For example, one set of modality might be one wheel and a luggage boot as obligatory and windows and additional wheels as optional; another set might be wheel and hood as obligatory and windows and additional wheels as optional.

The complete definition of each part and concrete relationship is done by the substructure *link description*. Different from most similar knowledge representation languages, the "dimension" item inside a link description causes no differentiation for the corresponding "role" in the instances. However, the role in an instance has to be filled with just as many instances as the dimension interval requires. In order to facilitate an analysis process it is useful to avoid certain cycles in the knowledge base. This is done by imposing some constraints on the introduction of links between concepts. If a concept A is related to a concept B by a specialization link directed from A to B (or a chain of specialization links), then it must not occur that there is also a sequence of specialization links from B to A, nor a sequence of part or concrete links from A to B or vice versa. An analogous condition is imposed if A and B are related by a chain of part or concrete links.

With these restrictions the concepts of a network form a three-dimensional hierarchy along the specialization, part, and concrete links. There is one *most general concept* G such that there is no concept having G as a specialization, a part, or a concrete. Following the part and concrete links one finally reaches *minimal concepts* which do not have parts or concretes (and do not inherit them from more general concepts). The 3D hierarchy may be viewed as a 3D coordinate grid having a concept or a set of concepts in the grid points. The origin of the coordinate system may be chosen arbitrarily, for example, in the most general minimal concept. The grid point of any concept has a "degree" (i, j, k) giving the position of the grid point relative to the origin of the coordinate grid. It is assumed that instances of minimal concepts can be computed from the initial segmentation I^r. In this sense the minimal concepts are the *interface level* between segmentation or low-level methods and knowledge based processing.

Among the substructures is also an *adjacency* which allows a compact notation of required time or spatial neighborhoods of parts. Because it is possible to describe the facts of a modality description also by specialized concepts and the adjacency in the substructure "relation", both are not epistemological primitives. But they give more compact knowledge bases and therefore support the ergonomic adequacy.

For later reference we denote the set of parts and concretes of a concept K by PA(K) and CO(K), respectively. According to the above discussion the set PA(K) is partitioned into the sets of context-dependent and context-independent parts, denoted by CD-PA(K), and CI-PA(K), respectively. These two sets and the set CO(K) may form different subsets of obligatory and optional parts and concretes. The pth subsets of obligatory components are denoted by OB-CD-PA$_p$(K), OB-CI-PA$_p$(K), and OP-CO$_p$(K); their union gives the pth obligatory modality set OBM$_p$(K). The pth subsets of optional components are OP-CD-PA$_p$(K), OP-CI-PA$_p$(K), and OP-CO$_p$(K); their union gives the pth optional modality set OPM$_p$(K) which corresponds to the pth obligatory set OBM$_p$(K). This is illustrated in Fig. 7.14. In addition to the part and concrete links the inverse links part-of and concrete-of are also provided in the net. The set of concepts having K as part is PA-OF(K), and similarly the set CO-OF(K) is defined.

A concept also has a set of attributes which describe physical or other properties of the class, for example, length, color, location, or prize. An attribute is described by the data structure *attribute description*. In addition, structural relations between parts or concretes can be defined in the structure *relation description*. For example, a relation might be that the wheel is below a hood or a luggage boot and that the four wheels are in rectangular arrangement. The main item in the substructures defining an attribute description or a relation description is the *function* to calculate the value of the attribute or to judge the relation when building up an instance. A function description includes the explicit notation of the *arguments*, and it is possible to refer also to the inverse. While the activation of the function itself results in actual values if all arguments are known, the inverse is able to restrict the domain of attributes in modified concepts. In a similar way the domain of an attribute may be restricted, if not all the argument values are known. The complete domain description of an attribute is given by the three items "domain", "selection", and "dimension". The "domain" defines the type of attribute, for example, integer, real, concept, or set; the "selection" defines the range, for example $0 \leq$ integer < 100; the "dimension" gives the size, for example, $32 \cdot 32$ integers. Arguments are described by a role or a pair of roles, respectively. If only one role is referred to, the argument is taken from the instance or modified concept itself. If two roles are specified for an attribute of a concept K, the first role denotes a concept A which is a part or concrete of K, the second role denotes an attribute of A.

Whereas the attributes show the features of a concept, "analysis parameters" accept quantitative parameters which are necessary for the analysis process but

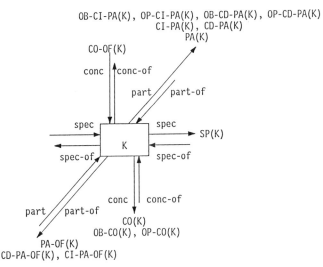

Fig. 7.14. The sets of concepts related to a concept K

$A \in PA(K)$: K has A as a part; $X \in PA\text{-}OF(K)$: K is a part of X

$OBM_p(K) = OB\text{-}CO_p(K) \cup OB\text{-}CD\text{-}PA_p(K) \cup OB\text{-}CI\text{-}PA_p(K)$: the pth *obligatory* set of modality related to concept K

$OPM_p(K) = OP\text{-}CO_p(K) \cup OP\text{-}CI\text{-}PA_p(K) \cup OP\text{-}CD\text{-}PA_p(K)$: the pth *optional* set of modality related to concept K

(CO: concrete; CO-OF: concrete-of; PA: part; PA-OF: part-of; OB: obligatory; OP: optional; CD: context-dependent; CI: context-independent)

do not contribute to the intensional description. The same situation distinguishes structural relations from "analysis relations".

The function for computing a score of an instance with respect to the conception modeled by the corresponding concept is referred to in the slot "judgement" of the concept. Arguments of this function are the judgements of the parts and concretes, the judgement of link and attribute descriptions, and the judgement of both kinds of relations. All arguments are referred to by the role of the corresponding substructure. The scheme for judging an instance is not fixed. In the different applications, for example, fuzzy logic, distance measurements, probabilities, or also vectors (quality, certainty, priority) are used.

In the course of instantiating a concept there will be intermediate and partial results of processing. For example, some elliptical shapes may be detected after preprocessing and initial segmentation. These intermediate results allow one to constrain the knowledge base of the system because the location of an elliptical shape may indicate the location of a wheel, this in turn indicates that there may be additional wheels within the neighborhood, and from this it becomes likely that a car will be found. These constraints result in more restricted domains of attribute values of the corresponding concept which then causes the creation of a *modified concept*. Thus a set of concepts may be viewed as a knowledge base valid for all elements of a task domain. A set of modified concepts is a knowledge base which is constrained by partial results computed from observations and therefore

302   7. Knowledge

is valid only for a subset of observations in the task domain. The data structure of a modified concept is similar to that for a concept. A constraint occurring in one concept will usually have effects on concepts related to it, that is, the constraint will propagate in the net. The details of this constraint propagation are defined in the rules stated in the next section.

Finally, an instantiation of a concept may already be started even if complete information for instantiation is not available. In this case, which is defined precisely by the rules given in the next section, a *partial instance* is computed. The data structure of a partial instance is the same as that of an instance.

To summarize, a system for image or speech understanding using a semantic net as knowledge base represents *declarative knowledge* by the concept hierarchy and *procedural knowledge* by attached functions. It represents knowledge about the whole task domain in concepts, about a restricted subset in modified concepts computed from actual observations, and about one particular element in an instance of a concept. This definition of a semantic net allows a direct realization of the structures (3.1.2, 4.5.1, and 7.1.1).

In order to work with such a system it is necessary to control appropriately the instantiation and modification of concepts. The basis for this control consists of two points. The first is a set of rules specifying the conditions when modified concepts and instances can be computed. The second point is a scoring and searching algorithm selecting alternatives for modifying and instantiating concepts. The basic principles of scoring and searching are discussed in Chap. 6.

7.4.3 Instantiation

The fundamental activity of a semantic net is the *instantiation* of concepts, that is, the computation of results using the initial description I^r of a pattern f^r as input and the semantic net as the explicit representation of structural and logical constraints. The conditions for the instantiation of a concept can be formulated as task independent. This means that general rules can be stated which only depend on the syntax of the network, but not on its content. If one has the general rules for instantiation of concepts then it is possible to combine them with some appropriate control strategy in order to achieve a "good" or "optimal" sequence of instantiations. This latter point will be discussed in the next section.

Five rules are sufficient to specify the conditions for the instantiation of concepts. First one builds *partial instances* for concept K or a modified concept $\text{mod}_i(K)$ derived from K. A partial instance may not contain values of all attributes due to context-dependent parts. However, it contains as many results as possible at some stage of processing.

Rule 1. (Creating partial instances of a concept K or a modified concept $\text{mod}_i(K)$)

IF   (for a concept K or a modified concept $\text{mod}_i(K)$ there is
      one obligatory set of modality $\text{OBM}_p(K)$ of K such that
      1. there are instances for all concepts in $\text{OB-CO}_p(K)$,
      2. there are instances for all concepts in $\text{OB-CI-PA}_p(K)$,

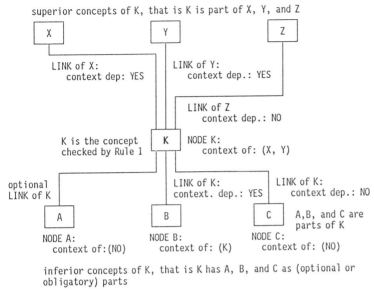

Fig. 7.15. Example illustrating the use of Rule 1. NODE K refers to the concept node K which has a slot "context of"; it points to a list of concepts in CD-PA-OF(K). The LINK referring to a part has an item "context depending"; it has entry "YES" if $B \in$ CD-PA(K)

    3. there is at least one instance of a concept in CD-PA-OF(K))
THEN (build up a partial instance $\text{inp}_j(K)$ by
    1. constructing an empty instance of K,
    2. connecting $\text{inp}_j(K)$ with the instance required
       in the condition of this rule,
    3. activating the functions for judgements
       of links, calculation of attributes, judgement of attributes,
       judgement of relations, and judgement (in this order)).   (7.4.2)

Figure 7.15 gives a small example of the effect of this rule. For this example the rule requires that there must be an instance of C, there need not be an instance of B, there must be an instance of X or Y, and there need not be an instance of Z. Because the part A of K is optional, no instance of the concept A is required. If the required instances exist, one or more new objects can be created. These objects are instances of a subtype called partial instances.

The second rule completes an instance when all the context-dependent parts of the corresponding concept have been instantiated. The third rule extends an instance by optional components. It is assumed that the effect of these two rules is obvious.

Rule 2. (Creating instances of a concept K)
IF   (1. a partial instance $\text{inp}_j(K)$ of a concept K has been
      computed from the obligatory set of modality $\text{OBM}_p(K)$,

2. there are instances of all concepts in OB-CD-PA$_p(K)$)
THEN (build up new instances in$_k(K)$ from inp$_j(K)$). (7.4.3)

Rule 3. (Extending instances)
IF (1. an instance in$_l(K)$ of a concept K exists,
2. there is at least one instance of a concept which is in the *optional* set of modality OPM$_p(K)$ corresponding to the obligatory set OBM$_p(K)$ used for constructing in$_k(K)$),
THEN (build up extended instances in$_l(K)$ from in$_k(K)$). (7.4.4)

Having built a certain number of instances it may be possible to compute better restrictions on attribute values or relations in other related concepts. This amounts to a propagation of constraints in the network. Propagation can be top-down or bottom-up and causes the computation of modified concepts by the last two rules.

Rule 4. (Bottom-up creation of modified concepts)
IF (for a concept K or a modified concept mod$_i(K)$ a new modified concept or a new instance was created for a concept in the set PA(K) or CO(K) or CD-PA-OF(K)),
THEN (create a new modified concept mod$_m(K)$ out of K or mod$_i(K)$). (7.4.5)

Rule 5. (Top-down creation of modified concepts)
IF (for a concept K or a modified concept mod$_i(K)$ a new modified concept or a new partial instance was created for a concept in the set PA-OF(K) or CO-OF(K))
THEN (create a new modified concept mod$_n(K)$ from K or mod$_i(K)$). (7.4.6)

These five rules define the use of a semantic net for the analysis of patterns and in this sense define the *pragmatics* of a net. Obviously, they are completely task independent since they use the syntactic properties of the net, but not the actual objects defined in a concept. The task dependence is introduced via the functions attached to a concept. The rules only activate functions without needing any knowledge of the effect or meaning of a function.

7.4.4 Control

The above rules state general conditions for the computation of instances, partial instances, and so on. Usually in a complex task domain there will be several alternatives for instantiating concepts and computing modified concepts. These alternatives arise, for example, from competing results of initial segmentation. The situation is in complete analogy to a production system where the conflict set of the rule set was introduced. A particular control strategy in a semantic net has to choose among the available alternatives and has to carry out the computations in a certain order. The basic idea is to combine the A^*-algorithm introduced in Sect. 6.5 with the above rules. Several different control strategies are conceivable. In this section two basic algorithms are outlined.

A simple, yet powerful control algorithm can be summarized by "top-down model expansion and bottom-up concept instantiation" or *top-down control* for short. The idea is to start with some goal concept K which may, for example, be the most general concept or which is stated by the user. The instantiation of K requires instances of parts and concretes as defined in the condition of rule 1 in (7.4.2). Instantiation of the parts and concretes in turn requires instances of their parts and concretes, unless they are *minimal*, that is, unless they do not have parts or concretes. So the first step is an expansion of that part of the model required by the goal concept K in order to determine all concepts which have to be instantiated prior to K. This expansion can be done whenever an instance of K is required; alternatively it can be done only once before any instances are computed and the list of required concepts is stored as auxiliary information of K. After this expansion the second step is a successive bottom-up instantiation of concepts until K has been instantiated. If there are competing instances, the instances are scored and only the most promising instances in the sense of the A^*-algorithm are considered further. The algorithm is given in Fig. 7.16.

If the model is only of moderate complexity (say 100 to 300 concepts), the above strategy is acceptable. If there is a large number of concepts, the strict top-down strategy may not be feasible because the algorithm always starts with the most general concept or requests a goal concept from a human operator. In this case it is necessary to modify the strategy. It is assumed that during structuring

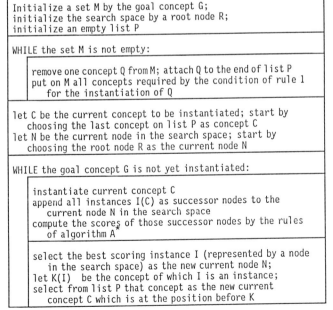

Fig. 7.16. An algorithm for top-down control in a semantic net and instantiation guided by the A^*-algorithm

and representing knowledge in the semantic net a number of attributes can be found which can only take special values. These attribute values can be defined as fixed sets, and can be used to direct analysis. The fixed attributes can be gathered, for instance, from a number of given colors (red, blue, green, ...) or shapes of a surface (plane, spherical, cylindrical, ...), or something similar. Such values will be defined as a fixed set in the domain of the specific attribute. For example, one can determine that a concept has an attribute "COLOR", which can take only the values "red", "blue", and "green". A concept having one or more fixed attributes can be characterized by them (e.g.: concept "red_ball"; fixed attributes "COLOR: red", "FORM: spherical"). After the semantic net has been developed, a procedure is searching for definitions of fixed attributes throughout the whole net. A table is then created containing all fixed attributes and values. Furthermore, a number of *concept productions* and *subnet productions* is created. The term "concept production" denotes a production rule which has a condition term consisting of a connection (logical AND) of fixed attribute values, and a list of concepts for conclusion. Moreover, for the condition only fixed attributes are allowed which are directly linked to the concept or inherited by a more general concept. The general form of a concept production is

IF (fixed attributes) THEN (list of concepts having those attributes), (7.4.7)

Example:
IF ((COLOR: red) (FORM: spherical) (MATERIAL: plastic))
THEN (red ball, frisbee, ...) .

The list given in the conclusion of the production contains the names of all concepts (OR connection), possessing attributes fulfilling the condition. Given a number of fixed attributes resulting from input data (preprocessing, segmentation, and so on), the use of the productions yields a number of concepts having the ability to match such attribute values. In other words, a number of possible goal concepts has been selected in a data-driven manner.

An even stronger selection is achieved by using not only attributes directly linked to a concept, but also attributes within the part hierarchy of this concept. This results in productions which have premises selecting a subnet of the total semantic network. Productions like these have been termed "subnet productions". The subnet production is

IF (logical expression of (fixed attributes in the parts hierarchy of concepts))
THEN (list of concepts having those attributes). (7.4.8)

The effect of this production is similar to (7.4.7), except that due to the more severe restrictions the number of concepts selected is smaller. The attribute table and the two types of production rules can be extracted automatically from a given semantic net before using the net for analysis. Finally the control algorithm has four knowledge sources to direct the analysis and limit the search space: attribute tables, concept productions, subnet productions, and the semantic net.

```
initialize empty lists GOAL and ACTIVE;
determine by concept and subnet productions a set of goal
   concepts which is as small as possible;
put goal concepts on list GOAL
```

```
WHILE list ACTIVE or list GOAL is not empty:

   IF list GOAL is empty:
      remove from list ACTIVE that concept B having the best
         scoring instance
      determine according to prespecified strategy a set of
         concepts superior to B and put them on list GOAL;
      IF no concept in ACTIVE has a superior concept OR IF B
         is on the final level of abstraction, THEN take
         instances of B as result and END
   END IF

   WHILE list GOAL is not empty:

      choose the goal concept being on the highest level of
         abstraction; if there are several, choose concept
         being most general in the generalization hierarchy;
         if there are several choose concept being most
         general in the parts hierarchy; if there are
         several, select an arbitrary one; the result of
         this choice is the current concept C;
      remove C from list GOAL;
      initialize local expanded model and a local search
         space for the current concept C

      instantiate C by top-down expansion and $A^*$ search
         and consider modality set

      IF during instantiation of C a concept D being in
         GOAL is instantiated, THEN remove D from GOAL and
         put D together with the maximal score of its
         instances on list ACTIVE;
      compare the best state of the local search space to
         those of previous search spaces and if appropriate
         continue with a previous state
      store local expanded model and local search space,
         put C together with maximal score of its instances
         on list ACTIVE
```

Fig. 7.17. An algorithm for bidirectional control in a semantic net

Control alternates between bottom-up and top-down phases, or between hypothesize and test phases. A general sketch of such a *bidirectional control algorithm* is given in Fig. 7.17. At first a small number of goal concepts by use of concept or subnet productions is selected. Starting with the highest level of abstraction according to the hierarchies concrete-of, specialization-of, and part-of, the selected goal concepts are processed sequentially. A local search space and a local expanded model are built up for the actual goal concept. Instantiation of this concept is done according to Fig. 7.16. Furthermore, concepts instantiated during this process are removed from the set of possible goal concepts (GOAL) and together with the maximal score of their instances entered in the list ACTIVE. In this list all concepts are collected which can possibly be expanded bottom-up.

After this phase has been terminated, all preselected goal concepts have been instantiated (first processing step).

The concepts on the list ACTIVE can be expanded bottom-up. The concept with the most promising instances (maximal score) is chosen. According to a desired strategy a number of "superior concepts" will be determined, that is concepts on a higher level of abstraction. A possible strategy would be to take concepts referred by "part-of". Concepts that have been found are stored in the list GOAL. If there is no superior concept for all concepts in ACTIVE, the analysis stops. The results are then the best scoring instances of the concept. Otherwise the concepts of the list GOAL are processed. In addition the score of search space states is compared with the score of concepts in ACTIVE. If the score of an actual state becomes larger than the score of one of the elements of ACTIVE, the goal concept will be changed and control will go on with the new concept. Concepts already processed are also saved in ACTIVE, because they may be expanded one at a time.

As discussed in Sect. 6.5 the A^*-algorithm requires an estimate $\hat{\varphi}(v_i)$ of the costs of an optimal solution path. The information available are the instances generated so far (intermediate results), the semantic net (task dependent knowledge), and perhaps the general heuristics of the designer. The estimate $\hat{\psi}$ is the sum of the costs of the path found from the goal v_g to v_i. To compute the estimate $\hat{\chi}$ it is useful to remember the meaning of a node v_i in the search graph. This node conceptually consists of the list L_g of concepts necessary for the instantiation of v_g and of the instances generated so far; alternative results give rise to different nodes in the search graph. A concept in L_g which is not yet instantiated is given cost 0 or certainty 1 (optimistic estimate) in order to assure the condition of Theorem 6.5.1. It is assumed that a score of an instance of every concept can be computed by an appropriate function. The arguments of this function in general will be the results of attribute judgement and relation judgement as well as the scores of parts and concretes of the concept. Since v_i itself will have such a scoring function, the score of v_i can be estimated by using either the optimistic estimate of concepts not yet instantiated in L_g or the scores of available instances.

It is obviously possible to include a certain part of segmentation as concepts in the net. This in turn allows one to include segmentation into the process of control in a natural manner and to alternate between high-level symbolic processing and low-level numeric processing.

7.4.5 An Application in Image Understanding

In this section an example of knowledge representation in a semantic net is described which is from *image understanding* and concerns the diagnostic interpretation of scintigraphic image sequences from the human heart. The knowledge base is organized into 9 levels of abstraction as shown in Fig. 7.18. The interface between low-level processing methods and objects contained in the images of one sequence, in this case contours of the heart, the left ventricle, and possibly other organs visible in an image. Figure 2.18 gives an example of such an image. The only preprocessing operation is median filtering. The contours are obtained by a dynamic programming approach to contour following.

The results of contour detection are represented in the "interface" of the knowledge base. Additionally, angles and areas are computed. For one particular image sequence special reference systems are given, the "space reference", given by the center of gravity of the left ventricle and angles of the anatomical segments and sectors, and the "time reference" (images per heart beat), which is fixed by the length of one image sequence containing 12 to 32 images. The reference systems and each object are represented by concepts as shown in Fig. 7.18. The next level "time complete objects" represents an image sequence. Now it is

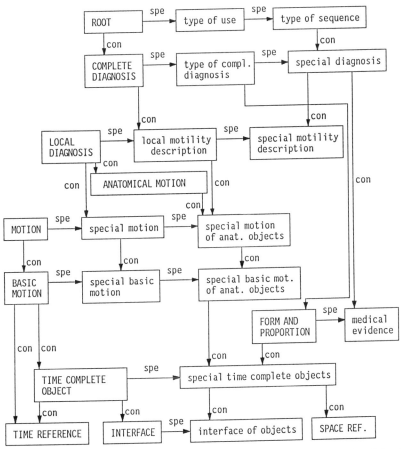

Fig. 7.18. A condensed view of the semantic net representing the task specific knowledge for the analysis of scintigraphic image sequences. It consists of nine levels of abstraction. The "part" hierarchy is omitted in the figure. (CAPITAL LETTERS: one general concept, lower case letters: a set of specialized concepts)

possible to talk about a modified volume curve of the heart, or the left ventricle, for example.

The shape and size of the left ventricle and its segments are modeled in the level "form and proportion". Relations between different parts of the ventricle are checked, as well as proportions, and the analysis of the elliptical approximation is performed. This static information is needed as evidence for the final medical diagnosis such as "the inferioapical segment of the left ventricle is enlarged". The next two levels in the knowledge base describe motion. The level "basic motion" computes changes in area from image i to image $i+1$, classifies them as contraction, stagnation, or expansion, and scores them by fuzzy functions. The results are combined in the next level to obtain larger phases of one motion type, for example, a contraction from image 1 to image 5. In this way the motion of the left ventricle and its segments are determined.

Motional behavior is represented in common medical terms at the next level of abstraction. The link between a description by general terms like "contraction phase" and medical terms like "pre-ejection period" is drawn in the level "anatomical motion". An anatomical interpretation of a motion cycle of the left ventricle consists of a systolic and diastolic phase, which are further subdivided. Attributes are start and end times of those phases. The next two levels give a medical interpretation based on the results obtained above. First, a "local diagnosis" is obtained, for example, "the inferioapical segment shows weak motion". Then effects of one segment on a neighboring one are analyzed and combined with the results of shape analysis to derive a "complete diagnosis" at the last level of abstraction.

In order to compute actual values concerning one special image sequence, adequate procedural knowledge has to be provided. This is activated by a control module of the type shown in Fig. 7.16. Most of the procedures in the procedural knowledge are straightforward like computation of areas, angles, and center of gravity. There are also more sophisticated ones. The motion phases are determined by syntactic pattern recognition and dynamic programming. Medical terms are added by pattern matching algorithms. A rule-based approach is adopted for deriving a diagnosis. Certainty factors assess the result of the application of a rule as shown in (7.3.28).

This knowledge is sufficient for the analysis of one image sequence. The last level with the most general concept "root" contains knowledge about different types of sequences for the comparison of different image sequences. Therefore, concepts exist to describe the comparison of two sequences taken under the same mode or under different modes. This example shows that all aspects of a reasonably complex task domain can be represented using the same framework, that is the semantic net formalism. Procedural knowledge can be attached which is best suited to the problem, for example, dynamic programming algorithms for waveform parsing or rules for diagnostic inferences.

7.4.6 An Application in Speech Understanding

In this section an example from *speech understanding* is treated. Relevant processing steps are acoustic segmentation to obtain subword units, word hypothesization, syntactic, semantic, and pragmatic analysis, dialog strategies, and answer generation. Syntactics, semantics, and pragmatics are defined as follows:

1. *Syntactics* studies the formal relations of signs to one another. It deals with the combination of signs without regard to their specific significations or their relation to the behavior in which they occur.
2. *Semantics* studies the relation of signs to the objects to which the signs are applicable. It deals with the signification of signs in all modes of signifying.
3. *Pragmatics* studies the relation of signs to interpreters. It deals with the origin, use, and effect of signs within the behavior in which they occur.
Incidentally, *Morris* [7.1], in his treatment of semiotics and the definition of syntactics, semantics, and pragmatics, warned that "these terms have already taken on an ambiguity which threatens to cloud rather than illumine the problems of this field ...".

In the approach described here syntax does not contain any knowledge about the meaning of words. The meaning in a general task independent way is treated in the semantic processing stage. The specialized task dependent knowledge is represented in the pragmatic module. By this separation of syntax, semantics, and pragmatics a clear separation of competence is achieved. A problem is that the border between syntax, semantics, and pragmatics is not well-defined from a linguistic standpoint. Instead, the border is defined by the specifications of the processing levels. We first give an example of the representation of semantic knowledge using the semantic net formalism, then demonstrate that pragmatic knowledge can be modeled in a semantic net, and finally show that different levels of speech understanding can be represented in a semantic net.

The representation of the general task-independent meaning of words is based on case and valency theory. The basic idea of *valency theory* is outlined for the example of verb valency; it may be used in a similar way also for noun, adjective, and adverbial valency. A verb used in a particular meaning will require certain elements having well-defined properties. Those elements constitute the "valency frame". The same verb used in another meaning will usually require some other elements resulting in a different valency frame. Three types of elements are distinguished, the obligatory elements, the optional elements, and the free elements. Obligatory elements are those necessary to obtain a grammatically correct sentence. Optional elements are those which in addition to the obligatory ones fully define a certain verb meaning. Free elements, finally, are those which may be used fairly independently of a certain verb meaning. It is seen that the emphasis of valency theory is on the syntactic structure defined by a certain verb. On the other hand, *case theory* emphasizes the "functional role" or "(deep) case" of a constituent in a sentence. It emphasizes the logical relations of words with respect to the verb.

Some examples of cases are the following, with sample sentences in quotation marks:

1. Agent: The agent is the thing causing an action and often is the surface subject as in "*Tom* washed the car for his father".
2. Beneficiary: This is the person for whom an action is performed, that is in the above sample sentence "for his *father*".
3. Coagent: This is a partner of the agent as in "Tom washed the car with *Jim*".
4. Conveyance: This is the means of transportation as in "Tom went to Chicago by *train*".
5. Destination: The destination is the result of a movement or state change, for instance, to *Chicago*. Similarly the source case is the origin or initial state.
6. Instrument: This is something used by the agent as in "Tom washed the car with a *spoon*".
7. Location: The location indicates where an action occurs as in "Tom washed the car near the *house*".
8. Object: The object is the thing the sentence is concerned with, and it usually undergoes a state change as in "Tom washed the *car*".
9. Raw Material: A final product is made out of something which is the raw material of this product as in "Tom built a doghouse out of *wood*".
10. Time: It indicates when an action is performed as in "Tom washed the car *before dinner*".

This list of cases is not exhaustive, but it is supposed to illustrate the idea. There is no general agreement about a set of cases necessary or sufficient for a natural language or a limited task domain. Knowing the case of a noun group helps the understanding of a sentence and answering questions about it. For instance, in the last sample sentence "Tom washed the car before dinner", there are three noun groups belonging to the cases

Agent → Tom,
Object → the car,
Time → before dinner.

This allows one to answer the questions

Who washed the car? → Agent → Tom,
What was washed? → Object → the car,
When did Tom wash it? → Time → before dinner.

If, for instance, the above sentence is spoken in continuous speech into a machine, if afterwards one of the above questions is also spoken into the machine, and if then the machine can answer this question correctly, then it seems reasonable to assume that the machine "understood" the sentence and the question.

Determination of the case of a noun group is supported by various constraints such as prepositions preceding a noun group (for instance, before, after, or during, indicate the time case), position of the noun group within the sentence (for instance, the agent often is followed by the verb), meaning of the verb (for instance, to build may be accompanied by an object, instrument, or raw material case, but not a conveyance case), meaning of the noun (for instance, the noun train indicates a conveyance, but not a raw material), other cases in the sentence (for instance, if one noun group was identified as a conveyance, it is unlikely to have another case of conveyance but likely to have a case of destination, source, or time), and pragmatics or the dialog context. As an example of the use of these contraints consider the two sentences

Tom washed the car with a spoon,
Tom washed the car with his sister.

Both have three noun groups. Without any constraints the ten cases mentioned above would allow 10^3 combinations of different cases. The meaning of the verb is apparent – although often the meaning of a verb may depend on a noun as in "take a picture, take a plane, take a seat, take a tablet" – and indicates an action. Similarly in these simple sentences the agent and the object are easily determined to precede and follow the verb, respectively. The last noun group contains the preposition "with" in both sentences, so from the ten cases only coagent and instrument remain. But a coagent must be a person, and an instrument must not be a person. The preposition and the noun thus constrain the ten cases to just one.

The raw semantic knowledge about words is usually contained in the entries of a lexicon. From the lexicon a preprocessor extracts the relevant semantic knowledge. For each verb a *case frame* is constructed according to valency and case theory as outlined above. An example of two case frames of a verb is given in Fig. 7.19. The reader should note that the formalism is designed for German, and translations have to cope with the problem that different meanings of one German word may be mapped to one English word but also to different English words.

It was shown in Fig. 7.18 that different levels of a system for image analysis can be represented in a semantic net; the same also holds for different levels of a speech understanding system as shown in Fig. 7.20. It contains the linguistic processing starting at the level of words, the *interface level*, and proceeding via syntactic and semantic analysis to pragmatic analysis. Obviously, an extension to lower levels, for example the level of parametric representation of speech, and to higher levels, for example, to the level of dialog and answer generation, is conceivable. Incidentally, time relations of syntactic constituents are represented by the "adjacency" substructure shown in Fig. 7.13. It is fairly straightforward to represent an ATN representation of the syntax to a semantic net representation, and similarly semantic and pragmatic constraints can be represented. However, it is beyond the scope of this book to go into the details of these steps.

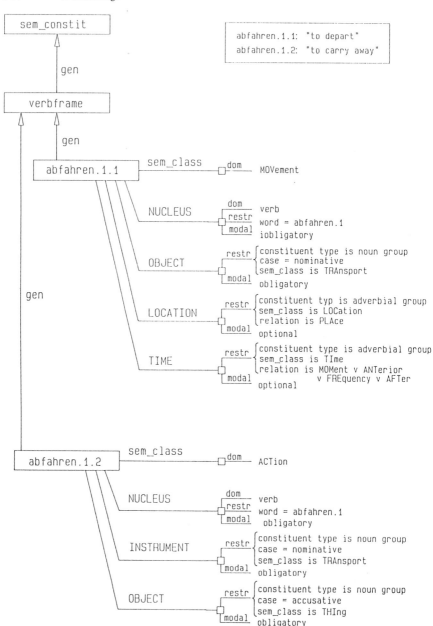

Fig. 7.19. A semantic net representation of two case frames of a verb, representing two different meanings of the verb

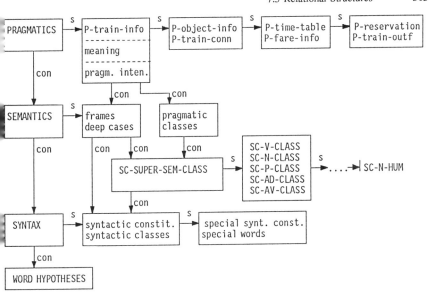

Fig. 7.20. The linguistic processing levels of a speech understanding system represented in a semantic net with part links omitted; spe and con denote a specialization and concrete link, respectively. The prefixes P, PC, and SC stand for pragmatics, pragmatic classification, and semantic classes, respectively. Capital letters denote a single concept, lower case letters summarize a set of concepts

7.5 Relational Structures

Since a graph or a net may also be represented by tables of nodes and edges, as indicated by Fig. 5.4e, an alternative but equivalent representation of a net, for example, a net representing an object model, is possible by a relational data model in the sense of Sect. 5.2.4. It consists of two tables (or sets), one table SC for the simple constituents or segmentation objects and their properties, the other table IR for the interrelations between constituents and the properties of these interrelations (see also Fig. 5.14). A generalization would be to include n tables IR$(i), i = 1, \ldots, n$, where IR(i) is an i-ary relation. This allows one to include n-ary, not just binary, relations between constituents in a natural manner and to use relational data base techniques. It should be noted that each line (or element) of SC has a unique identification which corresponds to the primary key (level, number, name) in Fig. 5.15, and a set of values of certain properties, attributes, or parameters. Similarly, if IR represents interrelations between two simple constituents which are called the components of the relation, each line of IR contains the identification of the two simple constituents being related and an identification of the interrelation, the primary key (result), as in Fig. 5.15. If also non-normal forms are allowed, it is possible to use a table as an attribute of a relation. This in turn results in hierachical relational descriptions which in principle allows the representation of structured task dependent knowledge.

It is possible to use relational descriptions directly for structural matching. Let $\{SC_m, IR_m\}$ and $\{SC_o, IR_o\}$ be the tables of an object model and an observation, respectively. An *inexact match* should tolerate certain differences of property values, some missing segmentation objects, and some missing relations between constituents. By assigning weights to properties and relations an inexact match is defined as a mapping $SC_m \rightarrow SC_o \cup \phi$ meeting the following requirements:

1. If an element of SC_m is mapped to an element of SC_o, then these two elements have the same properties and similar values, for example the difference of the values is less than a threshold.
2. If an element of SC_m having no corresponding element in SC_o is called a missing constituent (that is an element of SC_m mapped to ϕ), then the sum of the weights of all missing constituents is less than a threshold.
3. If the mapping is applied to the components of IR_m, then the result is either an element of IR_o or not, and the sum of weights of elements of IR_m not mapping to elements of IR_o is less than a threshold.

In general there will be several mappings between object and scene model. In this case the mapping giving smallest differences in steps 1–3 above is to be determined. Such a mapping may be determined by search procedures of the type discussed in Sect. 6.5 but details are omitted here.

7.6 Grammars

7.6.1 Augmented Transition Networks

Grammars defining formal languages have been applied to various problems of pattern classification and analysis. Although in a critical assessment of syntactic methods it was pointed out that they are limited to specification of geometrical and topological constraints and allow no consideration of the "use" of an object, there is no doubt that these constraints are of extreme value and that there are practical problems where syntactic methods are well suited. Furthermore, an extension to include semantic information is possible. String grammars were introduced in Sect. 4.3.

The use of formal grammars to represent syntactic relations has turned out to be very useful in natural language and speech understanding systems. The following discussion is limited to this aspect. Syntactic constraints occurring in natural language or speech can be represented by a graph. This graph can be generalized to allow recursion and additional structure building rules. Recursion is handled by a *recursive transition network (RTN)*, which is a directed graph with labeled nodes or states and edges. An example is given in Fig. 7.21 for generation or acceptance of sentences like "The boy goes to school" or "The boy saw a car with a flat tire", but not sentences like "Did the boy see a car". To accept, for instance, the first sample sentence above, the RTN starts in state S and may only go to state $Q1$ if a state or nonterminal NP is found in the input. This is checked by calling state NP of the RTN. A transition from state NP to $Q5$ is possible in this example because the article "the" occurs in the sentence,

7.6 Grammars   317

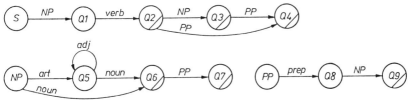

Fig. 7.21. A recursive transition network for generating or accepting simple sentences

and from $Q5$ a transition to the accepting state $Q6$ occurs because the next word is the noun "boy". Thus an instance of NP was found, a return to the S level is made, and a transition to $Q1$ is possible. Next the verb "goes" allows a transition to $Q2$. From $Q2$ a call of NP — which would end unsuccessfully — or of PP is possible. Calling PP results in recognizing "to school" via states PP, $Q8$, NP, $Q6$, $Q9$, and this results in acceptance of the sentence by going from $Q2$ to the final state $Q4$. Incidentally, the RTN of Fig. 7.21 is equivalent to the h graph in Fig. 7.22, but the role of nodes and edges is interchanged.

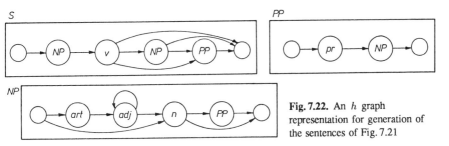

Fig. 7.22. An h graph representation for generation of the sentences of Fig. 7.21

An RTN is equivalent to a pushdown store acceptor which is known to recognize context-free languages. The power of an RTN is increased by adding tests and actions to the edges. Only if a test gives a positive result, may an edge be followed, and if it is followed, the specified action is taken. This extension gives an *augmented transition network (ATN)*. The actions are used to build structures and store them in registers. By specifying suitable tests and actions it is possible to give an ATN the power of a *Turing machine*.

An ATN may be specified by a list of *edge* sets, where each edge set is, in turn, a list containing a *state* and an arbitrary number of edges leaving this state. This is defined formally by

⟨ATN⟩ → (⟨edge set⟩ ⟨edge set⟩*) ;
⟨edge set⟩ → (⟨state⟩ ⟨edge⟩*) ;
⟨edge⟩ → (CAT ⟨category⟩ ⟨test⟩ ⟨action⟩* ⟨next state⟩)|
         (WRD ⟨word⟩ ⟨test⟩ ⟨action⟩* ⟨next state⟩)|
         (VIR ⟨const − type⟩⟨test⟩⟨action⟩* ⟨next state⟩)|
         PUSH ⟨state⟩⟨test⟩ ⟨pre − action⟩* ⟨action⟩* ⟨next state⟩)|
         JUMP ⟨test⟩ ⟨action⟩* ⟨next state⟩)|
         (POP ⟨form⟩ ⟨test⟩) .   (7.6.1)

There are six types of edges named CAT, WRD, VIR, PUSH, JUMP, and POP. Common to all of them is the term ⟨test⟩ meaning that the edge only can be followed if the condition specified in ⟨test⟩ is met. The term ⟨next state⟩ in all edge types except POP indicates the next state, resulting from following the edge. The first five edge types contain an arbitrary number of ⟨actions⟩; the effect of an action is outlined below. A CAT edge may be followed if the symbol currently pointed to in the input string is a member of ⟨category⟩. For instance, an edge "CAT verb ..." may be followed if the pointer is on "saw", but not if it is on "boy". A WRD edge can be followed if precisely the word requested in ⟨word⟩ occurs at the present location of the input pointer. A PUSH edge causes an exit from the present subnet of the ATN and a transition to ⟨state⟩ of a new subnet; the current state, register contents, and input pointer are saved or pushed onto a stack. Usually a subnet will correspond to some syntactic constituent, for example, a verbal group. The result of this processing step is saved in a special register RP. Tests and actions are carried out only after returning from the subnet. Actions which have to be performed before following this edge are specified in ⟨pre-action⟩. When following a PUSH edge execution is continued at ⟨state⟩, and if this execution was successful, it ends with a POP edge. This returns the computed value specified in ⟨form⟩ to a special register TEMP and causes execution to continue at the place of the activating PUSH edge. A jump edge does not consume a word from the input string, it allows optional constituents to be treated by going to another state. The VIR edge allows the testing of a constituent which was analyzed before and stored in a special HOLD register. The edge may be followed if the stored constituent type is that requested by ⟨const-type⟩. Part of the RTN of Fig. 7.21 is given in this notation which specifies ⟨action⟩ only informally, and does so also for ⟨form⟩; T is a test which is always true.

```
((S   (Push NP T (store value of NP) Q1))
 (Q1  (Cat verb T (store verb found) Q2))
 (Q2  (POP (store structure found) T)
      (PUSH NP T (store value of NP) Q3)
      (PUSH PP T (store value of PP) Q4))
 (Q3  (POP (store structure found) T)
      (PUSH PP T (store value of PP) Q4))
 (Q4  (POP (store structure found) T))
 (PP  (CAT prep T (store prep found) Q8))
 (Q8  (PUSH NP T (store value of NP) Q9))
 (Q9  (POP (store value of PP) T))).                       (7.6.2)
```

This portion of an ATN suffices for the first sample sentence mentioned above to be accepted, and the processing should be obvious.

The ⟨test⟩ of an ATN edge may contain arbitrary expressions. They test, for example, the register contents or features of an input word. This allows context information at arbitrary locations in the input to be checked. The ⟨action⟩ of an ATN edge allows the building of syntactic structures and the storing of results in registers. Examples of actions are the storage of a ⟨form⟩ in a register which may be attributed to the same, lower level, or higher level subnet. In addition a global register, the HOLD register, is available and can be used by actions. This allows, in combination with the VIR edge, the processing of discontinuous constituents which are common, for example, in German verbal groups. The ⟨form⟩ may be an arbitrary function of a register content. For example, the value stored in a register or the value of some syntactic feature of a word may be recovered. In addition there are operations for generating structures. This is done by providing a template for part of a parse tree; the template contains certain positions which have to be filled by register contents. Therefore, an ATN allows the representation and parsing of *context-sensitive languages*.

Figure 7.23 gives a section of an ATN grammar where tests and actions are omitted in the graphical representation. It is designed for a subset of German and it is used in the design of a speech understanding system. Figure 7.24 gives the result of parsing an example sentence. The complete ATN has 4 subnets and 16 states, 19 word classes, and 1 word corresponding to a WRD edge. In general, syntax should be task independent; however, to date, speech understanding systems are only able to accept a certain subset of a language, and the relevant subset may be task dependent to some extent. For example, in the task domain "train connections" the time of the day and the date are very important, but may be unimportant in other applications. Therefore, the grammar contains subnets handling the time of the day (UHRZ in Fig. 7.23) and the date (DATUM in Fig. 7.23). On the other hand, there are syntactic constructions which will occur in any application, for example, the verbal groups or noun groups.

7.6.2 Unification-Based Grammar

Before considering a unification-based grammar it is useful to introduce some definitions and notations. The goal of syntactic analysis is to obtain a (syntactic) description d of an entity e which is a string of words or a sentence in a natural language. Let E be the set of sentences and D the set of descriptions. If an element $e \in E$ is described by some $d \in D$ this is denoted by

$$e \,|\!\rightarrow d \,. \tag{7.6.3}$$

The *description* is constructed from a set A of attribute names and a set W of attribute values. A *simple description* consists of a sequence of (attribute, value) pairs

$$d_1 = \left[(a_1, w_1), (a_2, w_2), \ldots, (a_n, w_n)\right], \quad a_i \in A, \quad w_i \in W \,. \tag{7.6.4}$$

But a description may also contain a description as a value of an attribute; this is called a *complex description*, for example,

$$d_2 = \left[(a_2, w_2), \quad (a_5, w_5)\right],$$
$$d_3 = \left[(a_1, w_1), \quad (a_4, d_2)\right]. \tag{7.6.5}$$

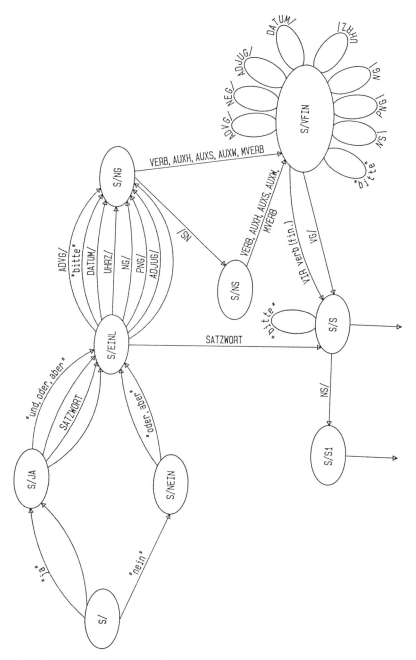

Fig. 7.23. The sentence net of an ATN grammar for German speech

```
Sentence: Die Abfahrt    des      nächsten  Zuges  ist  um  fünf  Uhr.
          <The departure of the   next      train  is   at  five  o" clock.>

Parse structure:
(S (STRING   (die Abfahrt des nächsten Zuges ist um fünf Uhr))
   (NUCLEUS  (verb sein.1))
   (TYPE     (declarative))
   (STRUCTURE
     ((NG   (STRING   (die Abfahrt))
            (NUCLEUS  (n Abfahrt.1))
            (CASE     nom)
            (GENDER   fem)
            (NUMBER   3)
            (STRUCTURE
              ((det  die.1)
               (n    Abfahrt.1))))
      (NG   (STRING   (des nächsten Zuges))
            (NUCLEUS  (n Zug.1))
            (KASUS    gen)
            (GENDER   mas)
            (NUMBER   3)
            (STRUCTURE
              ((det  der.1)
               (adj  nächster.1)
               (n    Zug.1))))
      (UHRZ (STRING   (um fünf Uhr))
            (NUCLEUS  (n Uhr.1))
            (HOURS    5)
            (MINUTES  0)
            (STRUCTURE
              ((preap   um.1)
               (NUMERAL (STRING  fünf)
                        (NUCLEUS 5)
                        (STRUCTURE ((number fünf.1))))
               (n Uhr.1)
      (VG   (STRING   (ist))
            (NUCLEUS  (verb sein.1))
            (NUMBER   3)
            (MODE     ind)
            (TENSE    praes)
            (STRUCTURE
              ((verb sein.1)))))))    Fig. 7.24. Parse of an example sentence
```

The following discussion will be illustrated by simple examples. Therefore, the following sets of attributes, values, and their abbreviations are introduced:

A = (category <cat>, agreement <agr>, member <mem>,
       person <per>, subject <sub>, number <num>,
W = (noun group <NG>, singular <SG>, three <3>, plural <PL>). (7.6.6)

These sets allow one to construct various simple and complex descriptions, for example,

d_ϕ = [ ],   the empty description

d_{NG} = [(cat, NG)] ,

d_{SG} = [(num, SG)] ,

d_{PL} = [(num, PL)] ,

d_3 = [(per, 3)] ,

$d_{\text{NGSG}} = [(\text{cat, NG}), (\text{agr}, d_{\text{SG}})]$, (7.6.7)

$d_{\text{NGPL}} = [(\text{cat, NG}), (\text{agr}, d_{\text{PL}})]$,

$d_{\text{NG3}} = [(\text{cat, NG}), (\text{agr}, d_3)]$,

$d_{\text{3SG}} = [(\text{num, SG}), (\text{agr}, d_3)]$,

$d_{\text{NG3SG}} = [(\text{cat, NG}), (\text{agr}, d_{\text{3SG}})]$,

$d_{\text{NG3SGsub}} = [(\text{cat, NG}), (\text{agr}, d_{\text{3SG}}), (\text{sub}, d_{\text{3SG}})]$.

It is assumed that the above notation is self-explanatory in the natural language context.

If e is an object meeting the requirements of d_{NGSG}, then e also meets the requirements of d_{NG}. This is called a *subsumption* of d_{NGSG} by d_{NG}, or $\text{subs}(d_{\text{NG}}, d_{\text{NGSG}})$. In general, for two descriptions d and d',

$$\text{subs}(d, d') \leftrightarrow \wedge_e (e \,|\!\to d' \to e \,|\!\to d) \ . \qquad (7.6.8)$$

So d subsumes d' if d is more general than d'. No simple description can subsume another different simple description. The empty description subsumes any description. For example,

$$\text{subs}(d_\phi, d_{\text{NG}}) \ , \quad \text{subs}(d_{\text{NG}}, d_{\text{NGSG}}) \ , \quad \neg\text{subs}(d_{\text{NG}}, d_{\text{SG}}) \ , \qquad (7.6.9)$$

where $\neg\text{subs}(,)$ means no subsumption. A formal definition of subsumption requires the notion of a path in d. By $d(a), a \in A$, the value of attribute a in the description d is selected. It may be reasonable to take a sequence of attributes $p = (a_1, \ldots, a_n) \in A^*$ or a *path* $p = A^*$. The path p defines a sequence of attributes in a complex description. For example,

$d_{\text{NGSG}}(\text{cat}) = \text{NG}$,

$d_{\text{NGSG}}(\text{agr}, \text{num}) = \text{SG}$, (7.6.10)

$d_{\text{NGSG}}(\text{cat}, \text{num}) = \text{NIL}$,

where NIL is the undefined value. By $\text{dom}(d)$ the domain of description d is denoted, for instance, $\text{dom}(d_{\text{3SG}}) = \{\text{num, per}\}$. If two descriptions d and d' are simple and if $d = d'$, d subsumes d' and vice versa. If d and d' are complex,

$$\begin{aligned}\text{subs}(d, d') &\leftrightarrow \wedge_{a \in \text{dom}(d)} \text{subs}(d(a), d'(a)) \\ \text{AND } &\wedge_{p,q \in A^*} d(p) = d(q) \to d'(p) = d'(q) \ .\end{aligned} \qquad (7.6.11)$$

Two descriptions d' and d'' are *compatible*, denoted $\text{comp}(d', d'')$, if there is a more special description d such that d is subsumed by both d' and d'', that is,

$$\text{comp}(d', d'') \leftrightarrow \vee_d (\text{subs}(d', d) \text{ AND } \text{subs}(d'', d)) \ . \qquad (7.6.12)$$

With the above definitions of compatiblity and subsumption we have the relations

$$\text{comp}(d_{\text{NG3SG}}, d_{\text{NGSG}}), \quad \text{subs}(d_{\text{NGSG}}, d_{\text{NG3SG}}),$$

$$\text{comp}(d_{\text{NGSG}}, d_{\text{NG3}}), \quad \neg\text{subs}(d_{\text{NGSG}}, d_{\text{NG3}}), \tag{7.6.13}$$

$$\neg\text{comp}(d_{\text{NGSG}}, d_{\text{NGPL}}), \quad \neg\text{subs}(d_{\text{NGSG}}, d_{\text{NGPL}}).$$

There are no incompatible descriptions which can be in a subsumption relation. The descriptions d_{NGSG}, d_{NG3} are an example of two descriptions which are compatible but not subsumptive. They are not subsumptive since the attribute "agr" is different; they are compatible since there is a more special description which is subsumed by both d_{NGSG} and d_{NG3}, for example, d_{NG3SG}. The construction of d_{NG3SG} may be viewed as a merging of the attribute values of "agr" occurring in the other two descriptions. This is an example of the *unification* of two descriptions.

Unification was introduced in Sect. 7.2 in connection with predicate logic. It meant the replacement of variable symbols by terms such that two WFFs become equal. The analogous operation for two descriptions d' and d'' is called *unification* of d' and d'', denoted by

$$d = d' \sqcup d''. \tag{7.6.14}$$

It yields a new description d which is the most general description meeting the requirements

$$\text{subs}(d', d) \text{ AND } \text{subs}(d'', d). \tag{7.6.15}$$

Incompatible descriptions cannot be unified. For the description relation $\mid \rightarrow$ introduced above it holds that

$$\wedge_e e \mid \rightarrow (d' \sqcup d'') \rightarrow e \mid \rightarrow d' \wedge e \mid \rightarrow d'', \tag{7.6.16}$$

$$\wedge_e e \mid \rightarrow d' \wedge e \mid \rightarrow d'' \rightarrow e \mid \rightarrow (d' \sqcup d''). \tag{7.6.17}$$

To compute the unification d the following operations have to be performed for all attribute-value pairs in d' and in d'':

1. IF $\quad ((a', w') \in d' \text{ AND } (a'', w'') \in d'' \text{ AND } a' \neq a'')$,
   THEN $([(a', w'), (a'', w'')]$ are in $d)$. $\tag{7.6.18}$
2. IF $\quad (\{(a, w) \in d' \text{ AND there is no pair } (a, u) \text{ such that } (a, u) \in d''\}$
   OR $\{(a, w) \in d'' \text{ AND there is no pair } (a, u) \text{ such that } (a, u) \in d'\})$,
   THEN $((a, w)$ is in $d)$. $\tag{7.6.19}$
3. IF $\quad ((a, w') \in d' \text{ AND } (a, w'') \in d'')$,
   THEN (3.1. IF $\quad (w' \in W \text{ AND } w'' \in W \text{ AND } w' \neq w'')$,
   THEN $(d = \text{NIL} -$ no unification is possible$)$, $\tag{7.6.20}$
   3.2. IF $\quad ((w = w' \text{ AND } w'' = [\,] \text{ AND } w' \neq [\,])$
   OR $(w = w'' \text{ AND } w' = [\,] \text{ AND } w'' \neq [\,])$
   OR $(w = w' = w'') \text{ OR } (w = w' \sqcup w'')),$

THEN $((a, w)$ is in $d)$. (7.6.21)

The result of unification is *invariant* with respect to the order of the above steps, it is *additive* in the sense that $[(a, u)] \sqcup [(b, w)] = [(a, u), (b, w)]$, and it is *idempotent* in the sense that $[(a, u)] \sqcup [(a, u), (b, w)] = [(a, u), (b, w)]$.

A description can be represented by a *directed acyclic graph* (DAG). The nodes are attribute values and they are labeled by the attribute value if this is simple, that is, if it has a value $w \in W$. The edges are attribute names and are labeled by attribute names.

Having introduced the "description" d we now define how sentences in a language can be characterized. It is useful to have a finite scheme for defining a possibly infinite set of sentences; such a finite scheme is a suitable grammar. There are many (slightly) different definitions for a unification-based grammar. In the following a *unification-based grammar* is defined by the tuple

$$G = (V_T, V_N, R, s, U, L) \,.$$ (7.6.22)

In this definition V_T and V_N are finite, disjoint, non-empty sets of terminal and nonterminal symbols, respectively. R is a finite set of *annotated rules* consisting of a *context-free part*

$$r_i \in R : t_j \to \beta_k \,, \quad t_j \in V_N \,, \quad \beta_k \in (V_N \cup V_T)^*$$ (7.6.23)

and a set of equations where each equation has the form

$$\texttt{<path>} = \texttt{<path>} \text{ or } \texttt{<path>} = t_j \,.$$ (7.6.24)

The start symbol is $s \in V_N$, L is a set of labels, and U a relation in $V_T \times D$, where D is the set of descriptions (or DAGs) over L.

As an example consider the rule

$$\begin{aligned}
& t_0 \to t_1 t_2 \,, \\
& \langle t_0 \, \texttt{cat} \rangle = \text{S} \,, \\
& \langle t_1 \, \texttt{cat} \rangle = \text{NG} \,, \\
& \langle t_2 \, \texttt{cat} \rangle = \text{VG} \,, \\
& \langle t_0 \, \texttt{head} \rangle = \langle t_2 \, \texttt{head} \rangle \,, \\
& \langle t_0 \, \texttt{head sub} \rangle = \langle t_1 \, \texttt{head} \rangle \,.
\end{aligned}$$ (7.6.25)

The above rule is applicable if the restrictions imposed by the equations are met. For example, the attribute "category" (cat) of t_0, t_1, and t_2 must be S (sentence), NG, and VG, respectively; the attribute "head" of t_0 must coincide with that of t_2, and so on. This gives an example of the two fundamental operations of combining structures or descriptions to larger structures and of defining the relations of the combined structures. The equations may be viewed as the order to replace the substructures by their unification. Obviously, (7.6.25) can be summarized by

$$S \to NG\ VG\ ,$$

$$\langle S\ \text{head}\rangle = \langle VG\ \text{head}\rangle\ , \qquad (7.6.26)$$

$$\langle S\ \text{head sub}\rangle = \langle NG\ \text{head}\rangle\ .$$

The above rule may be written with little modification in the formalism of a *lexical functional grammar* (LFG). The names of constituents are replaced by the *syntactic meta-variables* "↑" and "↓" denoting the description in the father and son node, respectively, of a parse tree. The *lexicon* becomes a central data structure and provides a schema for each word. The annotated rules of an LFG determine allowed configurations of words represented in the lexicon. So viewing S as the father and NG, VG as the sons, the rule (7.6.26) in this notation becomes the first rule shown in Fig. 7.25. To present a simple example two more rules are given in this figure and also the lexicon representation of the words in the sentence "Peter kauft die Fahrkarte" (Peter buys the ticket).

a)
```
S →    NG          VG         .
      (↑subj)=↓    ↑=↓

NG → (def)    N          ,
      ↑=↓    ↑=↓

VG →  V       (NG)       (S')
     ↑=↓    (↑obj)=↓   (↑scomp)=↓
```

b)
```
Peter       N    (↑ pred) = "Peter"
                 (↑ num)  = SG
                 (↑ gen)  = M
                 (↑ per)  = 3

kauft       V    (↑ pred)    = "kaufen <(↑ sub)(↑ obj)>"
(buys)           (↑ sub num) = SG
                 (↑ sub per) = 3
                 (↑ ten)     = PR

die        def   (↑ spec) = def
(the)            (↑ num)  = SG
                 (↑ gen)  = F

Fahrkarte   N    (↑ pred) = "Fahrkarte"
(ticket)         (↑ num)  = SG
                 (↑ gen)  = F
                 (↑ per)  = 3
```

Fig. 7.25. An example of three rules in a lexical functional grammar and lexicon entries of a sentence

Since the rules have a context-free part, it is possible to determine the parse tree of the sentence by standard techniques. In this tree the equations given on the right side of a rule are also noted at the corresponding node of the tree as shown in Fig. 7.26a. The description is derived from this representation. At first, every node is given a unique function name d_i by numbering the nodes from the upper left to the lower right. The function names then replace the syntactic meta-variables, where "↑" is replaced by the function name of the father node and "↓" by that of the son node. This results in the tree shown in Fig. 7.26b. In

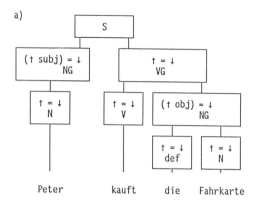

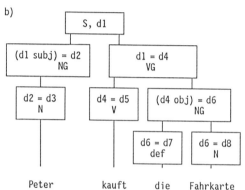

Fig. 7.26a,b. The context-free parse tree of the example sentence in Fig. 7.25

addition the syntactic meta-variables in the lexicon entries of the corresponding words are replaced by the function names. The solution of the resulting system of equations by unification gives the description of the sentence. This is indicated for the example of the NG "die Fahrkarte" only. From the lexicon it follows that

$$d7 = [(\text{spec, def}), (\text{num, SG}), (\text{gen, F})],$$

$$d8 = [(\text{pred, "Fahrkarte"}), (\text{num, SG}), (\text{gen, F}), (\text{per, 3})]. \quad (7.6.27)$$

From the context-free parse tree it follows that $d6 = d7 = d8$, which means that the corresponding descriptions have to be unified. In this case unification is possible, resulting in

$$d7 \cup d8 = [(\text{pred, "Fahrkarte"}), (\text{num, SG}), (\text{gen, F}), (\text{per, 3}), (\text{spec, def})]. \quad (7.6.28)$$

For incompatible structures the result of unification would have been NIL so that these structures would not be accepted. From the parse tree it also follows that $(d4 \text{ obj}) = d6 = d7 \cup d8$. This process is repeated until either the description $d1$ of S results, in which case the sentence is accepted, or the result NIL is obtained, in which case the sentence is rejected. The rule defining NG in Fig. 7.25 and the tree in Fig. 7.26 may be interpreted as "the element "die" having description $d7$

and the element "Fahrkarte" having description $d8$ may be combined to a new structure "NG" having description $d6$, where $d6$ is subsumed by both $d7$ and $d8$"

The above discussion only gives a basic introduction to the formalism of unification-based grammar and lexical funtional grammar. In addition to the so-called *defining equations* associated with the rule in Fig. 7.25 the *constraining equations* may be used. If a defining equation occurs, an attribute is transferred to the result of unification if it occurs in at least one description, see (7.6.19); if a constraining equation is employed, unification is allowed only if the attribute occurs in both descriptions with the same value. Finally, existentially constraining equations can be used which do not require a special value of an attribute, but only that some value is attached to the attribute.

7.6.3 Lexical Organization

The *lexicon* is a central knowledge base in a speech understanding system. It has to represent information about pronunciation, i.e. the *phonetic word*, and about syntax, semantics, and pragmatics. For readability it is also useful to provide the spelling, i.e. the *graphemic word*. The unit of a lexical entry is called a *word* for short. This unit may be the phonetic word or the graphemic word. We suggest using the graphemic word as the unit of a lexical entry or as the "word" in the lexicon. This facilitates readability and knowledge acquisition from standard sources. A particular meaning of a word is invariant from different syntactic or graphemic variants and is represented in a *lexeme*. This usually is a "normal form" of a word, for example, the infinitive of a verb or the nominative singular of a noun. At least in German the inflexions of a word are numerous and may be fairly different from the normal form. Therefore, all inflexions of a word are also represented in the lexicon. It is sufficient to represent information about the meaning in the normal form only and to provide a pointer to this normal form from the inflexions. Finally, for bookkeeping it is useful to provide a unique number (integer) for each lexical entry.

The various properties of a word are described by attribute-value pairs. The value of an attribute may be several attribute-value pairs as in (7.6.5). The format of a lexical entry then is

```
lexical entry = (graphemic word, number, phonetic word(s),
    IF    (the word is a lexeme),
    THEN ([attribute-value pairs for syntax],
          [attribute-value pairs for semantics],
          [attribute-value pairs for pragmatics]),
    ELSE ([attribute-value pairs for syntax], (pointer to lexeme)).   (7.6.29)
```

It should be noted that this only provides the general outline of a lexical entry. For example, there may be several syntactic descriptions of the same graphemic word, there may be several different semantic descriptions for the same syntactic description, and there may be different pragmatic descriptions for the same

```
word ::= [ WORD          string
          WORDNUMBER     integer
          PRONUNCIATION ( phonetic_word + )

          SYNTACTIC_DESCRIPTION ( syntactic_description + ) ]

phonetic_word ::= phoneme_symbol * '. phoneme_symbol +
phoneme_symbol /* out of a set of 64 phonemes */

syntactic_description ::= [ SYNTACTIC_WORD_NUMBER integer.integer
                            WORD_CLASS word_class
                            morpho_syntactic_description ]

morpho_syntactic_description ::= BASE_FORM (base_form +)
                    { syntactic_description_dependent_on_word_class,
                      inflectible_word_class, inflected_form }
                    OR
                    { syntactic_description_dependent_on_word_class,
                      word_class_with_semantic_description, BF }
                      SEMANTIC_DESCRIPTION
                         ( semantic_description + )
                    OR
                    { syntactic_description_dependent_on_word_class,
                      word_class_without_semantic_description, BF }

base_form ::= [ WORD string
                SYNTACTIC_WORD_NUMBER (integer.integer + )
                { PREFIX_OF_VERB
                  [ WORD string
                    SYNTACTIC_WORD_NUMBER integer.integer ] } 0,1

semantic_description ::=
        [ SEMANTIC_WORD_NUMBER integer.integer.integer
          MEANING "string"
          { semantic_description_dependent_on_word_class, word_class } 0,1
          { pragmatic_description } * ]

pragmatic_description ::=
        [ CONCEPT pragmatic_concept
          ATTRIBUTES [ pragmatic_attr_to_semantic_attr * ]]

Notation:
  [ ... ]  → description by attribute-value pairs
  ( ... )  → list of alternatives
  X +      → arbitrary number of repetitions of X, but at least one
  {X} 0,1  → X exactly once, or not once
  {X, Y}   → X with variable Y
```

Fig. 7.27. A detailed format of a lexicon entry in a speech understanding system

semantic description. An example of a detailed format of a lexical entry is given in Fig. 7.27.

A lexicon compiled according to the above description is a source of raw linguistic knowledge. By means of a preprocessor it is possible to extract, for example, the (phonetic) word models mentioned in Sect. 4.4, the case frames of words mentioned in Sect. 7.4, or the syntactic word representation mentioned above. If the speech understanding system is a dialog system which gives an answer to a user question, the lexicon may also be used for answer generation.

7.7 Acquisition of Knowledge (Learning)

7.7.1 Introductory Remarks

In the previous sections we suggested representing knowledge in a frame, predicate calculus, production rule, semantic net, grammar, and so on. But no comments have been given about how to find (or learn, construct, acquire, ...), the task specific knowledge. In this section a short account of approaches to *knowledge acquisition* or *learning* is given. Two definitions of learning are:

"Learning is the process of gathering experiences in order to optimally fit the behavior of humans, animals, or machines to the properties of the environment" (quotation translated from "Meyers Enzyklopaedisches Lexikon", edition 1975, see entry *lernen*).

"Learning denotes changes in the system that are adaptive in the sense that they enable the system to do the same task or tasks drawn from the same population more efficiently and more effectively the next time" (quotation from H. A. Simon in [7.2]).

There are basically two approaches to knowledge acquisition: leave it to the designer, that is, the manual approach, or leave it to the machine, that is, the (automatic) learning approach. And there are an arbitrary number of related approaches resulting from some kind of mixture of the manual and the learning approach. In any case, the relevant knowledge is inferred from a sample of patterns (see Postulate 1 in Sect. 1.3), even if the designer assumes he does not need it because of his familiarity with the task domain.

In the manual approach a sufficiently experienced, intelligent, and patient person looks at the problem and tries to write down, in some of the aforementioned representations and at the necessary level of detail, the required knowledge. Often one person is not sufficient because one needs one or more persons experienced in the task domain, and one or more persons experienced in the computer representation of knowledge. It is safe to say that there would be no pattern analysis without this approach; it is tempting to foretell that sooner or later the learning approach will replace it; and probably it is reasonable to assume that in the mixture approach the importance of automatic learning will increase.

At the present state of the art it is impossible (and would be inefficient) to start a learning process from "zero". Usually the system designer will influence the system behavior by some of the following initial design choices:

1. Define a formalism for knowledge representation.
2. Supply procedures with which the system can operate on observations.
3. Decide upon a strategy for (internal) representation of (external) observations.
4. Define criteria for the generalization of observations.
5. Limit the complexity of the learned model.
6. Supply a certain corpus of initial or *a priori* knowledge.

Machine learning usually does not change any of the initial choices in steps 1–6 above. It concentrates on the automatic generation of models of the observed environment. Step 5 above usually results in a limitation of learning to one level of abstraction, for example, the learning of a limited number of visible objects, or motion types, or object groups, but not learning to "see" in general, or also, learning about words, or syntax, or phonemes, but not learning to "hear" in general.

In the above sense learning is limited to the construction or inference of a concept or a class of patterns. More generally, learning means automatic improvement of system performance. This certainly includes improvement of the representations, of the procedures, and of the criteria for generalization. All this is excluded since at present there seem to be no feasible approaches to solving these problems. It is evident from this discussion that a learning system in this restricted sense is implicitly equipped with a good deal of *a priori* knowledge, experience, and intuition of the designer who only leaves a strictly limited task to the system. However, since this task may be laborious and tedious, its automation is nevertheless rewarding.

There are different situations in learning and different sources of information for a learning system. It may be that the system is required to learn a *single situation*, such that a model contains precisely one element, for example, when learning the floor-plan of a building by exploration of that building. It may also be required to learn a *class of situations* where a model contains a set of elements, for example, when learning the pronunciations of the word "car" or a 3D-model of a "car". The two main sources of information for learning are by *demonstration*, that is by presenting typical image or sound signals to the learning system (a sample ω in the sense of Sect. 1.3), or by *description*, that is by presenting a symbolic representation or a natural language description to the learning system.

The system may learn *supervised*, in which case the correct name of a concept is stated with every demonstration or description; learning may also be *unsupervised*, that is new demonstrations or descriptions have to be grouped or clustered into classes of similar elements. Finally, the result of learning may be a set of parameters or weights and/or a symbolic knowledge structure, for example, a semantic net.

7.7.2 Generalization

A fundamental problem of (automatic) learning of a class containing a set of elements is *generalization*, that is, the inference of a class of patterns or a concept with a possibly infinite number of members from a finite number of observed patterns. A trivial approach with no generalization would be to use just the set of observations as a definition of the concept. The basic approach to generalization is *similarity*, and the basic problem is a useful definition of similarity. An example is depicted in Fig. 7.28, where for simplicity it is assumed that observations may be represented by points in a plane. If a positive observation is made – that is, a pattern is known to belong to the class – then it is generalized by assuming that patterns *near* or *similar* to the observation also belong to the class; and if a negative observation is made, then it is assumed that patterns near to this also do not belong to the class. This was stated as Postulate 6 in Sect. 1.3. The vast number of possibilities of defining "similarity" is only indicated in Fig. 7.28.

Different representations of knowledge about patterns will not be treated further although, of course, the choice of representation is essential. Further procedures will be quite different if a pattern is represented by a vector $c \in R^n$, a string $v \in L$, a frame, a graph, or something else. Perhaps it is appropriate here to mention that an "observation" need not be restricted to a single object, but it may also consist of an ordered pair (f_1, f_2) of objects (or patterns or situations). The pair (f_1, f_2) indicates that if f_1 occurs, it can be replaced or it is followed by f_2. This allows the learning of production rules of type (7.3.1).

Similarity is obviously related to a distance in the case where the observations are represented by vectors of real numbers. Since representations by vectors were not a point in this chapter, only some remarks on similarity and generalization for nonnumeric representations are given. For strings of symbols the *Levenshtein distance* may be used to define similarity; it is simply the minimum number of deletions, insertions, and substitutions necessary to convert one string into the other. A refinement is obtained by consideration of transition probabilities (Sect. 4.4) or transition weights. This principle can also be applied to define a distance between graph structures. At first a set of transformations of nodes and links is defined, for example, insertion, deletion, and substitution of nodes and links as well as the relabeling of nodes and links; these transformations are given weights and the weight of a sequence of transformations is defined, for example, as the sum of individual weights. Then the set of transformations having minimal weight and transforming one graph structure into the other is defined as the "distance" of the two graphs.

If an observation is represented by some kind of symbolic description, it can be generalized by mitigating constraints. Some special rules for this are given for the special case where the descriptions are a conjunction of conditions, or where a concept consists of a set of obligatory parts and concretes. Rule 1 is the dropping of a condition

Rule 1. (dropping of a condition)

IF (a representation has several conditions),
THEN (a generalization is obtained by elimination of one
or more attributes). (7.7.1)

Rule 2. (turning constants into variables)

IF (there are two or more representations concerning the same
concept AND in corresponding conditions an attribute
has different constant values),
THEN (a generalization is obtained by replacing the constants
by a variable). (7.7.2)

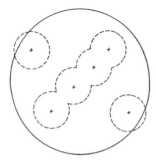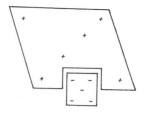

Fig. 7.28. Examples of generalizing positive and negative (−) observations

Rule 3. (introduction of internal disjunctions)

    IF   (in two representations an attribute has particular values),
THEN (a generalization is obtained by extending the range of the values according to the following special cases): (7.7.3)
    IF   (the range of values are real (or integer) numbers),
THEN (generalize to the interval given by the attribute values). (7.7.4)
    IF   (the range of values is a set),
THEN (generalize to the set). (7.7.5)
    IF   (the set of values may be ordered in a hierarchy),
THEN (generalize to the next higher level in the hierarchy). (7.7.6)

The first two rules are assumed to be self-explanatory although it is not specified in rule 1 which attribute to drop. The third is illustrated by brief examples. Assume two descriptions of objects known to belong to the same class to be "the shape of object 1 is square" and "the shape of object 2 is triangular". Possible generalizations by internal disjunctions are "the shape of objects is square, or triangular, or rectangular", or also "the shape of objects is polygonal" provided that the concept of a polygon is available to the system for shape description. Another example are the descriptions "the size of object 1 is l_1" and "the size of object 2 is l_2", which generalizes to "the size of objects is between l_1 and l_2". In view of Fig. 7.28 another generalization might be "the size of objects is between $l_1 \pm \Delta l$ or $l_2 \pm \Delta l$" which requires a priori or interactive selection of the parameter Δl, but avoids the possibility of over-generalization. Additional rules are the following:

Rule 4. (turning of a conjunction to a disjunction)

    IF   (a representation consists of a conjunction of conditions),
THEN (a generalization is obtained by turning a conjunction
        to a disjunction). (7.7.7)

Rule 5 concerns the case that positive and negative observations are given.

Rule 5. (positive and negative observations)

    IF   (there are two representations, one of which belongs to a
        particular concept and the other one does not, AND
        if they have an attribute with disjoint values),

THEN (a generalization is obtained by extending the attribute value
with the exception of those not belonging to the concept). (7.7.8)

The above generalization rules are "simple" in the sense that they do *not* introduce any *new* attributes. This is possible if additional knowledge is available. An example is the last rule.

Rule 6. (generating new attributes)
IF (a representation contains a component A and it is known
that A implies B),
THEN (a generalization is obtained by replacing A by B). (7.7.9)

In any case one should remember that a useful generalization cannot be obtained from just one or two observations. As the results about confidence intervals in statistics indicate, reliable inferences about a set of elements can only be made from observing a large number of elements. Additional rules are necessary if, for example, one 3D model is to be obtained from several 2D views. The remarks on generalization may be summarized as

a description $D1$ is a *generalization* of another description $D2$
if $D2$ is valid only for a proper subset of the objects
for which $D1$ is valid. (7.7.10)

In Sect. 1.3 it was noted that if the machine representations of two observations or patterns have a *small distance*, then they are *similar*. This only makes a statement about two patterns, not about class formation. However, similarity is a strong indication that the observations or patterns belong to the same concept or class. Two possible rules for class formation are:

IF (in a set of observations there are always two observations
with small distance),
THEN (the set contains observations belonging to the same class). (7.7.11)

IF (in a set of observations all observations have
small distance from each other),
THEN (the set contains observations belonging to the same class). (7.7.12)

A problem with the first rule is the possible formation of bridges between "dissimilar" observations, and a problem with the second rule is the possible separation of adjacent observations. Point 4 in the last section may now be refined to the following subpoints:

4.1. Definition of a distance measure for observations.
4.2. Choice of a procedure for generalization.
4.3. Definition of a class or concept.
4.4. Construction of a generalization of several observations belonging to the same class.

7.7.3 Outline of Some Algorithms

Some general remarks on algorithms for constructing generalizations are possible. The model-driven and the data-driven strategies can be distinguished. The former use *a priori* knowledge about the problem to generate hypothetical concept descriptions which are then tested against the observed patterns. The latter use the observations (or the data) to construct new hypotheses such that discrepancies between the current hypothesis and a new observation are removed. Usually the model-driven strategies use all observations in each test, whereas data-driven strategies evaluate observations one at a time to improve the description of the class. A basic requirement for the test or construction of a hypothesis is consistency with observations defined as follows:

A description of a class or concept is *consistent* with the observations or the data if it is valid for all positive observations and for no negative observations. A weaker requirement is to replace "all" and "no" by "almost all" and "almost no", respectively. (7.7.13)

With (7.7.10, 13) learning as understood here may be viewed as a search problem. Among the possible hypotheses or descriptions a search is made for one which suitably generalizes the data according to (7.7.10) and is consistent with the data according to (7.7.13). Usually there will be many hypotheses meeting these requirements, thus necessitating the selection of a "good" hypothesis. This may be done implicitly by the type of generalization rules, or explicitly by a criterion of goodness. Such criteria may be derived from the likelihood of a hypothesis or from its complexity. The results produced by a learning system – in particular, the patterns included in the description via generalization, but not included in the observations – will be judged by a human observer who compares the result using his intuition, or by evaluation of an independent test sample.

In this section short sketches of the main steps of two types of learning algorithms are given. The first is an algorithm for the supervised generation of disjoint concepts. The input is a set of observations belonging to a concept K_1, and possibly a set of observations belonging to a concept K_0; K_1 and K_0 are disjoint. A representation is selected for the observations and a quality score Q for generated descriptions. The observed patterns form the disjoint sample sets ω_1, ω_0 where

$$K_1 \supseteq \omega_1 = (b_{11}, b_{12}, \ldots, b_{1N_1}) ,$$

$$K_0 \supseteq \omega_0 = (b_{01}, b_{02}, \ldots, b_{0N_0}) . \qquad (7.7.14)$$

A starting description $B_1(0)$ of the sample is computed and considered as the concept K_1 with

$$b_{1i} \in B_1(0) , \quad i = 1, \ldots, N_1 ,$$
$$b_{0i} \notin B_1(0) , \quad i = 1, \ldots, N_0 . \qquad (7.7.15)$$

An example is the starting description

$$B_1(0) = \bigvee_{i=1}^{N_1} b_{1i}. \tag{7.7.16}$$

Then $B_1(i)$ is modified (generalized, simplified, expanded) such that the quality score Q improves and $B_1(i)$ does not contain observations from ω_0. The process stops if Q is maximal or after a certain number of iterations. The final description B_1 is a concept with the property

$$K_1 \cup K_0 - \omega_0 \supseteq B_1 \supseteq \omega_1. \tag{7.7.17}$$

A description of this type is *consistent* in the sense of (7.7.13). A modification of the above algorithm is possible such that iteratively a new observation is used to update an available description $B_1(i)$.

The second example is an outline of an algorithm for concept formation from a (nearly) natural language (NL) description. Examples of such descriptions are

The "white house" is a building where the president of the USA is living and which is located in Washington DC.

The "wanted car" is a red sedan, probably of type ABC or XYZ, with a damaged right headlight.

The general idea is to create a new concept by means of other known concepts and to introduce appropriate relations, restrictions, and modifications in order to meet the description. This results in the following processing steps:

1. A set of *known* concepts is given, together with attributes, relations, and parts. An NL definition of a *new* concept in terms of the known concepts is given.
2. Create a raw concept containing *a priori* knowledge if available.
3. Determine information which is to be transferred from the definition of the new concept to the raw concept.
4. If adequate, make a selection among the information in step 3.
5. Transfer the selected knowledge to the raw concept.

7.7.4 Learning Concepts of a Semantic Net

In this section we give some details of an approach to learning concepts in the semantic net structure defined in Sect. 7.4. The *a priori* knowledge is represented in one or more model-schemes in the sense of Fig. 7.4. A concept is generated by the acquisition process using a special concept from the model-scheme called concept-scheme. This is indicated in the network by a link named *model* from this concept-scheme to the generated concept. The inverse link is named model-of. A concept in the knowledge base of the analysis system can only have one model-scheme, while one model-scheme can have several model-concepts. There is no inheritance from a model-scheme to its model-concepts. Instead each model-scheme contains a description about how to construct such a model-concept.

336   7. Knowledge

Learning is done incrementally in three steps. The first step is observation. Any sensor data has to be transformed into an internal representation. A new sample description is constructed during the second step whenever a new observation is made. During the third step the model is constructed by combination of the information gained. The general outline of the acquisition process, the goals of each step, and the data flow is shown in Fig. 7.29.

The knowledge acquisition process may be structured according to Fig. 1.4. The knowledge base of this process contains procedural and declarative knowledge. The procedural part of the knowledge consists of the procedures referred to by the slots *test of arguments, computation of splitting, computation of dimension, computation of default*, and *selection of specialization* in Fig. 7.13 and of the rules used to compare intermediate results. The declarative part is the *model-scheme*. An example of the model-scheme of a "binary object", that is an object represented by a binary image, is shown in Fig. 7.30. The module "methods" is the same as for the analysis process. The data base of results contains acquired concepts instead of instances. Control is performed by a specialized algorithm.

During the first step of the acquisition process the existing parts, relations, and attributes of the concept-scheme may be split. For instance, the concept-scheme 3D-OBJECT may have a part "surface". Of course, the generated concepts of real objects will have several surfaces. An example of splitting of a link occurs when generating the concept of a "cylinder", which has three surfaces. To build up a model of a cylinder the concrete "surface" of the concept-scheme has to be split three times. The function referred to by the slot "computation of splitting" computes the actual number of splittings. It is controlled by a given interval defining the allowed number. The same problem occurs in the case of dimension. Different model-concepts of the same concept-scheme may have different dimensions of their network structures. The problem is solved in the same way

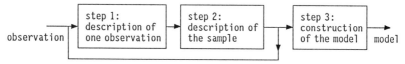

Fig. 7.29. The main steps of knowledge acquisition in a semantic net

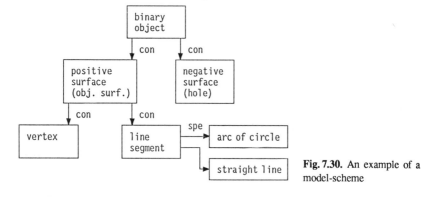

Fig. 7.30. An example of a model-scheme

7.7 Acquisition of Knowledge (Learning) 337

as is done in the case of splitting. An item "M-modifies" is used if a network structure of the concept-scheme has to be split. This item then contains the role of the part, relation, or attribute of the concept-scheme. Otherwise "M-modifies" is set to NO.

If a concept-scheme has specializations, the appropriate one has to be selected. This is done by a function which is referred to by the slot "selection of specialization". After having found the correct specialization the link model-of/model will no longer connect the concept-scheme and the model-concept but the chosen specialization and the model-concept. For example, the three surfaces of the model of a cylinder are specialized twice to "plane surface" and once to "cylindrical surface".

During the *first phase of step 1* the acquisition algorithm will build up a "concept-shell". This means that the new concept, its parts, attributes, and relations and the links between the parts and the concept are created, and the slots which later will contain special values are empty. During this phase, computation of splitting and computation of dimension takes place.

The goal of the *second phase* is to test the existence of the arguments of the analysis functions which are restricted by the syntax of the network language. Only a concept which can be reached along one link or an attribute description of such a concept can be an argument for an analysis function. In the first case the argument will consist only of the role of the link, in the second case it will consist of the role of the link and the role of the attribute description. The test of the existence is done by calling the routines referred to by the slots "test of argument". If network structures have been split, they will have new and different roles and they cannot have the role of the network structure from the concept-scheme. In this case the names of the arguments which are composed of the roles have to be determined again. For example, the judgement function of the concept-scheme "object" has only 1 argument, the role of the only concrete link. The judgement function of the generated concept of a cylinder in the above example has three arguments, the roles of the three links that have been constructed by splitting. It is required that the analysis functions are able to work with different numbers of arguments, but, on the other hand, as a consequence of the acquisition process, the generated concepts can be used for analysis directly without adaptation of the procedural knowledge. This is an important precondition for step 2 of the acquisition process.

During the *third phase* the default value of each attribute is computed and the domain may be restricted with respect to some inaccuracy. Such a restriction can improve the analysis process, but if, for example, the analysis is supposed to recognize an object regardless of its position, it would be fatal to restrict the domain of position dependent attributes. Ranges can be restricted by definition of an interval, a region, a distribution, or a fuzzy set.

The concepts created by the first step are the intermediate results of the acquisition process that are stored in the data base. A new sample description is constructed whenever a new observation has been transformed into the internal representation by the first step. The *first phase of step 2* consists of a search for correspondences between the network structures. This matching is done using the analysis functions referred to by the slots of the concepts and of the substructures. The same functions will later be used to analyze a whole scene with several objects. Indeed, matching in the case of knowledge acquisition is simpler than in the case of analysis. The new description does not have to be compared with the whole knowledge base, but only with the former sample description of the object that is to be learned. Of course, the preconditions described above for the analysis functions have to be taken into account.

During the *second phase*, the frequency counters of the network structures for which correspondences could not be found, are updated. The other network structures of the new observation are added to the new sample description with a frequency counter value of 1. This second step is done to combine several observations of the subject to be learned. Especially if one wants to build up a complete 3D-model of an object one has to examine several images from different views. The merging of the old sample description and a new observation is guided by the following rules:

Rule 1: (merging of two concepts K_1, K_2)

    IF   (there is a correspondence between K_1, K_2,
           AND K_1 and K_2 have been generated using the same concept-scheme,
           AND the arguments of the judgement procedure can be merged
           or redetermined),
    THEN (merge K_1 and K_2 as follows:
           1. increase the frequency counter of K_1,
           2. try to activate rules 2–4 for all substructures with
           equal roles,
           3. determine the best merge for those substructures that
           have been split, try to activate rules 2–4 respectively,
           4. create a new modality description for K_1 with respect
           to the links of K_2,
           5. copy all remaining substructures that cannot be
           merged from K_2 into K_1).   (7.7.18)

Rule 2: (merging of two link descriptions L_1 and L_2)

    IF   (there is a correspondence between the domains of L_1 and L_2,
           AND the concepts to which the link descriptions belong
           have been merged,
           AND L_1 and L_2 have been generated using the same link
           description from the model-scheme),
    THEN (merge L_1 and L_2 as follows:
           1. increase the frequency counter of L_1,
           2. append the domain of L_2 to the domain of L_1,
           3. append the preference of L_2 to the preference of L_1,
           4. recompute the allowed interval for dimension,

    5. for all those concepts from the domains for which there are correspondences, try to activate rule 1. If the concepts can be merged, redetermine the domain). (7.7.19)

Rule 3: (merging of two attribute descriptions A_1 and A_2)

    IF   (the concepts to which the attribute descriptions belong have been merged,
AND A_1 and A_2 have been generated using the same attribute description from the model-scheme,
AND the arguments of the computation of value function can be merged or redetermined),
THEN (merge A_1 and A_2, respectively, as follows:
    1. increase the frequency counter of A_1,
    2. append the preference of A_2 to the preference of A_1,
    3. recompute the allowed interval for dimension,
    4. recompute the restriction of the domain (selection) according to the selections of A_1 and A_2). (7.7.20)

Rule 4: (merging of two relations R_1 and R_2)

    IF   (the concepts to which the relations belong have been merged,
AND R_1 and R_2 have been generated using the same relation from the model-scheme,
AND the arguments of the judgement of relation function can be merged or redetermined),
THEN (merge R_1 and R_2 as follows:
    1. increase the frequency counter of R_1). (7.7.21)

The *third step* creates a new model by a rule based comparison of the sample description. Such a comparison of concepts may lead to a description of a given class of objects (generalization). On the other hand, given only examples of a single situation, segmentation faults may be corrected. Examples of the rules for step 3 are given in Sect. 7.7.2.

An *optional fourth step* may compare the created model with negative examples. The goal of this step is to verify and to specialize the model. It is known from parameter learning in pattern analysis that near misses will lead to better results in discriminating patterns. But what near misses are in learning complex structural knowledge is not easy to decide. For a given application one may know the answer, but in general it seems to be complicated to find these near misses. Therefore, our knowledge acquisition algorithm will build up complete models with respect to the input data from positive examples and is able to use near misses if there are any.

7.8 Explanation

In general, the purpose of an *explanation module* in a system for pattern analysis and understanding is to make the system (more) transparent to the user. This pertains in particular to the six main components of a knowledge based system for pattern understanding, which according to Chap. 1 are intermediate results, low-level methods, high-level knowledge, control, learning, and explanation itself. The two main tasks of this module are the explanation of *system resources* and the explanation of *system activities*. One reason for incorporating an explanation facility is to provide some kind of on-line documentation and a user support, in particular for the user who is not familiar with the implementation details of the system. Another reason is to provide a useful tool for system design and debugging. In this sense the explanation module and the user interface are closely linked.

Relevant to the explanation of system resources are the *knowledge* of the system, for example, the available set of rules and/or concepts in the knowledge base, the *procedures*, the stored *experience* about system performance, the *processing* alternatives, and *what* can and cannot be explained. This type of explanation also may cover system learning. In this case it should be pointed out what can be learned, what strategies for automatic knowledge acquisition are available, what are appropriate training sequences (instances), and what were the a priori design choices for the learning component.

It is also relevant to the explanation of system activities to point out to the user the chains of *actions and reasoning*, the *facts and results* used for processing, the order of *instantiation* of concepts or of *activation* of rules, perhaps the reasons for the *non-applicability* of concepts and rules, and the activated *procedures*, together with a display of relevant results and parameters. As with most types of explanations, one can give "immediate" reasons for certain actions, but no "final" reasons.

All types of explanation facilities should be provided through a *user interface* which does not require the reading of lengthy manuals.

It would be beyond the scope of this limited volume to go into the details of explanation modules and user interfaces here.

7.9 Bibliographical Remarks

Standard definitions of knowledge are obtained from [7.3, 4] and Fig. 7.1 is according to [7.5].

A three-dimensional hierarchy for knowledge representation as in Fig. 7.3 was described in [7.6, 7]. The frames were introduced in [7.8], a frame based programming language in [7.9].

Introductions to first order predicate calculus are given, for example, in [7.10, 11]. The use of FOPC for knowledge representation is discussed in [7.12–15]. The resolution principle is based on [7.16] and discussed in [7.17–19]; other

7.9 Bibliographical Remarks

approaches to logic inferences are [7.20–22]. Extensions of FOPC are given, for example, in [7.23–28]; the counterfactuals were introduced in [7.29]. A programming language designed for logic programming is described in [7.30].

Rule based systems are introduced in [7.31] and considered in the AI context in [7.32]; additional material on this topic is provided by [7.33–35]. The programmed production rule was discussed in [7.36] and the censor part of a rule in [7.37]. The fuzzy inference rules are treated, for example, in [7.38, 39]. The example of a general rule format (OPS5) is according to [7.40] where also a detailed description of the RETE algorithm is given. Conflict resolution is treated in [7.41, 42]. Examples of production systems in pattern analysis are [7.43–49].

Semantic nets were introduced in [7.50] for representing the meaning of words and are treated in [7.51, 52] – an often used name is also associative nets or networks. Apparently, the frames in [7.8] are very similar. The problem of what to represent in a node and in a link is addressed in [7.53–55]. The definition of a net structure given here and the rules for instantiation are based on [7.6, 7, 56]. The special control algorithms were developed in [7.6, 7, 57, 58]. The examples of applications in image and speech understanding are treated in more detail in [7.6, 56, 59, 60]; case theory is covered in [7.61, 62]. Other examples in pattern analysis and understanding are [7.63–65]. The use of semantic nets as a system architecture for pattern analysis systems is considered in [7.66]. Software environments for semantic nets are described in [7.67–69]. The object-oriented programming style employs techniques used also in semantic nets [7.70–72]. The relations between semantic nets and logic are discussed in [7.15, 7.73, 74].

Relational structures for knowledge representation in pattern analysis have been treated only very briefly in this text; additional material is available in [7.75–78].

The use of linguistic methods for pattern recognition was suggested in [7.79] and some references to standard string grammars were given in Sect. 4.7. The definition of syntax, semantics, and pragmatics is from [7.1]. A detailed treatment of syntactic methods for natural language processing is given in [7.80]. The ATN formalism was introduced in [7.81], the unification based grammars and lexical functional grammars are treated in [7.80, 82, 83]. The example of an ATN grammar and of lexicon organization is according to [7.84] and [7.85], respectively.

The section on automatic knowledge acquisition concentrated on approaches related to learning of symbolic structures. References on parameter or weight learning and grammatical inference are quoted in Sect. 4.7. Concept and rule learning are treated in [7.2, 86–93]. The generalization rules stated in Sect. 7.7 are developed in [7.88, 94]. The approach to concept learning in a semantic net is that of [7.95, 96]. Another approach to learning, the genetic learning algorithm, is developed in [7.97].

References

Chapter 1

1.1 L. Wittgenstein: *Tractatus Logico-Philosophicus* (Routledge and Kegan Paul, London 1961)
1.2 K.S. Fu: *Syntactic Methods in Pattern Recognition* (Academic, New York 1974)
1.3 K. Fukunaga: *Introduction to Statistical Pattern Recognition* (Academic, New York 1972)
1.4 K.S. Fu: *Sequential Methods in Pattern Recognition and Machine Learning* (Academic, New York 1968)
1.5 H. Niemann: *Klassifikation von Mustern* (Springer, Berlin, Heidelberg 1983)
1.6 J.T. Tou, R.C. Gonzalez: *Pattern Recognition Principles* (Addison-Wesley, Reading, MA 1974)
1.7 T.Y. Young, T.W. Calvert: *Classification, Estimation, and Pattern Recognition* (Elsevier, New York 1973)
1.8 W.A. Lea (ed.): *Trends in Speech Recognition* (Prentice Hall, Englewood Cliffs, NJ 1980)
1.9 R. De Mori: *Computer Models of Speech Using Fuzzy Algorithms* (Plenum, New York 1983)
1.10 D.R. Reddy (ed.): *Speech Recognition* (Academic, New York 1975)
1.11 E.C. Schwab, H.C. Nusbaum (eds.): *Pattern Recognition by Humans and Machines*, 1 Speech Perception (Academic, New York 1986)
1.12 D.H. Ballard, C.M. Brown: *Computer Vision* (Prentice Hall, Englewood Cliffs NJ 1982)
1.13 A.R. Hanson, E.M. Riseman (eds.): *Computer Vision Systems* (Academic, New York 1978)
1.14 B.K.P. Horn: *Robot Vision* (McGraw-Hill, New York 1986)
1.15 M.D. Levine: *Vision in Man and Machine* (McGraw-Hill, New York 1985)
1.16 D. Marr: *Computer Vision* (Freeman, San Francisco 1982)
1.17 H. Niemann: *Pattern Analysis* (Springer, Berlin, Heidelberg 1981)
1.18 W.K. Pratt: *Digital Image Processing* (Wiley, New York 1978)
1.19 Y. Shirai: *Three-Dimensional Computer Vision*, Springer Ser. Symb. Comp., Vol.6 (Springer, Berlin, Heidelberg 1987)
1.20 M. Brodie, J. Mylopoulos (eds.): *On Knowledge Base Management Systems* (Springer, Berlin, Heidelberg 1986)
1.21 B. Buchanan, E. Shortliffe: *Rule-Based Expert Systems* (Addison-Wesley, Reading, MA 1984)
1.22 N.V. Findler (ed.): *Associative Networks* (Academic, New York 1979)
1.23 N.J. Nilsson: *Principles of Artificial Intelligence* (Springer, Berlin, Heidelberg 1982)
1.24 J.F. Sowa: *Conceptual Structures: Information Processing in Mind and Machine* (Addison-Wesley, Reading, MA 1984)
1.25 P.H. Winston: *Artificial Intelligence*, 2nd edn. (Addison-Wesley, Reading, MA 1984)
1.26 T. Binford: Survey of model-based image analysis systems. Int'l Journ. Robotics Res., 1, 1-32 (1982)

1.27 R. Brooks: Symbolic reasoning among three-dimensional models and two-dimensional images. Artif. Intell. **17**, 285 (1981)
1.28 L. Erman: The HEARSAY II speech understanding system: Integrating knowledge to resolve uncertainty. ACM Comp. Survey **12**, 213 (1980)
1.29 D.H. Klatt: Review of the ARPA speech understanding project. J. Acoust. Soc. Am. **62**, 1345 (1977)
1.30 M. Nagao, T. Matsuyama: *A Structural Analysis of Complex Aerial Photographs* (Plenum, New York 1980)
1.31 H. Niemann, H. Bunke, I. Hofmann, G. Sagerer, F. Wolf, H. Feistel: A knowledge based system for analysis of gated blood pool studies. IEEE Trans. PAMI **7**, 246 (1985)
1.32 H. Niemann, A. Brietzmann, R. Mühlfeld, P. Regel, G. Schukat: The speech understanding and dialog system EVAR. In *New Systems and Architectures for Automatic Speech Recognition and Synthesis*, ed. by R. De Mori, C.Y. Suen, NATO ASI Ser. F16 (Springer, Berlin, Heidelberg 1985)
1.33 Y. Ohta: *Knowledge-Based Interpretation of Outdoor Natural Colour Scenes* (Pitman Books, Boston 1985)
1.34 G. Sagerer: *Darstellung und Nutzung von Expertenwissen für ein Bildanalysesystem*, Informatik Fachberichte, Vol.104 (Springer, Berlin, Heidelberg 1985)
1.35 W. Scherl: *Bildanalyse allgemeiner Dokumente*, Informatik Fachberichte, Vol.131 (Springer, Berlin, Heidelberg 1987)
1.36 G. Wyszecki, W.S. Stiles: *Color Science - Concepts and Methods, Quantitative Data and Formulae*, 2nd edn. (Wiley, New York 1982)
1.37 E. Zwicker, R. Feldtkeller: *Das Ohr als Nachrichtenempfänger* (Hirzel, Stuttgart 1967)

Chapter 2

2.1 D. Middleton: *An Introduction to Statistical Communication Theory* (McGraw-Hill, New York 1960)
2.2 G. Winkler: *Stochastische Systeme, Analyse und Synthese* (Akademische Verlagsgesellschaft, Wiesbaden 1977)
2.3 N.S. Jayant: Digital coding of speech waveforms, PCM, DPCM, and DM quantizers. Proc. IEEE **62**, 611 (1974)
2.4 J. Max: Quantizing for minimum distortion. IRE Trans. Inf. Theory **6**, 7 (1960)
2.5 A. Albert: *Regression and the Moore-Penrose Pseudoinverse* (Academic, New York 1972)
2.6 M.J. Lighthill: *Introduction to Fourier Analysis and Generalized Functions* (Cambridge Univ. Press, Cambridge 1958)
2.7 A. Papoulis: *Systems and Transforms with Applications in Optics* (McGraw-Hill, New York 1968)
2.8 A.V. Oppenheim, R.W. Schafer: *Digital Signal Processing* (Prentice Hall, Englewood Cliffs, NJ 1975)
2.9 A. Zygmund. *Trigonometric Series*, 2 (Cambridge Univ. Press, Cambridge 1968)
2.10 H. Babovsky, T. Beth, H. Neunzert, M. Schulz-Reese: *Mathematische Methoden in der Systemtheorie: Fourieranalysis* (Teubner, Stuttgart 1987) Kap.4
2.11 M.C. Pease: An adaptation of the fast Fourier transform to parallel processing. JACM **15**, 252 (1968)
2.12 H.C. Andrews, B.R. Hunt: *Digital Image Restoration* (Prentice Hall, Englewood Cliffs, NJ 1975)
2.13 N.S. Jayant (ed.): *Waveform Quantization and Coding* (IEEE Press, New York 1976)
2.14 N.S. Jayant, P. Noll: *Digital Coding of Waveforms* (Prentice Hall, Englewood Cliffs, NJ 1984)

2.15 P.A. Wintz: Transform picture coding. Proc. IEEE **60**, 809 (1972)
2.16 R. Zelinski, P. Noll: Adaptive transform coding of speech signals. IEEE Trans. ASSP-**25**, 299 (1977)
2.17 H. Freeman: On the encoding of arbitrary geometric configurations. IRE Trans. Electron. Computers **10**, 260 (1961)
2.18 H. Freeman: Computer processing of line drawing images. Comput. Surveys **6**, 57 (1974)
2.19 D.C. van Voorhis: An extended run-length encoder and decoder for compression of black/white images. IEEE Trans. IT-**22**, 190 (1976)
2.20 W.K. Pratt: *Digital Image Processing* (Wiley, New York 1978)
2.21 M. Kunt, M. Bénard, R. Leonardi: Recent results in high compression image coding. IEEE Trans. CAS-**34**,1306-1336 (1987)
2.22 A. Güdesen: Quantitative analysis of preprocessing techniques for the recognition of handprinted characters. Pattern Recognition **8**, 219 (1976)
2.23 F. Itakura: Minimum prediction residual principle applied to speech recognition. IEEE Trans. ASSP-**23**, 67 (1975)
2.24 L.R. Rabiner, A.E. Rosenberg, S.E. Levinson: Considerations in dynamic time warping algorithms for discrete word recognition. IEEE Trans. ASSP-**26**, 575 (1978)
2.25 A.T. Amin: An algorithm for grey-level transformation in digitized images. IEEE Trans. C-**26**, 1158 (1977)
2.26 R.B. Crane: Preprocessing techniques to reduce atmospheric and sensor variability. Proc. 7th. Int'l Symposium on Remote Sensing of Environment, Univ. of Michigan, Ann Arbor (1971) pp.1345-1350
2.27 W. Frei: Image enhancement by histogram hyperbolization. Comput. Graphics and Image Proc. **6**, 286 (1977)
2.28 E.L. Hall, R.P. Kruger, S.J. Dwyer, D.L. Hall, R.W. McLaren, G.S. Lodwick: A survey of preprocessing and feature extraction techniques for radiographic images. IEEE Trans. C-**20**, 1032 (1971)
2.29 R.A. Hummel: Histogram modification techniques. Comput. Graphics and Image Proc. **4**, 209 (1975)
2.30 G. Wyszecki, W.S. Stiles: *Color Science - Concepts and Methods, Quantitative Data and Formulae*, 2nd edn. (Wiley, New York 1982)
2.31 B.A. Wandell: The synthesis and analysis of color images: IEEE Trans. PAMI-**9**, 2 (1987)
2.32 J.F. Andurs, C.W.Campbell, R.R. Jayroe: Digital image registration method using boundary maps. IEEE Trans. C-**24**, 935 (1975)
2.33 D.I. Barnea, H.F. Silverman: A class of algorithms for fast digital image registration. IEEE Trans. C **21**, 179 (1972)
2.34 B.K.P. Horn, B.L. Bachman: Using synthetic images to register real images with surface models. Commun. ACM **21** 914 (1978)
2.35 P. Van Wie, M. Stein: A landsat digital image rectification system. IEEE Trans. GE-**15**, 130 (1977)
2.36 R. Lenz: Linsenfehlerkorrigierte Eichung von Halbleiterkameras mit Standardobjektiven für hochgenaue 3D-Messungen in Echtzeit. In *Mustererkennung*, ed. by E. Paulus (Springer, Berlin, Heidelberg 1987)
2.37 M. Meyer: Kalibrieren einer Stereokamera, Diplomarbeit, Lehrstuhl für Informatik 5 (Mustererkennung) Univ. Erlangen-Nürnberg (1987)
2.38 *Manual of Photogrammetry* (American Soc. Photogrammetry, Washington 1984)
2.39 R.Y. Tsai: A versatile camera calibration technique for high accuracy 3D vision metrology using off-the-shelf TV cameras and lenses. IBM Res. Rep. RC 11413 (1985)
2.40 A.V. Oppenheim (ed.): *Applications of Digital Signal Processing* (Prentice Hall, Englewood Cliffs, NJ 1978)

2.41 H.J. Nussbaumer: *Fast Fourier Transform and Convolution Algorithms*, 2nd edn., Springer Ser. Inf. Sci., Vol.2 (Springer, Berlin, Heidelberg 1981)
2.42 A.V. Oppenheim: Nonlinear filtering of multiplied and convolved signals. Proc. IEEE **56**, 1264 (1968)
2.43 D.G. Childers, D.P. Skinner, R.C. Kemerait: The cepstrum - a guide to processing. Proc. IEEE **65**, 1428 (1977)
2.44 C.H. Chen (ed.): *Computer Aided Seismic Analysis and Discrimination* (Elsevier, New York 1978) pp.55-96
2.45 O.R. Mitchell, E.J. Delp, P.L. Chen: Filtering to remove cloud cover in satellite imagery. IEEE Trans. GE-**15**, 137 (1977)
2.46 A.M. Noll: Cepstrum pitch determination. J. Acoust. Soc. Am. **41**, 293 (1967)
2.47 L.R. Rabiner, M.J. Cheng, A.E. Rosenberg, C.A. McGonegal: A comparative performance study of several pitch detection algorithms. IEEE Trans. ASSP-**24**, 399 (1976)
2.48 V.T. Rhyne: A comparison of coherent averaging techniques for repetitive biological signals. Med. Res. Eng. August-September, 22 (1969)
2.49 G.L. Anderson, A.N. Netravali: Image restoration based on a subjective criterion. IEEE Trans. SMC-**6**, 845 (1976)
2.50 D. Casasent, D. Psaltis: New optical transforms for pattern recognition. Proc. IEEE **65**, 77 (1977)
2.51 D. Casasent (ed.): *Optical Data Processing*, Topics Appl. Phys., Vol.23 (Springer, Berlin, Heidelberg 1978)
2.52 D. Casasent: Coherent optical computing. Computer **12**, 1-27 (1979)
2.53 P. Meer, S. Baugher, A. Rosenfeld: *Optimal Image Pyramid Generating Kernels*, CAR-TR-1665, Center for Automation Res., Univ. of Maryland, College Park (1986)
2.54 N. Levinson: The Wiener RMS error criterion in filter design and prediction. J. Math. Phys. N.Y. **25**, 261 (1947)
2.55 J. Makhoul: Linear prediction, a tutorial review. Proc. IEEE **63**, 561 (1975)
2.56 J.D. Markel, A.H. Gray, Jr.: *Linear Prediction of Speech*, Communications and Cybernetic, Vol.12 (Springer, Berlin, Heidelberg 1976)
2.57 T. Bohlin: Comparison of two methods of modeling stationary EEG signals. IBM J. Res. Dev. 194 (1973)
2.58 L.C. Wood, S. Treitel: Seismic signal processing. Proc. IEEE **63**, 649 (1975)
2.59 M.J.E. Golay: Hexagonal parallel pattern transformations. IEEE Trans. C-**18**, 733 (1969)
2.60 R.M. Haralick, S.R. Sternberg, X. Zhuang: Image analysis using mathematical morphology. IEEE Trans. PAMI-**9**, 532 (1987)
2.61 B. Jungmann: Segmentierung mit morphologischen Operationen. In *Mustererkennung*, ed. by W. Kropatsch (Springer, Berlin, Heidelberg 1984) pp.77-83
2.62 G. Matheron: *Random Sets and Integral Geometry* (Academic, New York 1975)
2.63 K. Preston: Feature extraction by Golay hexagonal pattern transformations. IEEE Trans. C-**20**, 1007 (1971)
2.64 J. Serra: *Image Analysis and Mathematical Morphology* (Academic, London 1982)
2.65 S.R. Sternberg: Biomedical image processing. Computer **16**, 1-22 (1983)
2.66 I. Pitas, A.N. Venetsanopoulos: Nonlinear order statistic filters for image filtering and edge detection. Signal Processing **10**, 395 (1986)
2.67 J. Jaschul: Adaption vorverarbeiteter Sprachsignale zum Erreichen der Sprecherunabhängigkeit automatischer Spracherkennungssysteme. Dissertation, Fakultät für Elektrotechnik, Techn. Universität, München (1982)
2.68 B. Kämmerer: *Sprecheradaption und Reduktion der Sprecherabhängigkeit für die automatische Spracherkennung*, Dissertation, Technische Fakultät, Universität Erlangen-Nürnberg (1989)

Chapter 3

3.1 H. Niemann: *Methoden der Mustererkennung* (Akademische Verlagsgesellschaft, Frankfurt 1974)
3.2 W.K. Pratt: *Digital Image Processing* (Wiley, New York 1978)
3.3 J.W. Strohbehn, C.H. Yates, B.H. Curran, E.S. Sternick: Image enhancement of conventional transverse-axis tomograms, IEEE Trans. BME-26, 253-262 (1979)
3.4 J.W. Modestino, R.W. Fries: Edge detection in noisy images using recursive digital filtering. Comput. Graph. Image Proc. **6**, 409 (1977)
3.5 J. Canny: A computational approach to edge detection. IEEE Trans. PAMI **8**, 679 (1986)
3.6 Y. Liu: Detektion von Konturpunkten in Bildern. Diplomarbeit, Lehrstuhl für Informatik 5 (Mustererkennung) Universität Erlangen-Nürnberg, Erlangen (1987)
3.7 N. Blasius: Extraktion elementarer Bestandteile des Herzens für ein nuklearmedizinisches Bildanalysesystem. Studienarbeit, Lehrstuhl für Informatik 5 (Mustererkennung) Universität Erlangen-Nürnberg (1988)
3.8 P. Brodatz: *Textures* (Dover, New York 1966)
3.9 W.I. Smirnov: *Lehrgang der höheren Mathematik*, Teil IV (VEB, Berlin 1966)
3.10 G. Schmid: Glattheitsbeschränkungen bei der Berechnung des optischen Flusses aus einem Stereobildpaar. Studienarbeit, Lehrstuhl für Informatik 5 (Mustererkennung), Universität Erlangen-Nürnberg (1988)
3.11 R.T. Frankot, R. Chellapa: A method for enforcing integrability in shape from shading algorithms. Proc. 1. Int'l Conf. on Computer Vision, London 1987, p.118-127
3.12 R. Prechtel: Extraktion dreidimensionaler Information aus Grauwertbildern durch "Local Shading Analysis". Diplomarbeit, Lehrstuhl für Informatik 5 (Mustererkennung), Universität Erlangen-Nürnberg (1988)
3.13 T.B. Martin: Acoustic recognition of a limited vocabulary in continuous speech. Ph.D. Thesis, Dept. Electr. Eng, Univ. of Pennsylvania (1970)
3.14 I. Biederman: Human image understanding: Recent research and theory. Computer Vision, Graphics, and Image Processing **32**, 29 (1985)
3.15 P. Regel: *Akustisch-Phonetische Transkription für die automatische Spracherkennung*. Fortschrittberichte VDI Reihe 10 **83** (VDI, Düsseldorf 1988)
3.16 D. Sahoa: A survey of thresholding techniques. Comp. Vision, Graphics and Image Processing **41**, 233-260 (1988)
3.17 H. Bley: Segmentation and preprocessing of electrical schematics using picture graphs. Comp. Vision, Graphics, and Image Processing **28**, 271 (1984)
3.18 H. Bunke, H. Feistel, H. Nieman, G. Sagerer, F, Wolf, G.X. Zhou: Smoothing, thresholding, and contour extraction in images from gated blood pool studies. Proc. First IEEE Comp. Soc. Int'l Symp. on Medical Imaging and Image Interpretation, Berlin (1982) pp.146-151
3.19 C.K. Chow, T. Kaneko: Boundary detection of radiographic images by a threshold method. In *Frontiers of Pattern Recognition*, ed. by S. Watanabe (Academic, New York 1972) pp.61-82
3.20 M. Ingram, K. Preston: Automatic analysis of blood cells. Sci. Am. **223**, 72 (1970)
3.21 R.S. Ledley: High-speed automatic analysis of biomedical pictures. Science **146**, 216 (1964)
3.22 J. Schürmann: Bildvorverarbeitung für die automatische Zeichenerkennung. Wiss. Ber. AEG Telefunken **47** Heft 3/4, 90 (1974)
3.23 J.R. Ullmen: Binarization using associative addressing. Pattern Recognition **6**, 127-135 (1974)

3.24 J.S. Weszka, R.N. Nagel, A. Rosenfeld: A threshold selection technique. IEEE Trans. C-23, 1322 (1974)
3.25 N. Otsu: Discriminant and least-squares threshold selection. Proc. 4th Int'l Joint Conf. on Pattern Recognition, Kyoto (1978) pp.592-596
3.26 W.A. Barrett: An iterative algorithm for multiple threshold detection. Proc. IEEE Comp. Soc. Conf. on Pattern Recognition and Image Processing, Dallas TX (1981) p.273
3.27 E.C. Greanias: The recognition of handwritten numerals by contour analysis. IBM J. Res. Dev. 7, 14 (1963)
3.28 S.J. Mason, J.K. Clemens: Character recognition in an experimental reading machine for the blind. In *Recognizing Patterns*, ed. by P.A. Kolers, M. Eden (MIT Press, Cambridge 1968) pp.155-167
3.29 C. Arcelli, S. Levialdi: Parallel shrinking in three dimensions. Comput. Graph. Image Proc. 1, 21 (1972)
3.30 Z. Kulpa: Area and perimeter measurements of blobs in discrete binary pictures. Comput. Graph. Imag. Proc. 6, 434 (1977)
3.31 B. Moayer, K.S. Fu: A syntactic approach to fingerprint pattern recognition. Pattern Recognition 7, 23 (1975)
3.32 I.S.N. Murthy, K.J. Udupa: A search algorithm for skeletonization of thick patterns. Comput. Graph. Image Proc. 3, 247 (1974)
3.33 R. Pavlidis: *Structural Pattern Recognition*, Springer Ser. Electrophysics, Vol. 1 (Springer, Berlin, Heidelberg 1977)
3.34 J. Sobel: Neighboorhood coding of binary images for fast contour following and general binary array processing. Comput. Graph. Image Proc. 8, 127 (1978)
3.35 A. Perez, R.C. Gonzalez: An iterative thresholding algorithm for image segmentation. IEEE Trans. PAMI-9, 742 (1987)
3.36 L.S. Davis: A survey of edge detection techniques. Comput. Graph. Image Proc. 4, 248 (1975)
3.37 E.M. Riseman, M.A. Arbib: Computational techniques in the visual segmentation of static scenes. Comput. Graph. Image Proc. 6, 221 (1977)
3.38 C.H. Chen: Note on a modified gradient method for image analysis. Pattern Recognition 10, 261 (1978)
3.39 K.K. Pingle: Visual perception by a computer. In *Automatic Interpretation and Classification of Images*, ed. by A. Grasselli (Academic, New York 1969) pp.277-284
3.40 L.G. Roberts: Machine perception of three-dimensional solids, in *Optical and Electro-Optical Information Processing*, ed. by J.T. Tippelt, D.A. Berkowitz, L.C. Clapp, C.J. Koester, A.V.D. Burgh (MIT Press, Cambridge 1965) pp.159-194
3.41 A. Rosenfeld: A nonlinear edge detection technique. Proc. IEEE 58, 814 (1970)
3.42 A. Rosenfeld, M. Thurston: Edge and curve detection for visual scene analysis. IEEE Trans. C-20, 562 (1971)
3.43 A. Rosenfeld, M. Thurston, Y.H. Lee: Edge and curve detection, further experiments. IEEE Trans. C 21, 677 (1972)
3.44 H. Wechsler, J. Sklansky: Finding the rib cage in chest radiographs. Pattern Recognition 9, 21 (1977)
3.45 D. Middleton: *An Introduction to Statistical Communication Theory* (McGraw Hill, New York 1960)
3.46 G. Winkler: *Stochastische Systeme, Analyse and Synthese* (Akademische Verlagsgesellschaft, Wiesbaden 1977)
3.47 B. Kruse, K. Rao: A matched filtering technique for corner detection. Proc. 4th Int'l Joint Conf. on Pattern Recognition, Kyoto (1978) pp.642-644
3.48 A.W. Lohmann, D.P. Paris: Computer generated spatial filters for coherent optical data processing. App. Opt. 7, 651 (1968)

3.49 B.J. Schachter, A. Rosenfeld: Some new methods of detecting step edges in digital pictures. Commun. ACM **21**, 172 (1978)
3.50 K.S. Shanmugam, F.M. Dickey, J.A. Green: An optimal frequency domain filter for edge detection in digital pictures. IEEE Trans. PAMI-1, 37 (1979)
3.51 D. Marr: *Computer Vision* (Freeman, San Francisco 1982)
3.52 D. Marr, E.C. Hildreth: Theory of edge detection. MIT AI Memo 518, MIT, Cambridge, MA (1979)
3.53 M. Hueckel: An opertor which locates edges in digitized pictures. J. Assoc. Comput. Mach. **18**, 113 (1971)
3.54 M. Hueckel: A local visual operator which recognizes edges and lines. J. Assoc. Comput. Mach. **20**, 634 (1973)
3.55 F. Holdermann, H. Kazmierczak: Preprocessing of gray-scale pictures. Comput. Graph. Image Proc. **1**, 66 (1972)
3.56 Y. Yakimovski: Boundary and object detection in real world images. J. Assoc. Comput. Mach. **23**, 599 (1976)
3.57 A.K. Griffith: Mathematical models for automatic line detection. J. Assoc. Comput. Mach. **20**, 62 (1973)
3.58 N.E. Nahi, S. Lopez-Mora: Estimation-detection of object boundaries in noisy images. IEEE Trans. AC-23, 834 (1978)
3.59 A. Kundu, S.K. Mitra: A new algorithm for image edge extraction using a statistical classifier approach. IEEE Trans. PAMI-9, 569-577 (1987)
3.60 H.P. Kramer, J.B. Bruckner: Iterations of a nonlinear transformation for enhancement of digital images. Pattern Recogn. **7**, 53 (1975)
3.61 A. Rosenfeld: Iterative methods in image analysis. Pattern Recognition **10**, 181 (1978)
3.62 S. Yokoi, T. Naruse, J. Toriwaki, T. Fukumura: A theoretical analysis of grey weighted distance transformations. Proc. 4th Int'l Joint Conf. on Pattern Recognition, Kyoto (1978) pp.573-575
3.63 S.W. Zucker, R.A. Hummel, A. Rosenfeld: An application of relaxation labeling to line and curve enhancement. IEEE Trans. C-26, 394 (1977)
3.64 G.J. Vander Brug: Experiments in iterative enhancement of linear features. Comput. Graph. Image Proc. **6**, 25 (1977)
3.65 R. Nevatia: A color edge detector. Proc. 3rd Int'l Joint Conf. on Pattern Recognition, Coronado, CA (1976) pp.826-832
3.66 J.B. Burns, A.R. Hanson, E.M. Riseman: Extracting straight lines. IEEE Trans. PAMI-8, 425 (1986)
3.67 S.A. Dudani, A.L. Luk: Locating straight-line edge segments on outdoor scenes. Pattern Recognition **10**, 145 (1978)
3.68 P. Gallinari, M. Milgram: A parallel edge following algorithm. Proc. 8th Int'l Conf. Pattern Recognition, Paris (1986) pp.907-909
3.69 T. Pavlidis, S.L. Horowitz: Segmentation of plane curves. IEEE Trans. C-23, 860 (1974)
3.70 M. Suk, O. Song: Curvilinear feature extraction using minimum spanning trees. Computer Vision, Graphics, and Image Processing **26**, 400 (1984)
3.71 C.M. Williams: An efficient algorithm for the piecewise linear approximation of planar curves. Comput. Graph. Image Proc. **8**, 286 (1978)
3.72 R.O. Duda, P.E. Hart: Use of the Hough transformation to detect lines and curves in pictures. Commun. ACM **15**, 11 (1972)
3.73 S.D. Shapiro: Properties of transforms for the detection of curves in noisy pictures. Comput. Graph. Image Proc. **8**, 219 (1978)
3.74 D.H. Ballard: Generalizing the Hough transform to detect arbitrary shapes. Pattern Recognition **13**, 111 (1981)
3.75 H. Bunke: *Modellgesteuerte Bildanalyse* (Teubner, Stuttgart 1985)

3.76 H. Elliot, L. Srinivasan: An application of dynamic programming to sequential boundary estimation. Comp. Graphics and Image Processing **17**, 291 (1981)

3.77 J.J. Gerbrands, E. Backer, W.A.G. v.d. Hoeven: Edge detection by dynamic programming. Proc. 6 Symp. on Information Theory in the Benelux, Mierlc (1985) pp.35-42

3.78 U. Montanari: On the optimal detection of curves in noisy pictures. CACM **14**, 335 (1971)

3.79 H. Niemann, H. Bunke, I. Hofmann, G. Sagerer, F. Wolf, H. Feistel: A knowledge based system for analysis of gated blood pool studies. IEEE Trans. PAMI-7, 246 (1985)

3.80 G.H. Granlund: In search of a general picture processing operator. Comput. Graph. Image Proc. **8**, 155 (1978)

3.81 M.K. Hu: Visual pattern recognition by moment invariants. IEEE Trans. IT-**18**, 179 (1962)

3.82 G. Nagy: Feature extraction on binary patterns. IEEE Trans. SSC-**5**, 273 (1969)

3.83 T. Pavlidis: A review of algorithms for shape analysis. Comput. Graph. Image Proc. **7**, 243 (1978)

3.84 E. Persoon, K.S. Fu: Shape discrimination using Fourier descriptors. IEEE Trans. SMC-**7**, 170 (1977)

3.85 R.Y. Wong, E.L. Hall: Scene matching with invariant moments. Comput. Graph. Image Proc. **8**, 16 (1978)

3.86 C.T. Zahn, R.Z. Roskies: Fourier descriptors for plane closed curves. IEEE Trans. C-**21**, 269 (1972)

3.87 H. Blum, R.N. Nagel: Shape description using weighted symmetric axis features. Pattern Recognition **10**, 167 (1978)

3.88 H.Y. Feng. T. Pavlidis: Decomposition of polygons into simpler components: feature extraction for syntactic pattern recognition. IEEE Trans. C-**24**, 636 (1975)

3.89 H. Freeman: Shape description via the use of critical points. Pattern Recognition **10**, 159 (1978)

3.90 Y. Nakimoto, Y. Nakano, K. Nakata, Y. Uchikara, A. Nakajima: Improvement of Chinese character recognition using projection profiles. Proc. 1st Int'l Joint Conf. on Pattern Recognition, Washington, DC (1973) pp.172-178

3.91 L.G. Shapiro, R.M. Haralick: Decomposition of two-dimensional shapes by graph-theoretical clustering. IEEE Trans. PAMI-**1**, 10 (1979)

3.92 E. Wong, J.A. Steppe: Invariant recognition of geometric shapes. In *Methodologies of Pattern Recognition*, ed. by S. Watanabe (Academic, New York 1969) pp.535-546

3.93 S.W. Zucker: Region growing: childhood and adolescence. Comput. Graph. Image Proc. **5**, 382 (1976)

3.94 C.R. Brice, C.L. Fennema: Scene analysis using regions. Artif. Intell. **1**, 205 (1970)

3.95 E.C. Freuder: Affinity, a relative approach to region finding. Comput. Graph. Image Proc. **5**, 254 (1976)

3.96 J.L. Muerle, D.C. Allen: Experimental evaluation of techniques for automatic segmentation of objects in a complex scene. In *Pictorial Pattern Recognition*, ed. by G.C. Cheng, R.S. Ledley, D.K. Pollock, A. Rosenfeld (Tompson, Washington 1968) pp.3-13

3.97 R. Ohlander, K. Price, D.R. Reddy: Picture segmentation using a recursive region splitting method. Comput. Graph. Image Proc. **8**, 313 (1978)

3.98 T.V. Robertson, P.H. Swain, K.S. Fu: Multispectral image partitioning. TR-WW 73-26, LARS Inf. Note 071373, School of Electr. Eng., Purdue Univ. (1973)

3.99 S.L. Horowitz, T. Pavlidis: Picture segmentation by a tree traversal algorithm. J. Assoc. Comput. Mach. **23**, 368 (1976)

3.100 F. Cheevasuvit, H. Maitre, D. Vidal-Madjar: A robust method for picture segmentation based on a split-and-merge procedure. Computer Vision, Graphics, and Image Processing **34**, 268 (1986)
3.101 C.H. Lee: Recursive region splitting at hierarchical scope views. Computer Vision, Graphics, and Image Processing **33**, 237 (1986)
3.102 A.M. Nazif, M.D. Levine: Low level image segmentation: an expert system. IEEE Trans. PAMI-**6**, 555 (1984)
3.103 P.A. Devijver: Probabilistic labeling in a hidden second order Markov mesh. In *Pattern Recognition in Practice* II, ed. by E.S. Gelsema, L.N. Kanal (North-Holland, Amsterdam 1986) p.113-123
3.104 P.A. Devijver, M.M. Dekessel: Real-time restoration and segmentation algorithms for hidden Markov mesh random field models. In *Real-Time Object Measurement and Classification*, ed. by A.K. Jain, NATO ASI Ser. F42 (Springer, Berlin, Heidelberg 1988) pp.293-307
3.105 B. Julesz: Experiments in the visual perception of texture. Sci. Am. **232**, 34 (April 1975)
3.106 B. Julesz: The role of terminators in preattentive preception of line textures. In *Recognition of Pattern and Form*, ed. by D.G. Albrecht (Springer, Berlin, Heidelberg 1982) pp.33-55
3.107 R.M. Haralick: Statistical and structural approaches to texture. Proc. 4th Int'l Joint Conf. on Pattern Recognition, Kyoto (1978) pp.45-69
3.108 J.K. Hawkins: Textural properties for pattern recognition. In *Picture Processing and Psychopictorics*, ed. by B.S. Lipkin, A. Rosenfeld (Academic, New York 1970) pp.347-370
3.109 G.R. Cross, A.K. Jain: Markov random field texture models. IEEE Trans. PAMI-**5**, 25 (1983)
3.110 A. Gagalowicz: Analysis of texture using a stochastic model, Proc. 4th Int'l Joint Conf. on Pattern Recognition, Kyoto (1978) pp.541-544
3.111 M. Hassner, J. Sklansky: Markov random field models of digitized image texture. Proc. 4th Int'l Joint Conf. on Pattern Recognition, Kyoto (1978) pp.538-540
3.112 O.D. Faugeras: Texture analysis and classification using a human visual model. Proc. 4th Int'l Joint Conf. on Pattern Recognition, Kyoto (1978) pp.549-552
3.113 M.M. Galloway: Texture analysis using gray level run lengths. Comput. Graph. Image Proc. **4**, 172 (1975)
3.114 R. Haralick, K. Shanmugan, I. Dinstein: Textural features for image classification. IEEE Trans. SMC-**3**, 610 (1973)
3.115 S.L. Tanimoto: An optimal algorithm for computing Fourier texture descriptors, IEEE Trans. C-**27**, 81 (1978)
3.116 S.W. Zucker, A. Rosenfeld, L.S. Davis: Picture segmentation by texture discrimination. IEEE Trans. C **24**, 1228 (1975)
3.117 S.Y. Lu, K.S. Fu: A syntactic approach to texture analysis. Comput. Graph. Image Proc. **7**, 303 (1977)
3.118 H.G. Musmann, P. Pirsch, H.-J. Gallert: Advances in picture coding. Proc. IEEE **73**, 523 (1985)
3.119 H.H. Nagel: Analyse und Interpretation von Bildfolgen. Informatik Spektrum **8**, 178, 312 (1985)
3.120 B.K.P. Horn, B.G. Schunk: Determining optical flow. Artif. Intelligence **17**, 185 (1981)
3.121 H.H. Nagel, W. Enkelmann: An investigation of smoothness constraints for the estimation of displacement vector fields from image sequences. IEEE Trans. PAMI-**8**, 565 (1986)
3.122 J. Hutchinson, C. Koch, J. Luo, C. Mead: Computing motion using analog and binary resistive networks. IEEE Computer 21 **3**, 52 (1988)

3.123 R.Y. Tsai, T.S. Huang: Uniquenses and estimation of 3-D motion parameters of rigid bodies with curved surfaces. IEEE Trans. PAMI-6, 13 (1984)
3.124 J. Weng, T.S. Huang, N. Ahuja: 3-D motion estimation, understanding, and prediction from noisy images. IEEE Trans. PAMI-9, 370 (1987)
3.125 A. v. Brandt, W. Tengler: Obtaining smoothed optical flow fields by modified block matching. Proc. 5th Scandinavian Conf. on Image Analysis, Stockholm (1987) pp.523-529
3.126 A. v. Brandt: Motion estimation and subband coding using quadrature mirror filters. Proc. EUSIPCO-86 2, 829 (1986)
3.127 D. Terzopoulos: Image analysis using multigrid relaxation methods. IEEE Trans. PAMI-8, 129 (1986)
3.128 S.T. Barnard, M.A. Fischer: Computational Stereo. Comp. Surveys 14, 553 (1982)
3.129 E.L. Hall, C.A. McPherson: Three-dimensional perception for robot vision. Proc. SPIE Conf. on Robotics and Robot Sensing Systems, San Diego 442, 117-143 (1983)
3.130 R.A. Jarvis: A perspective of range finding techniques for computer vision. IEEE Trans. PAMI 5, 122 (1983)
3.131 E.L. Hall, J.B.K. Tio, C.A. McPherson, C.S. Draper F.A. Sadjadi: Measuring curved surfaces for robot vision. Computer 15 12, 42 (1982)
3.132 A. Scheuing, H. Niemann: Computing depth from stereo images by using optical flow. Pattern Recognition Lett. 4, 205 (1986)
3.133 Y. Yakimovski, R. Cunningham: A system for extracting three-dimensional measurements from a pair of TV cameras. Comp. Graphics and Image Proc. 7, 195 (1978)
3.134 S. Barnard, W. Thompson: Disparity analysis of images: IEEE Trans. PAMI-2, 333 (1980)
3.135 G. Medioni, R. Nevatia: Segment based stereo matching. Comp. Vision, Graphics, and Image Processing 31, 2 (1985)
3.136 H. Baker, T. Binford: A system for automated stereo mapping. Proc. Image Understanding Workshop, Palo Alto CA, (1982) p.215
3.137 W. Grimson, D. Marr: A computer implementation of a theory of human stereo vision. Proc. Image Understanding Workshop, Palo Alto, CA (1979) pp.41-47
3.138 W. Hoff, N. Ahuja: Extracting surfaces from stereo images: An integrated approach. Proc. 1th Int'l Conf. on Computer Vision, London (1987) pp.284-294
3.139 S.A. Lloyd, E.R. Haddow, J.F. Boyce: A parallel binocular stereo algorithm utilizing dynamic programming and relaxation labeling. Comp. Vision, Graphics, and Image Processing 39, 202 (1987)
3.140 Y. Ohta, T. Kanade: Stero by intra- and inter-scanline search using dynamic programming. IEEE Trans. PAMI-7, 139-154 (1985)
3.141 S. Posch: Hierarchische linienbasierte Tiefenbestimmung in einem Stereobild, in Künstliche Intelligenz, Informatik-Fachberichte Vol.181 ed. by W. Hoeppner (Springer, Berlin, Heidelberg 1988) pp.275-285
3.142 B.K.P. Horn: Understanding image intensities: Artif. Intelligence 8, 201 (1977)
3.143 B.K.P. Horn, M.J. Brooks: The variational approach to shape from shading. Computer Vision, Graphics, and Image Processing 33, 174 (1986)
3.144 A.P. Pentland: Local shading analysis. IEEE Trans. PAMI-6, 170 (1984)
3.145 C.H. Lee, A. Rosenfeld: Improved methods of estimating shape from shading using the light source coordinate system. Artificial Intelligence 26, 125 (1985)
3.146 G. Healey, T.O. Binford: Local shape from specularity. Proc. 1. Int'l Conf. on Computer Vision, London (1987) pp.151-160
3.147 B.T. Phong: Illumination for computer generated pictures. CACM 18, 311 (1975)
3.148 J.E. Shoup: Phonological aspects of speech recognition. In [Ref.1.8, pp.125-138]

3.149 J.L. Flanagan: *Speech Analysis, Synthesis, and Perception, Kommunikation und Kybernetik in Einzeldarstellungen*, Vol. 3, 2nd edn. (Springer, Berlin, Heidelberg, New York 1972)
3.150 F. Itakura: Minimum prediction residual principle applied to speech recognition. IEEE Trans. ASSP 23, 67 (1975)
3.151 J.D. Markel, A.H. Gray, Jr.: *Linear Prediction of Speech, Communications and Cybernetics*, Vol.12 (Springer, Berlin, Heidelberg, 1976)
3.152 R.J. Niederjohn, P.F. Castelaz: Zero-crossing analysis methods for speech recognition. Proc. IEEE Conf. Pattern Recognition and Image Proc., Chicago (1978) pp.507-513
3.153 H.F. Silverman, N.R. Dixon: The 1976 modular acoustic processor (MAP). IEEE Trans. ASSP 25, 367 (1977)
3.154 G.M. White, R.B. Neeley: Speech recognition experiments with linear prediction, bandpass filtering, and dynamic programming. IEEE Trans. ASSP-24, 183 (1976)
3.155 S.S. McCandless: An algorithm for automatic formant extraction using linear prediction spectra. IEEE Trans. ASSP 22, 135 (1974)
3.156 J.D. Markel: Digital inverse filtering - a new tool for formant trajectory estimation. IEEE Trans. AU-20,129 (1972)
3.157 W. Hess: *Algorithms and Devices for Pitch Determination of Speech Signals*, Springer Ser. Inf. Sci., Vol.3 (Springer, Berlin, Heidelberg 1983)
3.158 J.D. Markel: The SIFT algorithm for fundamental frequency estimation. IEEE Trans. AU-20, 367 (1972)
3.159 S. Sennef: Real-time harmonic pitch detector: IEEE Trans ASSP-26, 358 (1978)
3.160 E. Nöth, H. Niemann, S. Schmölz: Prosodic features in German speech: stress assignment by man and machine. In *Recent Advances in Speech Understanding and Dialog Systems*, ed. by H. Niemann, M. Lang, G. Sagerer, NATO ASI Series F46 (Springer, Berlin, Heidelberg 1988) pp.101-106
3.161 A.M. Noll: Cepstrum pitch determination. J. Acoust. Soc. Am. 41, 293 (1967)
3.162 R.J. Niederjohn, I.B. Thomas: Computer recognition of the continuant phonemes in connected English speech. IEEE Trans. AU-21, 526 (1973)
3.163 W.A. Lea, M.F. Medress, T.E. Skinner: A prosodically guided speech understanding strategy. IEEE Trans. ASSP-23, 30 (1975)
3.164 A. Waibel: *Prosody and Speech Recognition*. PhD Thesis, Carnegie-Mellon Univ., Pittsburgh (1986)
3.165 J.Y. Cheung, A.D.C. Holden, F.D. Minifie: Computer recognition of linguistic stress patterns in connected speech. IEEE Trans. ASSP-25, 252-256 (1977)
3.166 J. Vaissiére: The use of prosodic parameters in automatic speech recognition: in *Recent Advances in Spech Understanding and Dialog Systems*, ed. by H. Niemann, M. Lang, G. Sagerer, NATO ASI Series F46 (Springer, Berlin, Heidelberg 1988) pp.71-100
3.167 M. Jalanko, S. Haltsonen, K.J. Bry, T. Kohonen: Application of orthogonal projection principles to simultaneous phonemic segmentation and labeling of continuous speech. Proc. 4th Int'l Joint Conf. on Pattern Recognition, Kyoto (1978) pp.1006-1008
3.168 R. Andre-Obrecht: Automatic segmentation of continuous speech signals. Proc. ICASSP Tokyo (1986) p.2275
3.169 M. Cravers, R. Pieraccini, F. Raineri: Definition and evaluation of phonetic units for speech recognition by hidden Markov models. Proc. ICASSP Tokyo (1986) pp.2235-2238
3.170 P. Demichelis, R. DeMori, P. Laface, M. O'Kane: Computer recognition of plosive sounds using contextual information. IEEE Trans. ASSP-31, 359 (1983)
3.171 F. Jelinek, L.R. Bahl, R.L. Mercer: Design of a linguistic statistical decoder for the recognition of continuous speech. IEEE Trans. IT-21, 250 (1975)

3.172 K. Mano, S. Ishige, K. Shirai: Phoneme recognition in connected speech using both static and dynamic properties of spectrum described by vector quantization. Proc. ICASSP Tokyo (1986) pp.2243-2246

3.173 G. Ruske, T. Schotola: The efficiency of demisyllable segmentation in the recognition of spoken words. Proc. ICASSP Atlanta (1981) pp.971-974

3.174 A. Tanaka, S. Kamiya: A speech processing based on syllable identification by using phonological patterns. Proc. ICASSP Tokyo (1986) pp.2231-2234

3.175 T. Watanabe: Syllable recognition for continuous Japanese speech recognition. Proc. ICASSP Tokyo 1986, p.2295-2298

3.176 W. Woods, M. Bates, G. Brown, B. Bruce, C. Cook, J. Klovstad, J. Makhoul, B. Nash-Webber, R. Schwartz, J. Wolf, V. Zue: Speech understanding systems Vol.2, Acoustic Front End, Final Report (Bolt, Beranek and Newman Inc. Cambridge, MA 1976)

3.177 V. Zue, L.F. Lamel: An expert spectrogram reader: A knowledge-based approach to speech recognition. Proc. ICASSP Tokyo (1986) pp.1197-1200

3.178 H. Niemann, M. Lang, G. Sagerer (eds.): *Recent Advances in Speech Understanding and Dialog Systems*. NATO ASI Ser. F46 (Springer, Berlin, Heidelberg 1988)

3.179 A. Abut, R.M. Gray, G.R. Robelledo: Vector quantization of speech and speech-like waveforms. IEEE Trans. ASSP-**30**, 423 (1982)

3.180 D. Wolf; H. Reininger: Recent advances in speech coding. In *Recent Advances in Speech Understanding and Dialog Systems*, ed. by H. Niemann, M. Lang, G. Sagerer, NATO ASI Ser. F46 (Springer, Berlin, Heidelberg 1988) pp.1-24

3.181 Y. Linde, A. Buzo, R.M. Gray: An algorithm for vector quantizer design. IEEE Trans. COM-**28**, 84 (1980)

Chapter 4

4.1 C.K. Chow: An optimum character recognition system using decision functions. IRE Trans. Electron. Comput. **6**, 247-254 (1957)

4.2 K.S. Fu, A.B. Whinston (eds.): *Pattern Recognition Theory and Application* (Nordhoff, Leyden 1977)

4.3 J. Schürmann: *Polynomklassifikatoren für die Zeichenerkennung* (Oldenbourg, München 1977)

4.4 H. Niemann: *Methoden der Mustererkennung* (Akademische, Frankfurt 1974)

4.5 H. Niemann: *Klassifikation von Mustern* (Springer, Berlin, Heidelberg 1983)

4.6 C.R. Rao: *Linear Statistical Inference and Its Applications* (Wiley, New York 1973)

4.7 T.M. Cover, P.E. Hart: Nearest neighbor pattern classification. IEEE Trans. IT-**13**, 21 (1967)

4.8 R.O. Duda, P.E. Hart: *Pattern Classification and Scene Analysis* (Wiley, New York 1972)

4.9 L.E. Baum: An inequality and associated maximization technique in statistical estimation for probabilistic functions of Markov processes. Inequalities **3**, 1 (1972)

4.10 E.G.Schukat-Talamazzini: *Generierung von Worthypothesen in kontinuierlicher Sprache*. Informatik Fachberichte, Vol.141 (Springer, Berlin, Heidelberg 1987)

4.11 C.A. Harlow: Image analysis and graphs. Comp. Graphics and Image Processing **2** 60-82 (1973)

4.12 M. Yachida, S. Tsuji: A versatile machine vision system for complex industrial parts. IEEE Trans. C-**26**, 882-894 (1977)

4.13 E.A. Patrick: *Fundamentals on Pattern Recognition* (Prentice Hall, Englewood Cliffs, NJ 1972)

4.14 K. Fukunaga: *Introduction to Statistical Pattern Recognition* (Academic, New York 1972)

4.15 P.A. Devijver, J. Kittler: *Pattern Recognition - A Statistical Approach* (Prentice Hall, Englewood Cliffs, NJ 1982)
4.16 P.E. Hart: The condensed nearest neighbor rule. IEEE Trans. IT-**14**, 515 (1968)
4.17 H. Niemann, R. Goppert: An efficient branch-and bound nearest neighbour classifier. Pattern Recognition Lett. **7**, 67 (1988)
4.18 A.K. Dewdney: Analysis of a steepest-descent image-matching algorithm. Pattern Recognition **10**, 31 (1978)
4.19 M.A. Fischler, R.A. Elschlager: The representation and matching of pictorial structures. IEEE Trans. C-**22**, 67 (1973)
4.20 R.Y. Wong, E.L. Hall: Sequential hierarchical scene matching. IEEE Trans. C-**27**, 359 (1978)
4.21 A. Rosenfeld, R.A. Hummel, S.W. Zucker: Scene labeling by relaxation operations. IEEE Trans. SMC-**6**, 420 (1976)
4.22 H. Yamamoto: A method of deriving compatibility coefficient for relaxation operators. Comput. Graph. Image Proc. **10**, 256 (1979)
4.23 J.M. Mendel, K.S. Fu (eds.): *Adaptive, Learning, and Pattern Recognition Systems* (Academic, New York 1970)
4.24 Y.Z. Tsypkin: *Foundations of the Theory of Learning Systems* (Academic, New York 1973)
4.25 R. Takiyama: A general method for training the committee machine. Pattern Recognition **10**, 255 (1978)
4.26 A.V. Kulkarni, L.N. Kanal: An optimization approach to hierarchical classifier design. Proc. 3rd Int'l Joint Conf. on Pattern Recognition, Coronado, CA (1976) pp.459-466
4.27 I.K. Sethi, B. Chatterjee: Efficient decision tree design for discrete variable pattern recognition problems. Pattern Recognition **9**, 197 (1977)
4.28 P.H. Swain, H. Hauska: The decision tree classifier - design and potential. IEEE Trans. GE-**15**, 142 (1977)
4.29 K.S. Fu: *Sequential Methods in Pattern Recognition and Machine Learning* (Academic, New York 1968)
4.30 G.T. Toussaint: The use of context in pattern recognition. Pattern Recognition **10**, 189 (1978)
4.31 R. Narasimhan: A linguistic approach to pattern recognition. RPT 121, Digital Comp. Lab., Univ. of Illinois, Urbana (1962)
4.32 K.S. Fu: *Syntactic Methods in Pattern Recognition* (Academic Press, New York 1974)
4.33 K.S. Fu (ed.): *Syntactic Pattern Recognition, Applications* (Springer, Berlin, Heidelberg 1977)
4.34 R.C. Gonzalez, MG. Thomason: *Syntactic Pattern Recognition, an Introduction* (Addison-Wesley, Reading, MA 1978)
4.35 R.H. Anderson: Syntax-directed recognition of hand-printed two-dimensional mathematics. Ph.D Thesis, Div. of Eng and Appl. Physics, Harvard Univ., Cambridge, MA (1968)
4.36 D.A. Inselberg: SAP: A model for the syntactic analysis of pictures. PhD Thesis, Washington Univ., St. Louis, Miss (1968)
4.37 A.C. Shaw: The formal description and parsing of pictures. PhD Thesis, Comp. Science Dep., Stanford Univ., Stanford, CA (1968)
4.38 G. Stockman: A problem reduction approach to the linguistic analysis of waveforms. TR 538, Univ. of Maryland, College Park, MY (1977)
4.39 E.M. Sussenguth: Structure matching in information processing. PhD Thesis, Appl. Math. Dep., Harvard Univ., Cambridge, MA (1964)
4.40 R.S. Ledley: High-speed automatic analysis of biomedical pictures. Science **146**, 216-223 (1964)

4.41 T.G. Evans: Grammatical inference techniques in pattern analysis. In *Software Engineering COINS III*, 2, ed. by J.T. Tou (Academic, New York 1971) pp.183-202
4.42 J.A. Feldmann: First Thoughts on Grammatical Inference. AIM-55, Comp. Science Dep., Stanford Univ., Stanford, CA (1967)
4.43 K.S. Fu, T.L. Booth: Grammatical inference, introduction and survey. IEEE Trans. SMC-5, 95, 409 (1975)
4.44 E.M. Gold: Language identification in the limit. Information and Control 10, 447 (1967)
4.45 J.J. Horning: A study of grammatical inference, TR-CS 139, Stanford Artificial Intelligence Project AIM-98, Computer Science Dept., Stanford Univ. (1969)
4.46 L.R. Rabiner: Mathematical foundations of hidden Markov models. In [Ref.4.49, pp.183-205]
4.47 L.R. Bahl, F. Jelinek, L.R. Mercer: A maximum likelihood approach to speech recognition. IEEE Trans. PAMI-5, 179 (1983)
4.48 R. Gemello, R. Pieraccini, F. Raineri: Diphone spotting with Markov chains. Proc. 7th ICPR, Montreal (1984) pp.176-178
4.49 H. Nieman, M. Lang, G. Sagerer, (eds.): *Recent Advances in Speech Understanding and Dialog Systems* NATO ASI Series F46 (Springer, Berlin, Heidelberg 1988)
4.50 G.D. Forney: The Viterbi algorithm. Proc. IEEE 61, 268 (1973)
4.51 M.S. Glassman: Hierarchical DP for word recognition. Proc. ICASSP (1985) pp.886-889
4.52 C.S. Myers, L.R. Rabiner, A.E. Rosenberg: Performance tradeoffs in dynamic time warping algorithms for isolated word recognition. IEEE Trans. ASSP-28, 623 (1980)
4.53 H. Ney: The use of a one-stage dynamic programming algorithm for connected word recognition. IEEE Trans. ASSP-32, 263 (1984)
4.54 G.M. White: Dynamic programming, the Viterbi algorithm, and low cost speech recognition. Proc. ICASSP (1978) pp.413-417
4.55 A. Guzman: Computer recognition of three-dimensional objects in a visual scene. PhD Thesis, AI TR 228, Artif. Int. Lab., MIT, Cambridge, MA (1968)
4.56 L.G. Roberts: Machine perception of three-dimensional solids, in *Optical and Electro-Optical Information Processing*, ed. by J.T. Tippelt, D. A. Berkowitz, L.C. Clark, C.J. Koester, A.v.d. Burgh (MIT Press, Cambridge 1965) pp.159-197
4.57 R.C. Bolles, P. Horaud, M.H. Hannah: 3DPO, a three-dimensional part orientation system. Proc. Int'l Joint Conf. on Artif. Intelligence, Karlsruhe (1983) pp.1116-1120
4.58 O.D. Faugeras: New steps towards a flexible 3d vision system for robotics. Proc. 7th Int'l Conf. on Pattern Recognition, Montreal (1984) pp.796-805
4.59 E. Grimson, T. Lozano-Perez: Model-based recognition and localization from sparse range or tactile data. Int'l J. of Robotics Res. 3, 3 (1984)
4.60 A. Rosenfeld (ed.): *Techniques for 3-D Machine Perception* (Elsevier, Amsterdam 1986)
4.61 R. Brooks: Symbolic reasoning among three-dimensional models and two-dimensional images. Artificial Intelligence 17, 285-348 (1981)
4.62 D.H. Ballard: Generalizing the Hough transform to detect arbitrary shapes. Pattern Recognition 13, 111-122 (1981)
4.63 J. Illingworth, J. Kittler: The adaptive Hough transform: IEEE Trans. PAMI-9, 690 (1987)
4.64 H. Li, M.A. Lavin, R.J. LeMaster: Fast Hough transform: A hierarchical approach. Computer Vision, Graphics, and Image Processing 36, 139 (1986)

4.65 D.G. Lowe: Three-dimensional object recognition from single two-dimensional images. AI-**31**, 355 (1987)
4.66 T.M. Silberberg, D. Harwood, L.S. Davis: Three-dimensional object recognition using oriented model points. In [Ref.4.60, pp.271-320]
4.67 M. Hermann, T. Kanade, S. Kuroe: Incremental acquisition of three-dimensional scene model from images. IEEE Trans. PAMI-**6**, 331 (1984)
4.68 H. Niemann, G. Sagerer, S. Schröder: Learning object models in associative networks. Proc. Int'l Symp. on El. Devices, Circuits, and Systems, Kharagpur, India (1987) pp.933-936
4.69 W.A. Perkins. INSPECTOR: A computer vision system that learns to inspect parts. IEEE Trans. PAMI-**5**, 584 (1983)
4.70 J.G. Augustson, J. Minker: An analysis of some graph theoretical cluster techniques. J. Assoc. Comput. Mach. **17**, 571 (1970)
4.71 C. Bron, J. Kerbosch: Finding all cliques of an undirected graph. Commun. ACM **16**, 575 (1973)
4.72 D.G. Corneil, C.C. Gotlieb: An efficient algorithm for graph isomorphism. J. Assoc. Comput. Mach. **17**, 51 (1970)
4.73 J.R. Ullman: An algorithm for subgraph isomorphism. J. Assoc. Comput. Mach. **23**, 31 (1976)
4.74 M. Clowes: On seeing things. Artif. Intell. **2**, 79 (1971)
4.75 T. Kanade: Recovery of the three-dimensional shape of an object from a single view. CMU-CS-79-153 (Dept. Comput. Sci., Carnegie-Mellon Univ., Pittsburgh, PA 1979)
4.76 K. Sugihara: Automatic construction of junction dictionaries and their exploitation for the analysis of range data. Proc. 6th Int'l Joint Conf. on Artif. Intell., Tokyo (1979) p.859-864; see also Artif. Intell. **12**, 41 (1979)
4.77 K. Sugihara: A necessary and sufficient condition for a picture to represent a polyhedral scene. IEEE Trans. PAMI-**6**, 578 (1984)
4.78 K. Sugihara: *Machine Interpretation of Line Drawings* (MIT Press, Cambridge, MA 1986)
4.79 D. Waltz: Understanding line drawings of scenes with shadows. In *The Psychology of Computer Vision*, ed. by P.H. Winston (McGraw-Hill, New York 1975) pp.19-91
4.80 M. Minsky, S. Papert: *Perceptrons: An Introduction to Computational Geometry* (MIT Press, Cambridge, MA 1969)
4.81 N.J. Nilsson: *Learning Machines* (McGraw-Hill, New York 1965)
4.82 F. Rosenblatt: *Principles of Neurodynamics: Perceptrons and the Theory of Brain Mechanisms* (Spartan, Washington, DC 1961)
4.83 S. Grossberg (ed.): *The Adaptive Brain*, Vols.1,2 (North-Holland, Amsterdam 1987)
4.84 S. Grossberg (ed.): *Neural Networks and Natural Intelligence* (MIT Press, Cambridge, MA 1988)
4.85 D.E. Rumelhart, J.L. McClelland: *Parallel Distributed Processing*, **1,2** (MIT Press, Cambridge, MA 1986)
4.86 Special Issue on Artificial Neural Systems. IEEE Computer **21**,3 (1988)
4.87 T. Kohonen: Clustering, taxonomy, and topological maps of patterns. Proc. 6th Int'l Conf. on Pattern Recognition, Munich (1982) pp.114-128
4.88 Y.T. Zhou, R. Chellapa: Neural network algorithms for motion stereo. Proc. Int'l J. Conf. Neural Networks, Washington, DC (1989) pp.II.251-258
4.89 B. Parvin, G. Medioni: A constraint satisfaction network for matching 3D-objects. Proc. Int'l J. Conf. Neural Networks, Washington, DC (1989) pp.II-281-286
4.90 L. Shastri: Semantic Networks: *An Evidential Formalisation and its Connectionist Realisation* (Pitman, London 1988)

Chapter 5

5.1 C.J. Date: *An Introduction to Database Systems*, 2nd edn. (Addison-Wesley, Reading; MA 1977)
5.2 E. Horowitz, S. Sahni: *Fundamentals of Data Structures* (Pitman, London 1976)
5.3 J.L. Pfaltz: *Computer Data Structures* (McGraw-Hill, New York 1977)
5.4 T.A. Standish: *Data Structure Techniques* (Addison Wesley, Reading, MA 1981)
5.5 N. Wirth; *Algorithmen und Datenstrukturen* (Teubner, Stuttgart 1975)
5.6 J. Piper, D. Rutovitz: Data structures for image processing in a C language and UNIX environment. Pattern Recognition Letters **3**, 119 (1985)
5.7 J. Guttag: Abstract data types and the development of data structures. Commun. ACM **20**, 396 (1977)
5.8 B.J. Cox: *Object Oriented Programming: An Evolutionary Approach* (Addison-Wesley, Reading, MA 1986)
5.9 C.J. Date: *An Introduction to Database Systems*, 4th ed. (Addison-Wesley, Reading, MA 1986)
5.10 P.C. Lockemann, E.J. Neuhold (eds.): *Systems for Large Data Bases* (North-Holland, Amsterdam 1977)
5.11 D. Maier: *The Theory of Relational Databases* (Computer Science, Rockville, MD 1983)
5.12 D. Tsichritzis, F. Lochovsky: *Data Models* (Prentice Hall, Englewood Cliffs, NJ 1982)
5.13 H. Wedekind; *Datenbanksysteme*, Bd.1 und 2 (Bibliographisches Inst., Mannheim 1974, 1976)
5.14 P. Dadam, K. Kuespert, F. Anderson, H. Blanken, R. Erbe, J. Guenauer, V. Lum, P. Pistor, G. Walch: A DBMS prototype to support extended 2NF relations. An integrated view on flat tables and hierarchies. Proc. ACM SIGMOD '86 Int'l Conf. on Management of Data, Washington, DC (1986), ACM SIGMOD Rec. 2, pp.356-367
5.15 K. Dittrich, U. Dayal (eds.): Proc. 1986 Int'l Workshop on Object-Oriented Database Systems (Computer Society, New York 1986)
5.16 G. Schlageter, R. Unland, W. Wilkes, R. Ziechang, G. Maul, M. Nagl, R. Meyer: OOPS - an object oriented programming system and integrated data management facility. Proc. 4th Int'l Conf. Data Engineering, Los Angeles, CA (1988)
5.17 W.D. Potter, R.P. Trueblood: Traditional, semantic, and hyper-semantic approaches to data modeling. Computer **21**, 53 (June 1988)
5.18 M. Brodie, J. Mylopoulos (eds.): *On Knowledge Base Management Systems* (Springer, Berlin, Heidelberg 1986)
5.19 A. Klinger: Data structures and pattern recognition. In *Advances in Information System Sciences*, ed. by J.T. Tou (Plenum, New York 1978) p.273-310
5.20 L.G. Shapiro: Data structures for picture processing. Comput. Graph. (ACM) **12**, No.3, 140 (1978)
5.21 N. Alexandridis, A. Klinger: Picture decomposition, tree data structures, and identifying directional symmetries as node combinations. Comput. Graph. Image Proc. **8**, 43 (1978)
5.22 R.D. Giustini, M.D. Levine, A.S. Malowany: Picture generating using semantic nets. Comput. Graph. Image Proc. 7, 1 (1978)
5.23 R.M. Haralick, L.G. Shapiro: The consistent labeling problem, part I. IEEE Trans. PAMI-1, 173 (1979)
5.24 S.L. Horowitz, T. Pavlidis: Picture segmentation by a tree traversal algorithm. J. Assoc. Comput. Mach. **23**, 368-388 (1976)
5.25 R.D. Merrill: Representation of contours and regions for efficient computer search. Commun. ACM **16**, 82 (1973)

5.26 B. Moayer, K.S. Fu: A syntactic approach to fingerprint pattern recognition. Pattern Recognition 7, 1-23 (1975)
5.27 T. Pavlidis, K. Steiglitz: The automatic counting of asbestos fibers in air samples. Proc. 3rd Int'l Joint Conf. on Pattern Recognition, Coronado, CA (1976) pp.789-792
5.28 U. Ramer: Extraction of line structures from photographs of curved objects. Comput. Graph. Image Proc. **4**, 81-103 (1975)
5.29 S. Tanimoto, T. Pavlidis: A hierarchical data structure for picture processing. Comput. Graph. Image Proc. **4**, 104 (1975)
5.30 S.L. Tanimoto: An iconic symbolic data structure scheme. In *Pattern Recognition and Artificial Intelligence*, ed. by C.H. Chen (Academic, New York 1976) pp.452-471
5.31 R.D. Fennell, V.R. Lesser: Parallelism in artificial intelligence problem solving, a case study of HEARSAY II. IEEE Trans. C-**26**, 98 (1977)
5.32 D.M. McKeown, D.R. Reddy: A hierarchical symbolic representation for an image database. Proc. Workshop on Picture Data Description and Management, Chicago (1977) pp.40-44
5.33 L.G. Shapiro, R.M. Haralick: A general spatial data structure. Proc. Conf. on Pattern Recognition and Image Proc., Chicago (1978) pp.238-249
5.34 M. Sties, B. Sanyal, K. Leist: Organization of object data for an image information system. Proc. 3rd Int'l Joint Conf. on Pattern Recognition, Coronado, CA (1976) pp.863-869
5.35 Y. Yakimovski, R. Cunningham: DABI - a data base for image analysis with nondeterministic inference capability. In *Pattern Recognition and Artificial Intelligence*, ed. by C.H. Chen (Academic, New York 1976) pp.554-592

Chapter 6

6.1 N.J. Nilsson: *Problem Solving Methods in Arificial Intelligence* (McGraw-Hill, New York 1971)
6.2 H. Niemann: *Control Strategies in Image and Speech Understanding*. Proc. GWAI, Informatik Fachberichte, Vol.76 (Springer, Berlin, Heidelberg 1983) pp.31-49
6.3 H. Niemann, H. Bunke: *Künstliche Intelligenz in Bild- und Sprachanalyse* (Teubner, Stuttgart 1987)
6.4 V.R. Lesser, R.D. Fennell, R.D. Erman, D.R. Reddy: Organisation of the HEARSAY II speech understanding system. IEEE Trans. ASSP-**23**, 11 (1975)
6.5 B.T. Lowerre: *The Harpy Speech Recognition System*. PhD Thesis, Dept. Comput. Sci., Carnegie-Mellon Univ., Pittsburgh, PA (1976)
6.6 C.A. Harlow: Image analysis and graphs. Comp. Graphics and Image Processing **2**, 60 (1973)
6.7 Z. Manna: *Mathematical Theory of Computation* (McGraw Hill, New York 1974)
6.8 T.W. Pratt: Hierarchical graph model of the semantics of programs. Proc. AFIPS, Summer Joint Conf. (1969) pp.813-825
6.9 C.A. Petri: Communication with automata. Suppl.1, Tech. Rep. RAD C-TR-65-337, Vol.1, Griffis Air Force Base, New York (1966) [transl. from Kommunikation mit Automaten, Univ. Bonn, Germany 1962]
6.10 A.C. Shaw: Parsing of graph-representable pictures. J. Assoc. Comput. Mach. **17**, 453 (1970)
6.11 C. Hewitt: Viewing control structures as patterns of passing messages. Artif. Intell. **8**, 323 (1977)
6.12 M. Georgeff: A framework for control in production systems. Rpt. No. STAN-CS-79-716, Comp. Science Dept., Stanford Univ. (1979)

6.13 H. Prade: A computational approach to approximate and plausible reasoning with applications to expert systems. IEEE Trans. PAMI-7, 260 (1985)
6.14 L.A. Zadeh: Fuzzy logic. Computer 21, No.4, 83 (1988)
6.15 R. De Mori, L. Saitta: Automatic learning of fuzzy naming relations over finite languages. Information Sci. 20, 93 (1980)
6.16 R. De Mori: *Computer Models of Speech Using Fuzzy Algorithms* (Plenum, New York 1983)
6.17 N.J. Nilsson: *Principles of Artificial Intelligence* (Springer, Berlihn, Heidelberg 1982)
6.18 R. Bellman, S. Dreyfus: *Applied Dynamic Programming* (Princeton, Univ. Press Princeton 1962)
6.19 Y. Ohta, T. Kanade, T. Sakai: An analysis system for scenes containing objects with substructures. Proc. 4th Int. Joint Conf. on Pattern Recognition, Kyoto (1978) pp.752-754
6.20 M. Nagao, T. Matsuyama: *A Structural Analysis of Complex Aerial Photographs* (Plenum, New York 1980)
6.21 M. Yachida, S. Tsuji: A versatile machine vision system for complex industrial parts. IEEE Trans. C-26, 882-894 (1977)
6.22 H. Nieman, H. Bunke, I. Hofmann, G. Sagerer, F. Wolf, H. Feistel: A knowledge based system for analysis of gated blood pool studies. IEEE Trans. PAMI-7 246-259 (1985)
6.23 S. Rubin: The ARGOS image understanding system. Tech. Rpt. (Dept. Comput Sci., Carnegie-Mellon Univ., Pittsburgh, PA (1978)
6.24 W. Woods, M. Bates, G. Brown, B. Bruce, C. Cook, J. Klovstad, J. Makhoul, B. Nash-Webber, R. Schwartz, J. Wolf, V. Zue: Speech understanding systems, Vol.1, Introduction and Overview, Final Report, Bolt Beranek and Newman Inc., Cambridge, MA (1976)
6.25 W.A. Woods: Optimal search strategies for speech understanding control. Artif. Intelligence 18, 295 (1982)
6.26 W.H. Paxton: Experiments in speech understanding system control. Tech. Note 134, Artif. Intell. Center, Stanford Res. Inst., Menlo Park, CA (1976)
6.27 G. Sagerer, F. Kummert: Knowledge based systems for speech understanding. In [Ref.3.178, pp.421-458]

Chapter 7

7.1 C. Morris: *Sign, Language, and Behavior* (Braziller, New York 1955)
7.2 R.S. Michalsky, J.G. Carbonell, T.M. Mitchell (eds.): *Machine Learning, An Artificial Intelligence Approach* (Tioga, Palo Alto, CA 1983)
7.3 Britannica World Language Edition of Funke and Wagnalls Standard Dictionary: "Knowledge" (Encyclopedia Britannica, Chicago 1962; Funk and Wagnalls, New York 1962)
7.4 Brockhaus Enzyklopädie: "Wissen" (Brockhaus, Wiesbaden 1974)
7.5 H. Niemann: Wissensbasierte Bildanalyse. Informatik Spektrum 8, 201 (1985)
7.6 H. Niemann, H. Bunke, I. Hofmann, G. Sagerer, F. Wolf, H. Feistel: A knowledge based system for analysis of gated blood pool studies. IEEE Trans. PAMI-7 246-259 (1985)
7.7 G. Sagerer: *Darstellung und Nutzung von Expertenwissen für ein Bildanalysesystem*, Informatik Fachberichte 104 (Springer, Berlin, Heidelberg 1985)
7.8 M. Minsky: A framework for representing knowledge, Artif. Int. Memo No. 306, A.I. Lab., MIT, Cambridge, MA (1974) also publ. in *The Psychology of Computer Vision*, ed. by P.H. Winston (McGraw-Hill, New York 1975)
7.9 I.P. Goldstein: The FRL primer, AI Memo 408, MIT, Cambridge, MA (1977)

7.10 A. Church: *Introduction to Mathematical Logic*, 1 (Princeton Univ. Press, Princeton, NJ 1956)
7.11 Z. Manna: *Mathematical Theory of Computation* (McGraw-Hill, New York 1974)
7.12 V. Dahl: Logic programming as a representation of knowledge. IEEE Computer **16**, No.10, 106 (1983)
7.13 A. Deliyanni, R.A. Kowalski: Logic and semantic networks. Comm. ACM **22**, 184 (1979)
7.14 J.P. Hayes: In defense of logic. Proc. IJCAI, Cambridge, MA (1977) pp.559-565
7.15 N.J. Nilsson: *Principles of Artificial Intelligence* (Springer, Berlin, Heidelberg 1982)
7.16 J. Herbrand. Investigations in proof theory (1930). In: *From Frege to Gödel*, ed. by J. Heijenoort (Harvard Univ. Press, Cambridge, MA 1967)
7.17 R. Kowalski: *Logic for Problem Solving* (North-Holland, Amsterdam 1979)
7.18 J.A. Robinson: A machine-oriented logic based on the resolution principle. JACM **12**, 23 (1965)
7.19 L. Wos, R. Overbeek, E. Lusk, J. Boyle: *Automated Reasoning, Introduction and Applications* (Prentice Hall, Englewood Cliffs, NJ 1984)
7.20 W. Bibel: *Automated Theorem Proving* (Vieweg, Braunschweig 1982)
7.21 W.W. Bledsoe: Non-resolution theorem proving. Artif. Intelligence **9**, 35 (1977)
7.22 H. Gentzen: Untersuchungen über das logische Schließen I. Math. Zeitschrift **30**, 176 (1935)
M.E. Szabo (ed.): *The collected papers of Gerhard Gentzen* (North-Holland, Amsterdam 1955) pp.68-132
7.23 Special issue on non-monotonic logic. Artif. Intelligence **13** (1980)
7.24 E. Charniak: Motivation analysis, abductive unification, and nonmonotonic equality. Artif. Intelligence **34**, 275 (1988)
7.25 S. Hanks, D. McDermott: Nonmonotonic logic and temporal projection. Artif. Intelligence **33**, 379 (1987)
7.26 R. Turner: *Logics for Artificial Intelligence* (Ellis Horwood, Chichester 1984)
7.27 D. Poole: A logical framework for default reasoning. Artificial Intelligence **36**, 27 (1988)
7.28 L.A. Zadeh: Knowledge representation in fuzzy logic. IEEE Trans. KDE-1, 89 (1989)
7.29 M.L. Ginsberg: Counterfactuals. Artif. Intelligence **30**, 35 (1986)
7.30 W.F. Clocksin, C.S. Mellish: *Programming in Prolog*, 2nd edn. (Springer, Berlin, Heidelberg 1984)
7.31 E. Post: Formal reductions of the general combinatorial problem. Am. J. Math. **65**, 197 (1943)
7.32 A. Newell: Production systems; models of control structures. In *Visual Information Processing*, ed. by W.C. Chase (Academic, New York 1973) p.463-526
7.33 R. Davis, J. King: An overview of production systems. In *Machine Intelligence*, Vol.8, ed. by E.W. Elcock, D. Michie (Ellis Horwood, Chichester 1977) pp.300-332
7.34 D.A. Waterman, F. Hayes-Roth (eds.): *Pattern Directed Inference Systems* (Academic, New York 1978)
7.35 B. Buchanan, T.M. Mitchell: Model Directed Learning of Production Rules, in [Ref.7.29, pp.297-312]
7.36 M. Georgeff: A framework for control in production systems, Rpt. No. STAN-CS-79-716, Comp. Sci. Dept., Stanford University (1979)
7.37 R.S. Michalsky, P.H. Winston: Variable precision logic. Aritf. Intelligence **29**, 121 (1986)
7.38 M. Togai, H. Watanabe: Expert system on a chip: an engine for real-time approximation reasoning. IEEE Expert **1** (3), 55 (1986)

7.39 L.A. Zadeh: Commonsense knowledge representation based on fuzzy logic. Computer **16** (10), 61-65 (1983)
7.40 C.L. Forgy: A fast algorithm for the many pattern/many object pattern match problem. Artif. Intelligence **19**, 17 (1982)
7.41 J. McDermott, C. Forgy: Production system conflict resolution strategies. In [Ref.7.34, pp.177-199]
7.42 A. Barr, E.A. Feigenbaum (eds.): *The Handbook of Artificial Intelligence*, Vols.1 and 2 (Pitman Books, London 1981 and 1982)
7.43 R. Engelmore, A. Terry: Structure and function of the CRYSALIS system. Proc. 6th Int'l Joint Conf. on Artificial Intelligence, Tokyo (1979) pp.250-256
7.44 C.L. Forgy: On the efficient implementation of production systems, Ph.D. Thesis, Dept. Comput. Sci., Carnegie-Melon Univ., Pittsburg, PA (1979)
7.45 D. McCracken: A Production System Version of the HEARSAY II Speech Understanding System. Ph.D. Thesis, Dept. Comput. Sci., Carnegie-Mellon Univ., Pittsburg, PA (1978)
7.46 D.J. Mostow, F. Hayes-Roth: A production system for speech undersanding. In [Ref.7.34, pp.471-481]
7.47 Y. Ohta, T. Kanade, T. Sakai: A production system for region analysis. Proc. 6th Int'l Joint Conf. on Artificial Intelligence, Tokyo (1979) pp.684-686
7.48 P. Szolovits, S.G. Pauker: Categorical and probabilistic reasoning in medical diagnosis. Artif. Intell. **11**, 115 (1978)
7.49 A.M. Nazif, M.D. Levine: Low level image segmentation: An expert system. IEEE Trans. PAMI-**6**, 555-577 (1984)
7.50 M.R. Quillian: Semantic memory, in *Semantic Information Processing*, ed. by M. Minsky (MIT Press, Cambridge, MA 1968) pp.227-270
7.51 N.V. Findler (ed.): *Associative Networks* (Academic, New York 1979)
7.52 J.F. Sowa: *Conceptual Structures: Information Processing in Mind and Machine* (Addison-Wesley, Reading, MA 1984)
7.53 R.J. Brachman. What's in a Concept: Structural Foundations for Semantic Networks. BBN (Bolt, Beranek, Newman) Report, Cambridge, MA (1977)
7.54 W. Woods: What's in a link: Foundations for semantic networks. In *Representation and Understanding*, ed. by D. Bobrow, A. Collins (Academic, New York 1975) pp.35-82
7.55 R.J. Brachman: What is-a is and isn't: An analysis of taxonomic links in semantic networks. IEEE Computer **16** No.10, 30-36 (1983)
7.56 G. Sagerer, H. Niemann: An expert system architecture and its application to the evaluation of scintigraphic image sequences. Proc. Symp. on the Eng. of Computer-Based Medical Systems, Minneapolis, MN (1988) pp.82-88
7.57 W. Eichhorn, H. Niemann: A bidirectional control strategy in a hierarchical knowledge structure. Proc. 8th Int'l Conf. Pattern Recognition, Paris (1986) pp.181-183
7.58 H. Niemann, G. Sagerer, W. Eichhorn: Control strategies in a hierarchical knowledge structure: Int'l J. Pattern Rec. and Artif. Intelligence **2**, 557-572 (1988)
7.59 H. Niemann, A. Brietzmann, U. Ehrlich, G. Sagerer: Representation of a continuous speech understanding and dialog system in a homogeneous semantic net architecture. Proc. ICASSP, Tokyo (1986)
7.60 G. Sagerer, F. Kummert: Knowledge based systems for speech understanding. In [Ref.3.178, pp.421-458]
7.61 C. Fillmore: The case for case. In *Universals in Linquistic Theory*, ed. by E. Bach, R. Harms (Holt, Rinehart, Winston, New York 1968) pp.1-90
7.62 B. Bruce: Case systems for natural language. Artif. Intell. **6**, 327 (1975)
7.63 D.H. Ballard, C.M. Brown, J.A. Feldman: An approach to knowledge-directed image analysis. In [Ref.1.13, pp.271-281]

7.64 J.K. Tsotsos: A framework for visual motion understanding. PhD Thesis, Dept. Comp. Sci., Univ. Toronto (1980)
7.65 J. Mylopoulus, T. Shibahara, J.K. Tsotsos: Building knowledge based systems: The PSN experience. IEEE Computer **16**, No. 10, 83 (1983)
7.66 H. Niemann: A homogeneous architecture for knowledge based image understanding. Proc. 2nd Conf. Artif. Intelligence Applications, Miami FL (1985) pp.88-93
7.67 D.G. Bobrow, T. Winograd: An overview of KRL, a knowledge representation language. Cognitive Science **1**, 3 (1977)
7.68 R.J. Brachman, J.G. Schmolze: An overview of the KL-ONE knowledge representation system. Cognitive Science **9**, 171 (1985)
7.69 F. Kummert, H. Niemann, G. Sagerer, S. Schröder: Werkzeuge zur Modellgesteuerten Bildanalyse und Wissensakquisition. In *GI 17. Jahrestagung*, ed. by M. Paul, Informatik Fachberichte, Bd.157 (Springer, Berlin, Heidelberg 1987)
7.70 D.G. Bobrow, M. Stefik: The LOOPS manual (Xerox Corp., Palo Alto, CA)
7.71 A. Goldberg, D. Robson: *SMALLTALK-80, The language and its Implementation* (Addison Wesley, Reading, MA 1983)
7.72 J.C. Kunz, T.P. Kehler, M.D. Williams: Applications development using a hybrid AI development system. The AI Magazine **5**, 41-54 (1984)
7.73 J.P. Hayes: The logic of frames. In *Frame Conceptions and Text Understanding*, ed. by D. Metzing (de Gruyter, Berlin 1979) pp.46-61
7.74 L.K. Schubert: Extending the expressive power of semantic networks. Artif. Intell. **7**, 163 (1976)
7.75 R. Bertelsmeier, B. Radig: Kontextunterstützte Analyse von Szenen mit bewegten Objekten. In *Digitale Bildverarbeitung, Digital Image Processing*, ed. by H.H. Nagel, Informatik Fachberichte, Bd.8 (Springer, Berlin, Heidelberg 1977) p.101-128
7.76 R.M. Haralick, L.G. Shapiro: The consistent labeling problem, Pt.I. IEEE Trans. PAMI-1, 173-184 (1979)
7.77 B. Radig: Image sequence analysis using relational structures. Pattern Recognition **17**, 161 (1984)
7.78 L.G. Shapiro, R.M. Haralick: Algorithms for inexact matching. Proc. 5th Int'l Joint Conf. on Pattern Recognition, Miami FL (1980) pp.202-207 [IEEE Catalog No.80 CH 1499-3]
7.79 R. Narasimhan: A linguistic approach to pattern recognition, Rpt.121, Digital Comp.Lab., Univ. of Illinois, Urbana (1962)
7.80 T. Winograd: *Language as a Cognitive Process*, Vol.1, Syntax (Addison Wesley, Reading, MA 1983)
7.81 W.A. Woods: Transition network grammars for natural language analysis. Commun. ACM **13**, 591 (1970)
7.82 G. Görz: Struturanalyse gesprochener Sprache - ein Verarbeitungsmodell. Dissertation, Univ. Erlangen-Nürnberg, Erlangen (1988)
7.83 S. Shieber: An introduction to unification-based approaches to grammar. CSLI Lecture Notes 4, Stanford Univ., Stanford, CA (1987)
7.84 A. Brietzmann: Eine ATN Grammatik des Deutschen für die automatische Spracherkennung. Sprache und Datenverarbeitung **8**, Heft 1/2, 54 (1984)
7.85 U. Ehrlich: Ein Lexikon für das natürlich-sprachliche Dialogsystem EVAR. Arbeitsberichte des Inst. für Mathematische Maschinen und Datenverarbeitung 19, No.3, Univ. Erlangen-Nürnberg, Erlangen (1986)
7.86 B.G. Buchanan, T.M. Mitchell: Model-directed learning of production rules. In [Ref.7.34, pp.297-312]
7.87 R. Davis: Knowledge acquisition in rule-based-systems - knowledge about representation as a basis for system construction and maintenance. In [Ref.7.34, pp.99-134]

7.88 T.G. Dietterich, R.S. Michalski: Learning and generalization of characteristic descriptions - evaluation criteria and comparative review of selected methods. Proc. 6th Int'l Joint Conf. on Artif. Intell., Tokyo (1979) pp.223-231
7.89 F. Hayes-Roth, J. McDermott: An interference matching technique for inducing abstractions. Commun. ACM **21**, 401 (1978)
7.90 C.L. Hedrick: Learning production systems from examples. Artif. Intelligence **7**, 21 (1976)
7.91 T.M. Mitchell: Analysis of generalization as a search problem. Proc. 6th Int'l Joint Conf. on Artif. Intel., Tokyo (1979) pp.577-582
7.92 R. J. Solomonoff: A formal theory of inductive inference. Information and Control **7** 1, 244 (1964)
7.93 P.H. Winston: Learning structural descriptions from examples. In *The Psychology of Computer Vision*, ed. by P.H. Winston (McGraw-Hill, New York 1975) pp.157-209
7.94 R.S. Michalsky: Pattern recognition as rule-guided inference. IEEE Trans. PAMI-2, 349 (1980)
7.95 H. Nieman, G. Sagerer, S. Schröder: Learning object models in associative networks. Proc. Int'l Symp. on El. Devices, Circuits and Systems, Kharagpur, India (1987) pp.933-936
7.96 S. Schröder, H. Niemann, G. Sagerer: Knowledge acquisition for a knowledge based image analysis system. Proc. Europ. Knowledge Acquisition Workshop, GMD Studien Nr. 143, Bonn (1988) pp.29.1-15
7.97 J.H. Holland: *Adaptation in Natural and Artificial Systems* (Univ. of Michigan Press, Ann Arbor 1975)

Subject Index

A^*-algorithm 256, 305
Absolute error 96
Abstraction 6, 8
Access control 222
Address reading 14
Adjacency matrix 210
Admissibility 257
Agent 312
Aliasing 27, 54
Allophone 139
Alphabet 243
Alternative 99, 140, 145
Analysis 5, 8
— system
Antecedent 286
Approximation 78, 95, 102, 126, 136, 143
—, piecewise 94, 105
Array 207
Aspect 276
Assignment 197
Association 221, 233
Attribute 75, 183, 274, 289, 297, 300, 306, 319
Autocorrelation approach 70
— function 114
Autonomous robot 16
Averaging 60

Background 76, 115, 118, 120, 130
Back-propagation training 202
Band limited 29
Bandwidth 112
Basic probability assignment 250
—, dissonant 251
—, cosonant 251
Baum-Welch algorithm 173
Beneficiary 312
Binary value 317
Bit rate 29
Border 76, 78, 103, 138
Boundary 80, 92, 97, 102, 104, 109, 140, 229, 233, 273
Branch 210

Camera calibration 46
— coordinates 47
— model 46
Canny operator 85
Cartesian product 225
Case 311
— frame 313
Cell 15, 105
— analysis 77
Cepstrum 59
—, complex 59
Certainty factor 252
Chain 233
— code 33
Change 6, 75, 116, 123, 278, 292, 329
—, gray level 81, 88, 93
Character 14
Chroma 41
Chromaticity coordinates 21
– diagram 21
Class 4, 152, 330, 333
—, reject 4, 154
Classification 4, 151, 183, 187
—, distribution free 158
—, nonparametric 160
—, statistical 154
—, syntactical 166
— system 10
Classifier 158, 160
—, hierarchical 166
—, sequential 166
Clause 280
Clique 197
Closing 67
Cloud transmission 60
— reflection 60
Cluster 98, 204
— analysis 165
Coagent 312
Code book 148
Coding 27
—, run-length 33
—, transform 31

Color 19, 40
— matching function 20
—, metameric 21
—, primary
Colorimetry 19
Compatibility 164
Component 11
Concept 274, 296
—, minimal 275, 299
—, modified 296
Concretization 273
Cone, generalized 229
Connectionist model 199
Consequent 286
Continuity constraint 181
Contour 80
— following 79
— line 94
Control 12, 236, 267, 304
— module 238, 254
— points 43
— strategy 304
— structure 230
Conveyance 312
Convolution 53
—, discrete 53
— integral 53
Coordinates, camera 47, 191
—, light source 131
—, object 47
—, viewer 131
—, world 47, 191
Correction, geometric 42
Corresponding image points 116, 123
Cost 154
— function 255
Covariance matrix 31, 157
— method 70
Crack 198
Credibility 250
Criterion of integrability 133
Crosscorrelation 162
Cutset 108

Data 206, 286
— independence 214
— model 215, 217
——, hierarchical 216
——, network
——, relational 219
——, semantic 227
— structure 206
——, iconic/symbolic 231

——, recursive 232
——, spatial 232
— sublanguage 224
Data base 12, 214
—, knowledge 274, 278
—, local 234
— machine 227
— management 215
—, results 214, 234
Decision rule 154
— tree 166
Decoding 27
Delete 224
Deletion 177
δ-function 52
Demisyllable 139
Density 39
—, conditional 154
—, multivariate normal 157
Dependence, functional 223
Depth 122, 136
Description 6, 319
—, initial 76
Designing 239, 247
Determinant 223
Destination 321
Dialog 311, 328
Dictionary 168
Difference 225
Digital methods 11
Dilation 65
Diphone 140
Discriminant function 158
Disparity 123
Displacement vector 117
Distance 153
Distortion-, geometric 42
— mapping 42
—, radial 47
Distribution, uniform 39
—, marginal 113
Divide 225
Document 14
Domain 219
Dynamic programming 27, 99, 180, 264

Edge 83, 85, 86, 90, 116, 210, 317
Eigenvalue 32, 46
Eigenvector 31, 46
Energy 37, 142
Enhancement 60, 61, 72, 82
Environment 2

Subject Index

Erosion 65
Error probability 157
Estimate 63, 85
Estimation 154
Euclidian distance 158, 160
Event 6, 269, 271
Execution cycle 287
Expansion 255
Expectation, conditional 160
Explanation 12, 340
Extension 297

Family, parametric 154, 158
Feature 9, 34, 100, 106, 109, 112, 114, 143, 145, 146
— map 204
File 208
—, image description 233
Film transmittance 39
Filter 52, 57
—, band-pass 112, 140
—, high-pass 61
—, inverse 69
—, linear 52, 57, 83
—, low-pass 60
—, matched 83
—, median 65
Find 225
First order predicate calculus 278
Flowchart schema 243
Fluctuation 76
Formant 71, 141
Formula 278
Forward-backward algorithm 171
Fourier descriptors 101
Fourier transform *see* transformation
Frame 38, 142, 276
Fundamental frequency *see* pitch
Fuzzy implication 290
— measure 250
— relation 290
— set 252

Gaussian function 86, 89
Generalization 330
Goal 237, 255
Grammar 166, 230, 316
—, lexical functional 325
—, unification based 319
Graph 195, 210, 238, 331
—, AND/OR 258
—, assignment 197
—, hierarchical (h) 244

—, line adjacency 229
—, Petri 245
—, region adjacency 229
Gray-level image *see* image
Gray-level run-length matrix 113

Hamming window 38
Hidden Markov model (HMM) 168, 169
Hierarchy 251, 273
Histogram 78
—, gray-level 39
Homogeneity 103
Hue 41
Hueckel operator 90
Hough transform *see* transformation
Hyperplane 160
Hypothesis 178, 293
Hypothesization 176

Illumination 18
Image 2, 16, 27, 29, 32
— analysis 29, 40, 256
—, binary 77, 107
—, calibration 46
—, color 23
—, distorted 46
—, gray-level 15, 33, 80
—, ideal 42
—, multispectral 5, 45
— pyramid 63, 229
—, radiographic 15, 61
—, reflection 16
— registration 45
—, remotely sensed 3, 15, 42
— segmentation 75
— sequence 2, 6, 120
—, stereo 49, 123
—, tomographic 17
—, transmission 17
— understanding 12, 284, 308
Importance 248
Impulse, unit 52
— response 53
Inconsistency of data 214
Industrial production 16
Inference rule 283
Insert 224
Insertion 177
Instance 273, 296
—, partial 302
Instantiation 288, 292, 302
Instrument 312

Intelligence 2
Intensity 37
Interactive system *see* system
Interpretation 6, 243, 279, 308
Interrelationship 220
Intersection 225
Isolated word recognition 14, 167, 169, 175
Iterative method 93

Join 226
Judgement 239, 248
Junction 191

Key, candidate 222
—, primary 222
Knowledge 6, 12, 227, 271
— acquisition 329
—, declarative 272, 275, 302
—, procedural 272, 275, 302
— representation 272

Label 163
Labeling, consistent 164
—, relaxation 163
Language 167
Laplacian operator 61, 82, 119
Lattice 43
LBG algorithm 148
Leaf 209
Learning 165, 329
—, supervised 165, 202, 330
—, unsupervised 165, 204, 330
Level 272
— of abstraction 6, 297
Levenshtein distance 331
Lexicon 168, 325, 327
— tree 176
Lightness 41
Line 99
— drawing 197
—, epipolar 50
Linear prediction 69
Link 219, 293, 297
List 208
Location 312
Lock 226
Loudness 24
Luminance 18
Luminous flux 17
— intensity 18

Markov model *see* hidden Markov model

Mapping, nonlinear 36
Mask 81
Matching, inexact 316
—, structural 195
Mean-square error 96
Mean vector 157
Median 65
Member 218
Merge 105
Method 12, 13
Model 169, 184, 189, 273, 335
—, connectionist 189
—, object 182
— scheme 273
—, stratified 12
— word 169
Modulation, delta 30
—, differential pulse code 30
—, pulse code 27
Module 11
Moment 100
Morphological operation 64
— algorithm 68
Motion 116

Nearest-neighbor rule 160
Necessity 251
Net, neural 199
—, Petri 245
—, semantic 227, 295, 335
Network
—, augmented transition (ATN) 317
— of states 242
—, recursive transition 316
Neural net *see* net
Node 208, 210, 258, 259
Normal form 222
Normalization 34, 50

Object 189, 211
— model 182
— oriented programming 211
—, polyhedral 197
— recognition 182
—, trihedral 197
Observation 152, 169
Oct tree 107
Opening 67
Optical flow 116, 124
Orthogonality principle 45
Owner 218

Part 273

Partition 4
Path 238
Pattern 3, 152
— analysis 207, 237, 247
—, complex 7
— recognition 2, 4
— restoration 62
—, simple 5
Perception 1
Perceptron 201
Perceptual grouping 190
Phone 139
Phoneme 139
Photographic density 39
Photometry 17
Picture *see* image
Pitch 23, 141
Place 245
Plane, epipolar 50
Planning 239, 247
Pointer 208
Possibility 251
Pragmatics 269, 304, 311
Predicate logic *see* first order predicat calculus
Predictor coefficients 70
Preprocessing 26, 59
Primaries 19
Primary color 19
Priority 249, 254
Probability 249
—, a posteriori 157
—, a priori 157
—, error 157
—, reject 155
Problem 237
—, primitive 258
— reduction 254, 258
—, solvable 258
—, unsolvable 258
Processing strategy 12
Production 166, 306
— rule 286
— system 286
Program 243
—, abstract 243
Project 225
Proof 282
Proposition, focal 251
Prototype 152, 167
Pruning 257, 261
Pseudocolor 45

Quad tree 107

Quantifier 280
Quantization, linear 29
— characteristic 28
— noise 29
—, uniform 30
Queue 208

Rank order operation 64
Raw material 312
Reading 14
Record 208
Reference white 21
Reflectance map 133
Reflectivity 18
Region 103
Relation 219, 231
—, binary 226
—, convolutional 61
—, fuzzy 290
—, multiplicative 60
—, unary 226
Relational structure 315
Reliability 248
Remotely sensed image *see* image
Resampling 43
Resolution 282
— hierarchy 63, 99, 107, 120
RETE algorithm 292
Risk 155
Robert's cross 81
Root 209
Rule 166, 230, 286, 291, 294, 302, 331, 338

Sample, representative 9
— value 27
Sampling period 28
— rate 28
— theorem 28
Saturation 41
Schema 215, 216, 275
Schema-schema 274
Scheme 273
Score 179, 248
—, shortfall density 268
Search, beam 269
—, state-space 254
Segmentation 228
— object 6, 10, 74, 75
Semantics 296, 311
Sentence 280, 319
Set, conflict 292
—, fuzzy 252
— of modality 299

Shape 100
— from shading 130
Signal, chrominance 23
— luminance 23
Signal-to-noise ratio 29, 83
Similarity 11, 330
Simple constituent 6
Skeletonization 79
Skolem function 280
Slant angle 131
Slot 277, 299
Smoothness constraint 117
Sound 23
Spatial co-occurrence matrix 112
Speaker identification 14
— verification 14
Specialization 273
Spectrum, logarithmic data 72
—, logarithmic model 72
—, model 71
Speech 17
—, connected 17, 175, 234
—, continuous 15, 17, 169, 175
— production 69
— recognition 14
— segmentation 138, 144
— understanding 12, 269, 293, 311
Split-and-merge 96, 107
Stack 208
State 169, 237, 317
—, goal 237
Statistical decision theory 152
Stereo image *see* image
Strategy 286
Structure, blackboard 11
—, heterarchical 11
—, hierarchical 11
—, model-directed 11
Substitution 281
Subword unit 139
Surface 130
Syllable 139, 168
Symbol, nonterminal 166
—, start 166
—, terminal 166
Syntactics 311
Syntax 278, 296, 311
System
—, characteristic 58
—, convolutional homomorphic 59
—, data-base-oriented 241
—, generative 240
—, heterarchical 241

—, hierarchical 241
—, homomorphic 58
—, interactive 239
—, inverse 69
—, linear 52
—, model-directed 240
—, multiplicative-homomorphic 58
—, shift invariant 53

Table 219
—, truth 279
Task domain 3, 304
Template 161, 167
— matching 161
Term 278
Test variable 155
Texture 110
Theorem 282
Thought 2
Threshold 77
— operation 77
Transformation
—, affine 43
—, discrete Karhunen-Loeve 31
—, eigenvector 31
—, Fourier 28, 53, 55, 101, 140
—, Hadamard 32
—, Hotelling 31
—, Hough 97
— —, generalized 188
—, medial axis 102
—, polynomial 43
—, principal axis 31
—, projective 43
Transition 245
Transmittance 39
Tree 208
—, AND/OR 258, 263
—, solution 259
Tristimulus value 19

Understanding 8
Unification 281, 323
Unifier 281
Union 225
User interface 12, 340
Update 224

Valency 311
Vector quantization 148, 204
Vertex 191, 210
Viterbi algorithm 172

Warping function 36
Window function 37
Word 168, 327
— hypothesis 176

— model 167
— recognition 169

Zero crossing 88

3459